INTRODUCTION TO MODERN
VIBRATIONAL SPECTROSCOPY

INTRODUCTION TO MODERN VIBRATIONAL SPECTROSCOPY

MAX DIEM
Department of Chemistry
City University of New York
Hunter College

A Wiley-Interscience Publication
JOHN WILEY & SONS
New York • Chichester • Brisbane • Toronto • Singapore

This text is printed on acid-free paper.

Copyright ©1993 by John Wiley & Sons, Inc.

All rights reserved. Published simultaneously in Canada.

Reproduction or translation of any part of this work beyond that permitted by Section 107 or 108 of the 1976 United States Copyright Act without the permission of the copyright owner is unlawful. Requests for permission or further information should be addressed to the Permission Department, John Wiley & Sons, Inc., 605 Third Avenue, New York, NY 10158-0012.

Library of Congress Cataloging in Publication Data:
Diem, Max, 1947–
    Introduction to modern vibrational spectroscopy / by Max Diem.
      p.   cm.
    "A Wiley-Interscience publication."
    Includes bibliographical references and index.
    ISBN 0-471-59584-5 (acid -free)
    1. Vibrational spectra.   I. Title.
QD96.V53D54   1993
543′.0858—dc20   93-1036

Printed in the United States of America

10 9 8 7 6 5 4 3 2

CONTENTS

Preface   x

1 Introduction   1

    References, 4

2 Results from Quantum Mechanics   5

    2.1 The Concepts of Quantum Mechanics, 5
    2.2 The Time-Independent Schrödinger Equation, 8
    2.3 Time-Dependent Description, 12
    2.4 Transition Moments for Absorption, 15
    2.5 Particle-in-a-Box: An Example for Stationary State Wavefunctions and Transition Moment, 17
    2.6 The Vibrational Schrödinger Equation, 21
    2.7 The Vibrational Transition Moment, 26
    2.8 Electromagnetic Radiation, 28
    2.9 Einstein Coefficients of Absorption and Emission, 31
    2.10 Rotational Energies of Rigid Molecules, 33
        2.10.1 Spherical Top Rotors, 35
        2.10.2 Linear Molecules, 36
        2.10.3 Symmetric Top Molecules, 36
        2.10.4 Rotational Spectra and Selection Rules for Linear Molecules, 37
        2.10.5 Rotational Energies and Spectra of Symmetric Top Rotors, 39
        2.10.6 Rotational–Vibrational Spectra, 39

2.11 Anharmonicity and Vibrational–Rotational Interaction, 43
References, 44

3 Polyatomic Molecules — 45

3.1 The Separation of Translational and Rotational Coordinates, 46
3.2 Classical Vibrations in Mass-Weighted Cartesian Coordinates, 47
3.3 Quantum Mechanical Treatment of the Vibrations of Polyatomic Molecules, 53
3.4 Accidental Degeneracy and Fermi Resonance, 58
3.5 Group Frequencies, 61
3.6 Isotopic Substitution and the Teller–Redlich Product Rule, 63
3.7 Normal Mode Calculations, 65
  3.7.1 Aim of Normal Mode Analysis, 65
  3.7.2 Internal Coordinates, 66
  3.7.3 Kinetic and Potential Energies in Internal Coordinates, 67
  3.7.4 Transformation Between Cartesian, Internal, and Normal Coordinates, 69
  3.7.5 Symmetry Coordinates, 72
3.8 Force Fields, 72
  3.8.1 The Generalized Valence Force Field (GVFF), 73
  3.8.2 Urey–Bradley Force Field (UBFF), 75
  3.8.3 *Ab Initio* Force Fields, 77
3.9 Example for the Computation of Normal Modes of Vibration, 78
3.10 Vibrational Intensities: Absorption, 86
References, 88

4 Symmetry of Molecular Vibrations — 90

4.2 Group Representations, 95
4.3 Representations of Molecular Vibrations, 100
4.4 Selection Rules for Normal Modes of Vibration, 103
4.5 Symmetry Coordinates, 106
References, 108

5 Introduction to Raman Spectroscopy — 109

5.1 General Considerations, 109
5.2 Classical Description of the Raman Effect, 112
5.3 The Polarizability Tensor, 114
5.4 Raman Selection Rules, 117
5.5 Polarization of Raman Scattering, 118
  5.5.1 Raman Scattering Geometries and Depolarization Ratios for Right Angle Scattering, 124

    5.5.2 Forward and Backscattering, 126
    5.5.3 Computation of Raman Scattered Intensities, 127
 5.6 Resonance Raman Spectroscopy, 128
 5.7 Time-Resolved Resonance Raman Spectroscopy, 130
 5.8 Nonlinear Raman Effects, 133
    5.8.1 Coherent Anti-Stokes–Raman Scattering (CARS), 135
    5.8.2 Stimulated Raman Scattering, 138
    5.8.3 Inverse Raman Effect, 139
    5.8.4 Raman-Induced Kerr Effects, 140
    5.8.5 The Hyper-Raman Effect, 140
 5.9 Surface Enhanced Raman Scattering (SERS), 142
 References, 143

6 Instrumentation for the Observation of Vibrational Spectra   145

 6.1 Dispersive Systems, 145
    6.1.1 Monochromators, 145
    6.1.2 Light Collection Optics for Monochromators, 148
 6.2 Interferometric Methods, 148
    6.2.1 General Aspects of Fourier Transform Infrared Spectroscopy, 149
    6.2.2 The Michelson Interferometer, 149
    6.2.3 Fourier Series and Fourier Transform, 152
    6.2.4 Discrete Fourier Transform and the Fast Fourier Transform Algorithm, 154
 6.3 Instrumentation for Infrared Spectroscopy, 157
    6.3.1 General Experimental Considerations, 157
    6.3.2 Dispersive Infrared Instrumentation, 159
    6.3.3 Principles of FT Spectrometers, 162
    6.3.4 Sampling Techniques in Infrared Spectroscopy, 163
        6.3.4.1 Transmission Spectroscopy, 163
        6.3.4.2 Attenuated Total Reflection and Diffuse Reflectance, 164
 6.4 Instrumentation for Raman Spectroscopy, 166
    6.4.1 Dispersive Single Detector Raman Instrumentation, 166
    6.4.2 Multichannel Dispersive Raman Spectrometers, 169
    6.4.3 Fourier Transform Raman Instrumentation, 172
 6.5 Hadamard Transform Spectroscopy, 173
 References, 174

7 Vibrational Spectra of Selected Small Molecules

 7.1 Triatomic Molecules, 176
 7.2 Tetratomic Molecules, 177
 7.3 Pentatomic Methane Derivatives, 179
    7.3.1 Pentatomic Molecules with T_d Symmetry, 180

7.3.2 Pentatomic Molecules with C_{3v} Symmetry, 183
7.3.3 Pentatomic Molecules with C_{2v} Symmetry, 187
7.3.4 Pentatomic Low Symmetry Species, 188
7.4 Ethane and Ethane Derivatives, 191
7.5 Example of an Amino Acid: Alanine, 194
7.6 Cyclic Molecules: Benzene, 198
7.7 Outlook, 202
References, 203

8 Biophysical Applications of Vibrational Spectroscopy 204

8.1 Peptide Vibrational Spectroscopy, 205
   8.1.1 Vibrations of the Peptide Linkage, 208
   8.1.2 The Conformational Sensitivity of the Amide I and III Modes, 215
8.2 Resonance Raman Spectroscopy of Peptides and Proteins, 220
   8.2.1 UV Resonance Raman Studies of the Peptide Linkage, 220
   8.2.2 Resonance Raman Spectroscopy of Prosthetic Groups, 221
      8.2.2.1 Heme Group Resonance Raman Studies, 221
      8.2.2.2 Heme Group Dynamic Studies, 224
      8.2.3.3 Other Prosthetic Groups, 225
8.3 Nucleic Acids: DNA and RNA, 226
   8.3.1 Phosphodiester Vibrations, 229
   8.3.2 Ribose Vibrations, 230
   8.3.3 Base Vibrations, 230
   8.3.4 Conformational Studies on Model Nucleotides, 230
8.4 Lipids, 232
References, 234

9 Vibrational Optical Activity 236

9.1 Infrared (Vibrational) Circular Dichroism, 237
   9.1.1 Phenomenological Description and Basic Equations, 237
   9.1.2 Observation of VCD, 238
9.2 Applications of VCD: Small Molecules, 240
9.3 Applications of VCD: Biomolecules, 243
   9.3.1 VCD of Peptides and Homo-oligoamino Acids, 243
   9.3.2 VCD of Proteins, 249
   9.3.3 VCD of Canonical Nucleic Acid Models, 250
   9.3.4 VCD Intensity Calculations for $d(CG)_5 \cdot d(CG)_5$, 252
   9.3.5 VCD of Small Oligomers, 253

9.4 Raman Optical Activity, 256
    9.4.1 Phenomenological Description, 256
    9.4.2 Observation of ROA, 258
9.5 Applications of ROA: Small Molecules, 259
9.6 Applications of ROA: Biological Molecules, 260
References, 261

Appendix I	**Vibrational Frequencies [cm^{-1}] for Some Common Groups**	**262**
Appendix II	**Refined Set of Urey–Bradley**	**264**
Appendix III	**Character Tables for Chemically Important Symmetry Groups**	**266**
Index		**279**

PREFACE

The aim of this book is to provide a text for a course in modern vibrational spectroscopy. The course is intended for advanced undergraduate students, who have had an introductory course in quantum chemistry and have been exposed to group theoretical concepts in an inorganic chemistry course, or for graduate students who have passed a graduate level course in quantum chemistry.

There are probably a dozen or so recent books on vibrational spectroscopy, and a few classic texts over three decades old. This large number seems to discourage any efforts to produce yet another book on the subject, unless one is willing to pursue a novel approach in presenting the material. This is, of course, exactly what was attempted with the present book, and the approach taken will be outlined in the next paragraphs.

There are two classic and comprehensive texts on vibrational spectroscopy, *Molecular Spectra and Molecular Structure II. Infrared and Raman Spectra of Polyatomic Molecules*, by G. Herzberg [Herzberg, 1945] and *Molecular Vibrations* by Wilson, Decius, and Cross [1955]. Both of these books are absolutely essential for the in-depth understanding of vibrational spectroscopy, and they devote hundreds of pages to theoretical derivations. However, due to the rapid progress in instrumental techniques and computational methods, and due to the fact that thousands of molecules have been studied since these two books were written, the practical aspects of these books are certainly limited. However, the value of these classic books for the serious vibrational spectroscopist is immeasurable, since they provide many of the fine points needed for the detailed understanding of the subject.

Among the more recent books, the reader will find either very specialized works dealing with one or a few specific topics of vibrational spectroscopy, or books that are more a compilation of data than a comprehensive text. In the former category, one might find, for example, a book entitled Biochemical Applications of Vibrational Spectroscopy, which treats the subject of the title exhaustively but offers little on instrumental aspects, calculations, and so on. The more practically oriented books often emphasize correlations of observed spectra with molecular structural features and may contain large compilations of spectra and group frequencies, and only a cursory treatment of theoretical principles. These books are essential for researchers who want to employ vibrational spectroscopy as a qualitative structural tool.

However, neither of these could be used as a textbook in a course, nor could they be used by a researcher who wants to gain insight into modern aspects of vibrational spectroscopy. Thus I was faced with the challenging task of writing a text that incorporates some theoretical background material that is necessary for the understanding of the principles of vibrational spectroscopy, in addition to computational methods, instrumental aspects, novel developments in vibrational spectroscopy, and a number of relatively detailed examples for the interpretations of vibrational spectra. Since the scope of this book is much broader than that of any of the aforementioned specialized texts, some of the theoretical material needed to be adjusted accordingly. Thus the quantum mechanics of molecular vibrations, time-dependent perturbations, transition moments, and many other topics are only summarized in this text, and detailed derivations are omitted. For the details, the reader is referred to specialized books, such as any of the many available textbooks on quantum chemistry [e.g., see *Quantum Chemistry*, Volumes I and II, Levine, 1970]. For the detailed theoretical background on symmetry aspects, the classic book by Cotton [1963], entitled *Chemical Applications of Group Theory*, is recommended, and the aforementioned books on vibrational spectroscopy for a more thorough treatment of theoretical aspects of molecular vibrations.

Thus the present book does not supersede any of the classic texts but is a further extension of them and intends to bring the reader to a more practical and up-to-date level of understanding in the field of vibrational spectroscopy. Subjects like Raman spectroscopy, which has become a major area of research in vibrational spectroscopy, is not treated as an afterthought, but experimental and theoretical aspects are discussed in detail. Items of historical significance, such as the Toronto arc for excitation of Raman spectra (which was actually mentioned as a viable source for Raman spectroscopy in a recent text) or the manual solution of the vibrational secular equation, have been banished from this book. Instead, modern experimental aspects, such as multichannel Raman instrumentation, time-resolved and resonance Raman techniques, nonlinear Raman effects, and Fourier transform infrared and Raman techniques are introduced. In addition, computational methods for the calculation of normal modes of vibration are treated in detail in this book.

One chapter is devoted to biological applications of vibrational spectroscopy. This is still a rapidly developing field and, perhaps, the most fascinating, for the molecules are often very large and difficult to study due to solvent interference and low solubility. It is in this area that the enormous progress of modern vibrational spectroscopy can best be gauged, since this field is not even discussed in the books of 30 or 40 years ago.

The final chapter is devoted to a new branch of vibrational spectroscopy carried out with circularly polarized light. The new techniques introduced here combine principles of vibrational spectroscopy with those of optical activity measurements of chiral molecules. Applications of these techniques to biological molecules, and to simple chiral systems, are presented.

I would like to thank my colleague, Prof. John Lombardi from the Department of Chemistry, City University of New York, City College, for his encouragement about this book and for correcting a large number of errors in the original manuscript. The Graduate Spectroscopy class (U761) at the City University of New York in Spring 1992 also was instrumental in pointing out inconsistencies in the manuscript, when an early version was used for the first time. I am grateful for the input received from these students.

M. DIEM

New York, New York
January 1993

INTRODUCTION TO MODERN
VIBRATIONAL SPECTROSCOPY

1

INTRODUCTION

Molecular spectroscopy is a branch of science in which the interactions of electromagnetic radiation and matter are studied. The aim and goal of these studies are to elucidate information on molecular structure and dynamics, the environment of the sample molecules and their state of association, interactions with solvents, and many other topics. Molecular (or atomic) spectroscopy is usually classified by the wavelength ranges (or energies) of the electromagnetic radiation (e.g., microwave or infrared spectroscopies), and the interaction of the radiation with the molecules or atoms is usually written in terms of a resonance condition, which implies that the energy difference between two stationary states in a molecule or atom must be matched exactly by the energy of the photon:

$$\Delta E_{\text{molecule}} = (h\nu)_{\text{photon}}$$

However, this view of the interaction between light and matter is rather cursory, since radiation interacts with matter even if its wavelength is far different from the specific wavelength at which an absorption occurs. Thus a classification of spectroscopy that is more general than that given by the wavelength range alone would be a resonance/off-resonance distinction. Many of the effects described and discussed in spectroscopy books are observed as resonance interactions where the incident light indeed possesses the exact energy of the molecular transition in question. Infrared and UV/visible absorption spectroscopy, microwave spectroscopy, or EPR are examples of such resonance interactions.

However, interaction of light and matter occurs, in a more subtle way, even if the wavelength of light is different from that of a molecular transition. These off-resonance interactions between electromagnetic radiation and matter give rise to well-known phenomena such as the refractive index of dielectric material and the anomalous dispersion of the refractive index with wavelength. Attenuated total reflection (ATR) spectroscopy, the optical rotatory dispersion in optically active molecules, and the normal (nonresonant) Raman effect are other phenomena that are best described in terms of off-resonance models. These effects are often ignored in spectroscopic books, but recent advances such as the ATR technique makes a discussion of these off-resonance effects necessary. Also, a discussion of nonresonance effects ties together many well-known aspects of classical optics and spectroscopy.

One of the numerous spectroscopic techniques employed is vibrational spectroscopy, which is a very useful technique indeed. It is a spectroscopic technique that has enormous pedagogical value, because vibrational spectra can be obtained easily by students in the laboratory, and the results are tangible and qualitatively interpretable. It can be used to identify compounds, explain eigenvector/eigenvalue problems during simple normal coordinate calculations, and demonstrate quantum mechanical principles such as allowed and forbidden transitions and breakdown of first-order approximations. Symmetry and group theory can be introduced logically when discussing vibrational spectroscopy, since the symmetry of atomic displacements during a normal mode of vibrations can be visualized easily and provides an intuitive approach for teaching the concepts of symmetry. Furthermore, vibrational spectroscopy is a useful probe for the structure of small molecules, because vibrational spectra can be predicted from symmetry considerations (group theory) and group frequencies. In fact, microwave (rotational) and vibrational spectroscopies were instrumental in determining the structure and symmetries of many small molecules. Rules to predict the structures and shapes of small molecules, such as the VSEPR (valence shell electron pair repulsion) model are partially based on vibrational spectroscopic results of the 1950s and 1960s. This aspect of the importance of vibrational spectroscopy becomes apparent when one reads the chapter on "Individual Molecules" in *Molecular Spectra and Molecular Structure* [Herzberg, 1945, Chapter III,3]. In this discussion, the emphasis is on the differentiation of the possible strucures of small molecules via their symmetry, which in turn determines their vibrational spectra. Since the structures of small molecules is now well established, the discussion of vibrational spectra, presented in Chapter 7, will emphasize the reverse process: how to assign and predict vibrational spectra from structural assumptions that can be derived from models such as VSEPR.

One other aspect of vibrational spectroscopy deserves further mention. Perhaps with the exception of nuclear magnetic resonance (NMR) techniques, it is probably safe to state that no other technique has grown at such a phenomenal rate over the past three decades than vibrational spectroscopy. Other spectroscopic methods, such as UV/visible, microwave, or EPR spectroscopy, have certainly profited from theoretical and technical advances; however, in vibrational spectroscopy, entire new branches have developed after the introduction of visible lasers. In the early 1960s, there were basically two techniques of vibrational spectroscopy: infrared absorption and Raman scattering. The latter of these was already somewhat esoteric, since large sample volumes and a lot of time were required to collect Raman data with the prevailing instrumentation.

After the introduction of high-power and pulsed lasers, not only were faster and more sensitive techniques developed (e.g., resonance Raman and time-resolved resonance Raman spectroscopies), but also entirely new spectroscopies were discovered. Hyper-Raman and CARS (coherent anti-Stokes–Raman scattering) are examples of such new techniques. In these, the nonlinear response of matter to extremely high field strengths produces interaction mechanisms that sample vibrational spectra via previously unknown molecular properties.

Dramatic progress was also achieved in infrared spectroscopy, due to the advent of infrared lasers and interferometric methods. The simplicity, specificity, and speed of infrared data acquisition make it one of the most generally applied spectroscopic measurements for quality control and qualitative and quantitative analysis. Thus the value of vibrational techniques in industrial laboratories for these routine tasks is enormous.

New techniques have been developed in infrared spectroscopy as well, such as infrared (vibrational) circular dichroism (VCD), a field that has enormously enhanced our ability to determine solution conformations of biological molecules. For these molecules in general, dramatic success has been achieved with modern infrared and Raman techniques. The structures of microgram quantities of molecules such as peptides, proteins, and DNA have been elucidated. Spatial resolution in Raman and infrared absorption microscopy has been sufficient to distinguish, by spectral features, parts of cells and tissue. Time-resolved techniques, which in the case of Raman scattering have reached subpicosecond time scales, have shed light on the dynamic behavior of biological molecules upon photoexcitation and oxidation/reduction. Some of these subjects will be discussed in Chapter 8. This is a novel feature of this particular text in vibrational spectroscopy, since these subject matters have emerged only during the past two decades.

Thus vibrational spectroscopy cannot be regarded as a static field that has outlived its usefulness. Quite to the contrary, it is a very dynamic and innovative field, and there is no reason to believe that the progress in this field is slowing down. Thus a book that emphasizes many of these new developments is necessary. As mentioned in the Preface, this volume is not intended to replace the previous book on vibrational spectroscopy, but to combine in one volume the necessary background to understand vibrational spectroscopy, and to introduce the reader to the many new and fascinating techniques and results.

REFERENCES

S. Califano, *Vibrational States*, Wiley, New York, 1976.

F. A. Cotton, *Chemical Applications of Group Theory*, Wiley-Interscience, New York, 1963.

G. Herzberg, *Molecular Spectra and Molecular Structure. II. Infrared and Raman Spectra of Polyatomic Molecules*, Van Nostrand Reinhold, New York, 1945.

I. N. Levine, *Quantum Chemistry, Volume I: Quantum Mechanics and Molecular Structure*, Allyn & Bacon, Boston, 1970.

I. N. Levine, *Quantum Chemistry, Volume II: Molecular Spectroscopy*, Allyn & Bacon, Boston, 1970.

E. B. Wilson, J. C. Decius, and P. C. Cross, *Molecular Vibrations: The Theory of Infrared and Raman Vibrational Spectra*, McGraw-Hill, New York, 1955.

2
RESULTS FROM QUANTUM MECHANICS

In this chapter, a relatively cursory review of the quantum mechanical principles necessary for the understanding of basic vibrational spectroscopy will be presented. In view of the large number of existing texts in introductory quantum chemistry, it was felt impossible to compete in detail with these books. On the other hand, spectroscopy relies heavily on quantum mechanics and is also one of the few methods that allows us to assess whether or not the quantum mechanical approach is valid: the energy eigenvalues predicted by the simplest quantum mechanical calculations for molecular vibration and rotation are, of course, the observables in infrared and microwave spectroscopies. Thus there is a synergy between quantum mechanics and spectroscopy, which necessitates a minimum discussion of this subject in any spectroscopy text. Since the mathematics of quantum mechanics can get rather complicated, the emphasis in this chapter will be more on the methods, approximations, and qualitative results, rather than the mathematical detail.

2.1 THE CONCEPTS OF QUANTUM MECHANICS

In classical (Newtonian) mechanics, the position and velocity (and hence the momentum) of any body in motion can be determined exactly, and, using the equations of motion, the exact trajetory of that body can be calculated. This, in turn, allows us to predict the exact position of that body at a future time. Classical mechanics holds for our everyday experiences with macroscopic objects. Any microscopic particle, however, is subject to different mechanical principles. The reason for this can be visualized conceptually to be due to the small mass of the microscopic particles, such as atoms and molecules, which makes the particles susceptible to perturbations when one tries to measure their positions or their momenta: the act of the measurement itself perturbs the momentum or position sufficiently that neither of these two quantities can be determined simultaneously. This is expressed in Heisenberg's uncertainty principle, which states that the product of the uncertainty in the momentum and the uncertainty in the position are larger than h:

$$\Delta x \, \Delta p_x \geqslant \hbar \qquad (2.1.1)$$

Mathematically, Eq. (2.1.1) follows from the fact that x and p_x are two operators that do not commute; that is, $[x, p_x] \neq 0$. The process of incorporating this uncertainty into a consistent physical picture of the motion of microscopic particles was one of the great merits of the physicists of the early twentieth century. Guided by the equations governing classical wave motion, such as the motion of a vibrating string or water waves in a trough, a set of equations was developed that govern the motions of microscopic particles. The major difference in the outcome of this "wave mechanical" or "quantum mechanical" description and the classical Newtonian picture is the quantization of energy, and the fact that the particle's position needs to be expressed as a probability of finding the particle at a given position—a direct consequence of the uncertainty principle.

Let us investigate the differences and similarities between classical and quantum mechanical wave equations. Macroscopic waves, such as the oscillation of a string or a water wave in a trough, obey a wave equation:

$$\frac{\partial^2 Z}{\partial t^2} = v^2 \frac{\partial^2 Z}{\partial x^2} \qquad (2.1.2)$$

Here, Z expresses an amplitude, perpendicular to the direction x, and the equation relates the propagation velocity v of the wave to the amplitudes at a given coordinate x. There are a multitude of functions that fulfill the differential equation 2.1.2:

$$Z(x, t) = F(x \pm vt) \qquad (2.1.3)$$

where F denotes any function that is twice differentiable. Commonly, the exact forms of the solutions of such differential equations are determined to a large extent by the boundary conditions. For example, if the differential equation describes an oscillating string for which the two ends are fixed, the boundary conditions impose that only certain trigonometric functions fulfill Eq. 2.1.3. Details of these arguments are given in Kauzmann's *Quantum Chemistry: An Introduction* [Kauzmann, 1957, Chapter 3].

The transition from classical to quantum mechanics proceeds as follows. For the quantum mechanical situation, a function is defined that contains all the information of the system we are interested in and that corresponds to the amplitude in the classical systems. This function is called the wavefunction $\Psi(x,t)$ of the system and is a function of time and space coordinates. For simplicity, let us restrict the space coordinate to just one dimension, x. The omission of two of three spatial coordinates simplifies the notation significantly, without changing the concepts and principles at all.

The square of this function $\Psi(x,t)$ (or the product of this function with its own complex conjugate in the case of complex wavefunctions), integrated over the spatial coordinate(s), gives a probability of finding the quantum mechanical particle at any given point in space and, therewith, is similar to the amplitude in Eq. 2.1.2. For the remainder of this text, we shall use the bold symbol $\boldsymbol{\Psi}$ for the total, time- and space-dependent wavefunction, Ψ for the product of various time-independent wave functions, and ψ for other, specific wavefunctions such as vibrational, rotational, or particle-in-a-box wavefunctions.

The quantum mechanical wave equation relating the temporal and spatial dependence of $\boldsymbol{\Psi}$ is given by

$$-\frac{\hbar}{i}\frac{\partial \boldsymbol{\Psi}(x,t)}{\partial t} = \left\{-\frac{\hbar^2}{2m}\frac{\partial^2}{\partial x^2} + V(x,t)\right\}\boldsymbol{\Psi}(x,t) \qquad (2.1.4)$$

Equation 2.1.4 is known as the time-dependent Schrödinger equation and contains, in addition to the terms in Eq. 2.1.2, a potential energy, $V(x,t)$. The first term in the brackets on the right-hand side is the kinetic energy T of the system, expressed as

$$T = -\frac{\hbar^2}{2m}\frac{\partial^2}{\partial x^2}\boldsymbol{\Psi}(x,t) \qquad (2.1.5)$$

Comparing the expression of the quantum mechanical kinetic energy to that of classical mechanics,

$$T = \frac{p^2}{2m} \qquad (2.1.6)$$

one finds that the classical momentum p needs to be substituted by

$$p(x) \rightarrow \frac{\hbar}{i}\frac{d}{dx}\boldsymbol{\Psi}(x,t) \qquad (2.1.7)$$

or

$$p^2 \rightarrow -\hbar^2\frac{d^2}{dx^2} \qquad (2.1.8)$$

Equation 2.1.7 is, in fact, one of the postulates of quantum mechanics. The substitution of the classical linear momentum by $(\hbar/i)(d/dx)$ cannot be proved per se, but the form can be made plausible by a number of procedures, which are discussed in textbooks on quantum mechanics [Levine (I), 1970, Chapter 7]. Other postulates of quantum mechanics are the existence of the wavefunction Ψ, the validity of the time-dependent Schrödinger equation (2.1.4), the existence of eigenvalues corresponding to the various quantum mechanical operators, and several others. It is interesting to note that these postulates, discussed in most texts on quantum chemistry, are true postulates, in that they cannot be proved or derived, although many of them make sense intuitively to those who are familiar with linear algebra. The ultimate proof of the form of the operators, functions, and the quantum mechanical formalism comes from experiment: if the static and dynamic properties, such as experimentally accessible transition energies and intensities of atomic and molecular systems subjected to an electromagnetic perturbation, are predicted properly by quantum mechanics, then it is quite likely that the formalism is correct. Although this sounds quite biased in favor of experimentalists (and particularly spectroscopists), it should be kept in mind that if the wave model proposed by Schrödinger had not fulfilled the experimentally well-established transition energies of the hydrogen atom (and many other spectroscopic measurements), quantum mechanics would have been dropped as a valid, alternative approach to describe microscopic systems.

2.2 THE TIME-INDEPENDENT SCHRÖDINGER EQUATION

We assume that the wavefunction $\Psi(x, t)$ can be separated into two independent functions of x and t according to

$$\Psi = \Psi(x) \cdot \varphi(t) = \Psi(x) \cdot e^{-i\omega t} \qquad (2.2.1)$$

Thus we can write a Schrödinger equation for time-independent, stationary systems (in one dimension) as follows:

$$\left\{-\frac{\hbar^2}{2m}\frac{\partial^2}{\partial x^2} + V(x)\right\}\Psi(x) = E\Psi(x) \qquad (2.2.2)$$

In Eq. 2.2.2, the expression on the left-hand side is referred to as the Hamilton operator H, or the *Hamiltonian* of the system, and the $\Psi(x)$ are a set of eigenfunctions such that the corresponding eigenvalues E are the energies of the operator H in the space of the eigenfunctions:

$$H\Psi(x) = E\Psi(x) \qquad (2.2.3)$$

The notation used in Eq. (2.2.3) is typical for eigenvalue/eigenvector problems in linear algebra.

To a first approximation, the functions Ψ can further be separated into products of wavefunctions, each of which is an eigenfunction of an independent energy operator only. For example, we assume that the energies of vibrational and rotational motion do not interact to a first approximation, and that the total energy is the sum of the vibrational and rotational energy eigenvalues. Thus we write independent Hamilton operators for rotational and vibrational (and electronic) energies in a molecule and assume that the wavefunction Ψ can be written as a product of electronic, vibrational, and rotational wavefunctions as follows:

$$\Psi = \psi_{elec} \cdot \psi_{vib} \cdot \psi_{rot} \tag{2.2.4}$$

The most basic (but by no means mathematically trivial) quantum mechanical problem arises from substituting appropriate functions for the potential energy into Eq. 2.2.2, writing the kinetic energy in a most convenient form, and solving the resulting Schrödinger equations for the energy eigenvalues and the eigenfunctions. For this, one assumes that the potential energy $V(x)$, which does not involve motion, is formally unchanged between classical and quantum mechanics.

For some of the more basic forms of spectroscopy, the following cases of potential functions and kinetic energy expressions are encountered.

In rotational spectroscopy, observed in the microwave spectral region, the potential energy V for a freely rotating molecule is assumed to be zero. The kinetic energy can be expressed in terms of the angular momentum, L:

$$T = L^2/2I \tag{2.2.5}$$

with

$$\mathbf{L} = \mathbf{r} \times \mathbf{p} \tag{2.2.6}$$

where the angular momentum is written as the cross product of the linear momentum $\mathbf{p}$ and the distance vector $\mathbf{r}$ from the center of mass. I is the moment of inertia. Equation 2.2.5 is the equivalent form, for rotational motion, of Eq. 2.1.6, which holds for linear motion:

$$T = \frac{p^2}{2m} \tag{2.1.6}$$

Thus there is a direct correspondence between mass and moment of inertia, and linear and angular momenta, for linear and circular motions, respectively.

10 RESULTS FROM QUANTUM MECHANICS

The x component of the angular momentum L can be expressed in terms of linear momentum components by

$$L_x = yp_z - zp_y \qquad (2.2.7)$$

with the other components of L following from Eq. 2.2.7 by cyclic permutation of the components. Substituting the quantum mechanical expression for the linear momentum (Eq. 2.1.7), one obtains

$$L_x = -i\hbar \left(y \frac{\partial}{\partial z} - z \frac{\partial}{\partial y} \right) \qquad (2.2.8)$$

When analogous expressions are written for the other components of the angular momentum operator, and the appropriate components are added and transformed to polar spherical coordinates, a Schrödinger equation is obtained, which can be solved in the simplest case of a two-body system in a lengthy, but well-documented procedure discussed in many of the elementary texts on quantum chemistry. The solutions of this rotational Schrödinger equation are the spherical harmonic functions, which were well established long before quantum mechanics to describe, for example, tidal waves on a spherical planet.

The energy eigenvalues obtained are (cf. Section 2.10)

$$E = BJ(J+1) \qquad (2.2.9)$$

for the case of a diatomic or any other linear molecule, where B is the rotational constant, inversely proportional to the moment of inertia, and J is a rotational quantum number, $J = 0, 1, 2, 3, \ldots$. These energy levels are relatively closely spaced, such that the transition energies in rotational spectroscopy occur generally in the subwavenumber range.

For vibrational spectroscopy, one defines, to a first approximation, the potential energy that holds the atoms in a molecule or ion in their equilibrium position by Hook's law:

$$V = \tfrac{1}{2}kx^2 \qquad (2.2.10)$$

where k is the restoring force (spring constant) counteracting nuclear displacements. This expression describes molecular systems approximately, ignoring effects of anharmonicity. The kinetic energy of the vibrational motion is that of the nuclear motion, which is given by

$$T_{\text{nuclear}} = p^2/2m \qquad (2.2.11)$$

Thus, in vibrational spectroscopy, one deals with the quantization of *nuclear* kinetic energy, with the nuclei being held in their equilibrium positions by the bonds formed by the electrons. The appropriate vibrational Schrödinger equation will be discussed in more detail in Section 2.6.

For a diatomic vibrating species, the obtained energy eigenvalues are

$$E \propto (v + \tfrac{1}{2})\nu \quad (2.2.12)$$

where v is the vibrational quantum number, $v = 0, 1, 2, 3, \ldots$, and ν is the frequency [in s^{-1}] of the vibrational motion, which for the simplest case of the diatomic oscillator is given by

$$\nu = (1/2\pi)\sqrt{k/M_R} \quad (2.2.13)$$

where M_R is the reduced mass of the oscillating atoms (Section 2.6). The spacing of vibrational energy levels is typically in the 100–4000 cm^{-1} (100–2.5 μm, or infrared) range. The units used here will be elaborated upon in Section 2.6.

The potential energy function used for molecular vibrations, expressed in Eq. 2.2.10, is a gross approximation of the real potential energy function, and much better approximations to the real potential energy functions are available, including anharmonic (cubic) force constants or expressions such as the Morse potential (cf. Section 2.6). The effects of the *nonharmonic* potential function will be discussed in Section 3.3; however, the quantum mechanical treatment of molecular vibration will use primarily the quadratic potential of Eq. 2.2.10, since the differential equations become very difficult or impossible to solve for any potential expression more involved than the quadratic function. Furthermore, for many transitions from the ground to the first excited state, the quadratic function provides an adequate description.

Finally, in atomic electronic spectroscopy, the potential energy is the nuclear potential acting on the electrons:

$$V = -e^2/r \quad (2.2.14)$$

and the kinetic energy is that of the electronic motion. Substitution of Eq. 2.2.14 into Eq. 2.2.2 leads, in the simplest case, to the well-known Schrödinger equation for the hydrogen atom, the solution of which proceeds in a similar fashion to that of the two-body rotational problem discussed above. The eigenvalues obtained here obey the Rydberg equation:

$$E = -\mathscr{R}/n^2 \quad (2.2.15)$$

where n is the primary hydrogen atom quantum number, and $\mathscr{R}$ is the Rydberg constant, $\mathscr{R} \approx 109{,}000$ cm^{-1}. The transition energies observed are typically in the 15,000–100,000 cm^{-1} (600–100 nm) range, which is the visible and ultraviolet spectral region. For atoms other than hydrogen, the energy expressions cannot be written in a simple form and will not be discussed here any further.

12 RESULTS FROM QUANTUM MECHANICS

Thus, for each of the cases above, a Schrödinger equation of the form

$$(T + V)\Psi = e\Psi \tag{2.2.16}$$

must be solved to give the particular eigenvalues and eigenfunctions. These solutions are the (stationary) eigenstates of the system and must agree with the observed electronic, vibrational, and rotational energy levels of the system.

2.3 TIME-DEPENDENT DESCRIPTION

Time-independent quantum mechanics introduced in the previous sections describes the energy expressions and wavefunctions of *stationary states*, that is, states that do not change with time. The stationary state wavefunctions and energies are obtained by solving the appropriate Schrödinger equations (Eq. 2.2.2) for the given system. This will be demonstrated in detail in Section 2.5 for the simplest quantum mechanical problem, the particle-in-a-box, and in Section 2.6 for the harmonic oscillator.

Next, one needs to describe a transition, that is, the process of how a system gets from one stationary state to another when a perturbation, typically electromagnetic radiation, is applied. For this, we need to invoke the time-dependent Schrödinger equation,

$$i\hbar \frac{\partial \Psi(x, t)}{dt} = H\Psi(x, t) \tag{2.1.4}$$

and solve it using the principles of time-dependent perturbation theory. As always in perturbation theory, one first defines the system in terms of a set of unperturbed, time-independent eigenfunctions, which describe the stationary states, and acts on this system with a perturbation operator. In this fashion, one obtains refined energy eigenvalues, and first-order corrections to the wavefunctions. This process may be repeated with a refined perturbation operator. Let the perturbed Hamiltonian be

$$[H + H'] \tag{2.3.1}$$

where H' is the perturbation imposed by the electromagnetic field for a dipole allowed transition. The principles of "dipole allowed" transitions will be discussed later on in more detail (Section 2.7 and Chapter 4), as will be the formalism used for electromagnetic radiation (Section 2.8). For the time being, let it suffice to state that the electric field vector along the x-direction, $\mathbf{E}_x$, of a light wave traveling in the z-direction is given by

$$\mathbf{E}_x = \mathbf{E}_x^0 [\cos(2\pi vt + 2\pi z/\lambda)] \tag{2.3.2}$$

and that an electrically charged particle, with charge e, experiences a force **F** in the electric field of the light given by

$$\mathbf{F} = e\mathbf{E} \tag{2.3.3}$$

where all symbols printed boldface are vector quantities. If the electric field is applied along the x-direction, we can write the force F_x as the derivative of the potential energy V:

$$\frac{dV}{dx} = -F_x = -eE_x \tag{2.3.4}$$

or, by integration,

$$V = -E_x(ex) \tag{2.3.5}$$

For a system of i charged particles with position vectors $\mathbf{r}_i$, the perturbation Hamiltonian is therefore written as (cf. Levine (I), 1970, Chapter 13]

$$H'(t) = -E^0 \sum_i (e_i \mathbf{r}_i)[\cos(2\pi v t + 2\pi z/\lambda)] \tag{2.3.6}$$
$$= -E^0 \boldsymbol{\mu}[\cos(2\pi v t + 2\pi z/\lambda)]$$

where $\boldsymbol{\mu} = \sum e_i \mathbf{r}_i$ is the dipole operator for the system.

We now return to the problem of solving the time-dependent Schrödinger equation

$$i\hbar \frac{d\Psi(x,t)}{dt} = [H + H']\Psi(x,t) \tag{2.3.7}$$

with the perturbation operator defined in Eq. 2.3.6. The strategy is to solve Eq. 2.3.7 by expanding the time-dependent wavefunctions $\Psi(x,t)$ in terms of the following summation:

$$\Psi(x,t) = \sum_k c_k(t)\varphi(t)\psi_k(x) \tag{2.3.8}$$

where $\varphi(t) = e^{i\omega t}$ (cf. Eq. 2.2.1), and the functions $\psi_k(x)$ are time-independent, stationary state wavefunctions appropriate for the system. $c_k(t)$ are time-dependent coefficients that determine the time evolution of the excited state population. In the absence of a perturbation, the system would be in a stationary state described by $\Psi(x,t) = \varphi(t)\psi_k(x)$. After a perturbation is applied for a certain time, the coefficients c_k change to account for the perturbation. For example, consider a hypothetical system with only two energy levels $\Psi(g)$

and $\Psi(e)$ for ground and excited states. Before a perturbation is applied, the system is completely in the ground state, $\Psi(g)$. Thus

$$\Psi(x, t) = 1\Psi(g) + 0\Psi(e) \tag{2.3.9}$$

At any time later, the coefficients 1 and 0 will change, and the system will be described by an equation

$$\Psi(x, t) = c_1\Psi(g) + c_2\Psi(e) \tag{2.3.10}$$

with $0 < c_i < 1$.

The time evolution of these c_i can be derived by classical perturbation methods and can be found in many quantum mechanical texts [Levine (II), 1970, Chapter 2]. This derivation proceeds by substituting Eq. (2.3.8) into Eq. (2.3.7) and taking the necessary derivatives, and integrating the resulting terms dc_m/dt between time 0 and the duration of the perturbation. Taking into account the explicit expression for the perturbation operator (Eq. 2.3.6), one arrives at an expression for the time evolution of the expansion coefficients:

$$c_m(t) = \delta_{nm} + \frac{iE^0}{2\hbar} \langle \psi_n|\mu|\psi_m\rangle \left[\frac{e^{i(\omega_{mn}+\omega)t} - 1}{(\omega_{mn} + \omega)} + \frac{e^{i(\omega_{mn}-\omega)t} - 1}{(\omega_{mn} - \omega)}\right] \tag{2.3.11}$$

where the wavefunctions ψ in Eq. 2.3.11 are time-independent (stationary state) wavefunctions, and the expression $\langle\psi_n|\mu|\psi_m\rangle$ is called the transition moment. The transition moment determines the selection rules, depending on the exact nature of the wavefunctions, the symmetry of the operator, and so on. This is the part to be discussed in Section 2.4.

The part in the square brackets in Eq. 2.3.11 describes the response of the system as the frequency of light is scanned through the absorption region. The second term in the square bracket becomes very large at the resonance condition, $\omega_{mn} = \omega$. When the frequency ω of the radiation is equal or very close to the energy difference ω_{mn} between states n and m, a transition between these states will occur and a photon with the corresponding energy will be absorbed.

The first term in the square bracket in Eq. 2.3.11 becomes very large if $\omega_{mn} = -\omega$. This case corresponds to the situation of stimulated emission, where a photon of the proper energy impinges onto a molecular or an atomic system in the excited state and causes this state to emit a second photon, thereby returning to the ground state. This time-dependent part of Eq. 2.3.11 also contains explicitly the expressions needed to treat certain off-resonance phenomena, such as molecular polarizabilities, to be discussed in Chapter 5. The magnitude of resonance versus off-resonance effects can be estimated from the expression in square brackets as well.

Both parts of this equation hold for one-photon absorption and emission situations; which include standard infrared (vibrational), microwave (rotational), and visible/ultraviolet (electronic) absorption spectroscopies.

2.4 TRANSITION MOMENTS FOR ABSORPTION

In the previous section, we have seen that two conditions are necessary for a transition to occur in an atomic or a molecular system under the influence of a perturbation by electromagnetic radiation. First, the radiation must possess the proper energy, or frequency, corresponding to the energy difference between the molecular or atomic states. This part is determined by the time-dependent part of Eq. 2.3.11. The second condition that must be fulfilled is that the transition moment must be nonzero. In Eq. 2.3.11, we called $\langle \psi_n | \mu | \psi_m \rangle$ the transition moment, which is, of course, just the expectation value of the dipole operator:

$$\boldsymbol{\mu} = \langle \psi_n | \mu | \psi_m \rangle \tag{2.4.1}$$

where ψ_n and ψ_m are the stationary state wavefunctions associated with energy levels n and m, and $\boldsymbol{\mu}$ is the dipole operator defined in the previous section. For a transition to occur, this transition moment must be nonzero. Whether or not the transition moment for a vibrational transition is zero depends on the geometry of the molecule and on the polarity of the atoms, as will be discussed later in detail. However, it is instructive even at this early point in this text to point out that certain vibrations have a zero transition moment and cannot be observed in the infrared absorption spectrum. A typical example for this is the symmetric stretching mode of CO_2, where both oxygen atoms move away from the central C atom in phase and with equal amplitudes. Although we know exactly what the energy of this transition is (from the Raman spectrum), irradiation of CO_2 molecules with infrared radiation of this energy does not produce a transition, since the transition moment is zero for symmetry reasons. The principle of zero and nonzero transition moments will be explained in the next section, using a simple quantum mechanical model.

The square of the transition moment is the actual observable in infrared absorption spectroscopy. We call the integrated intensity of an absorption band the dipole strength D of a transition, which can be determined experitally as the area under an infrared band, plotted in units of the molar extinction coefficient ϵ versus wavenumber:

$$D = \int (\epsilon/\tilde{\nu}) d\tilde{\nu} = |\langle \psi_n | \mu | \psi_m \rangle|^2 \tag{2.4.2}$$

The transition moment determines whether or not a transition will occur, and if it is allowed, how intensely it will appear in an absorption spectrum. The

16   RESULTS FROM QUANTUM MECHANICS

other piece of information from a vibrational spectrum is the energy of the transition, which is, of course, determined by the peak position (in wavenumbers, cf. Section 2.6).

It is generally true that at least two pieces of information may be extracted from a spectral band: its position (wavelength or wavenumber) and its intensity. The intensity is determined by the transition moment and the frequencies by the energy eigenvalues of the Hamiltonian. Both intensity and energy of a transition can be calculated from the quantum mechanical principles discussed so far in this chapter, if the wavefunctions of the system are known to a high degree of accuracy. In practice, the molecular wavefunctions are rarely known with sufficient accuracy to compute vibrational intensities accurately. *Ab initio* calculations of vibrational frequencies are even more difficult, since the second derivative of the molecular energy with respect to the normal coordinates needs to be calculated for the potential energy of the molecule. In fact, *ab initio* quantum mechanical calculations of vibrational frequencies and intensities are practical only for the smallest molecules (up to perhaps 8–10 atoms). However, there are a number of empirical calculations that have been utilized for the computation of vibrational eigenvalues; these methods will be discussed in Chapter 3.

2.5   PARTICLE-IN-A-BOX: AN EXAMPLE FOR STATIONARY STATE WAVEFUNCTIONS AND TRANSITION MOMENT

A simple hypothetical scenario will be examined next to illuminate the principles discussed in the last two sections: the example of the particle-in-a-box. Although this is, of course, an artificial situation, it is very instructive, since it shows in detail how the quantum mechanical formalism works in a situation that is sufficiently simple to carry out the calculations step-by-step. Furthermore, the vibrational wavefunctions to be discussed later and the particle-in-a-box wavefunctions have the same symmetry properties. Thus the principles of using symmetry arguments to determine whether or not a transition is allowed can be introduced here.

For any quantum mechanical calculation, we write the total energy of the system as the sum of the kinetic and potential energies, T and V:

$$(T + V)\psi = E\psi \tag{2.5.1}$$

The kinetic energy is given in Eq. 2.1.5 as

$$T = -\frac{\hbar^2}{2m} \frac{\partial^2}{\partial x^2} \psi \tag{2.2.2}$$

where m is the mass of the particle (typically an electron or a nucleus) and ψ

its wavefunction. For any real quantum mechanical system, the potential energy could be an expression such as Eq. 2.2.10. Using such an expression yields a complicated differential equation, which is not amenable to a simple solution. Thus we assume for the particle-in-a-box example that the potential well is not parabolic, but a simple square well, or box, defined such that (cf. Fig. 2.1)

$$V = 0 \quad \text{for } 0 < x < L \text{ (inside box)}$$
$$V = \infty \quad \text{for } x < 0 \text{ and}$$
$$x > L \text{ (outside box)}$$

Thus, for the region $0 < x < L$, the differential equation to solve is

$$-\left\{\frac{\hbar^2}{2m}\frac{d^2}{dx^2}\right\}\psi = (E - 0)\psi \tag{2.5.2}$$

whereas outside the box,

$$-\left\{\frac{\hbar^2}{2m}\frac{d^2}{dx^2}\right\}\psi = (E - \infty)\psi \tag{2.5.3}$$

Since $(E - \infty)$ is still infinite, Eq. 2.5.3 leads to the conclusion that outside the box, the wavefunction, and thus the probability of finding the particle, must be zero. Inside the box, the wavefunction must be nonzero (otherwise, the particle is not in the box). At the boundary, the wavefunctions must be continuous,

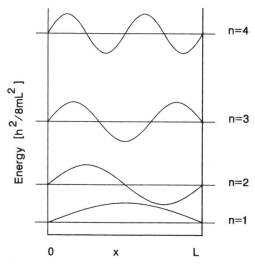

FIGURE 2.1. Wavefunctions and energy eigenvalues for the particle-in-a-box.

which leads to the boundary conditions

$$\psi(0) = 0 \quad \text{and} \quad \psi(L) = 0$$

These boundary conditions imply that the probability of finding the particle must be zero at the infinitely steep potential well, which means that the particle cannot escape from the well.

Equation 2.5.2 describes a typical eigenvalue problem: the Hamilton (total energy) operator is applied to a set of yet unknown eigenfunctions; the eigenvalues are the allowed, stationary energy states. The solutions of this differential equation, given in every elementary text on quantum chemistry, are basically any combinations of sine and cosine functions, since their second derivatives equal the original function, multiplied by a constant:

$$\psi_n = A \sin(Cx) + B \cos(Cx) \tag{2.5.4}$$

where A and B are amplitude constants, and C is given by

$$C = \frac{(2mE)^{1/2}}{\hbar} \tag{2.5.5}$$

However, it needs to be pointed out that the solutions of the differential equation depend to a great deal on the boundary conditions: the general solution of the differential equation may or may not describe the physical reality of the system, and it is the boundary conditions that force the solutions to be physically meaningful.

Since we required that the wavefunction is zero at the boundary of the box, the cosine terms in Eq. 2.5.4 cannot contribute to the wavefunctions since $\cos(0) \neq 0$. Thus the amplitude B of the cosine terms must be zero.

Because of the required continuity at $x = L$, the remaining term in Eq. 2.5.4 must be zero for $x = L$; thus

$$\psi_n = A \sin(CL) = 0 \tag{2.5.6}$$

which happens if $\sin(CL) = 0, \pm\pi, \pm 2\pi$, and so on. Substituting Eq. 2.5.5 into Eq. 2.5.6, one obtains

$$[(2mE)^{1/2}/\hbar]L = \pm\pi, \pm 2\pi \tag{2.5.7}$$

where the value 0 has been discarded because it would lead to a zero amplitude A, which is tantamount to saying that the particle is not in the box. Otherwise, Eq. 2.5.7 is true for any integer value n, and one obtains from it the energy eigenvalues

$$E(n) = \frac{n^2 h^2}{8mL^2} \tag{2.5.8}$$

Using these energy eigenvalues, and Eq. 2.5.5, the wavefunctions for the particle-in-a-box are obtained as

$$\psi_n = A \sin(n\pi x/L) \tag{2.5.9}$$

Here, n is an integer quantum number and A a still undefined amplitude factor. The choice of this amplitude is immaterial, since Eq. 2.5.9 fulfills Eq. 2.5.2 for any value of A. However, since the square of the wavefunction can be visualized as the probability of finding the particle, we postulate that the integral of ψ^2 over the length of the box must be unity, since we know that there is one particle in the box. Thus we adjust the amplitude A such that

$$\int (\psi_n)^2 \, dx = 1 \tag{2.5.10}$$

which leads to a normalization or amplitude factor

$$A = (2/L)^{1/2} \tag{2.5.11}$$

The resulting stationary state (time-independent) wavefunctions and energies are depicted in Fig. 2.1 and were obtained in a typical operator/eigenvalue fashion. They are orthonormal, as can be seen by evaluating the integral $\int \sin(n\pi x/L)\sin(m\pi x/L) \, dx$ over the length of the box. This integration yields

$$\int \psi_n \psi_m \, dx = \delta_{nm} \tag{2.5.12}$$

In order to study a transition from one stationary state to another, we apply the results of the time-dependent treatment of Section 2.3, which implies that we need to evaluate the transition moment $\langle \psi_n | \mu | \psi_m \rangle$. The electric dipole transition moment operator for a one-dimensional, one-particle system is

$$\mu = ex \tag{2.5.13}$$

where e is the particle's charge, and the **x** its position vector. Equation 2.5.13 follows from the general expression presented for the dipole operator in Section 2.3. The transition moments can be evaluated by integration of $\langle \psi_n | x | \psi_m \rangle$, since

20 RESULTS FROM QUANTUM MECHANICS

e is a constant. This leads to integrals of the form

$$\int \sin(n\pi x/L) x \sin(m\pi X/L) \, dx \qquad (2.5.14)$$

which can be integrated analytically. However, it is somewhat easier to determine whether or not the transition moment is zero by a symmetry argument, as follows.

One can shift the origin of the box such that it lies halfway between the potential energy walls; that is, the dimension of the box is from $-L/2$ to $L/2$. This does not change the shape of the wavefunctions; however, mathematically, they are expressed as cosine rather than sine functions. The wavefunctions can then be described as alternating odd and even functions with respect to the symmetry axis. In Fig. 2.2, for example, the lowest energy wavefunction (trace a) is an even (symmetric) function, whereas the first excited state is an odd function. The product of ψ_1 and ψ_2, for example, is shown in Fig. 2.2 (trace c). The integral of this product function is zero, since equal areas under the curve lie below and above the x-axis. This is, of course, a graphic representation of the orthogonality condition.

When the dipole moment operator is multiplied into the product of two consecutive wavefunctions, the resulting function is always even. This is shown in Fig. 2.2 (trace e). Since $\mu = ex$ is an odd function in this one-dimensional problem (which is the reason for shifting the origin into the middle of the box), the product of two consecutive wavefunctions and the dipole moment operator is always even, resulting in a nonzero transition moment.

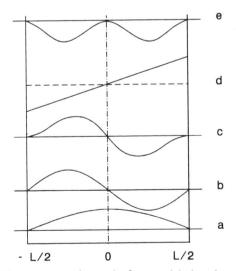

FIGURE 2.2. Transition moment integrals for particle-in-a-box: (a) ground state wavefunction, (b) first excited state wavefunction, (c) product of ground and first excited state wavefunctions, (d) transition dipole operator, and (e) product of parts (c) and (d).

For the particle-in-a-box, this leads to the following selection rules: transitions with $\Delta n = \pm 1, \pm 3, \pm 5$, and so on are allowed, whereas transitions with $\Delta n = \pm 2, \pm 4$, and so on are forbidden because the transition moment integrals are zero. Thus we encounter here for the first time the situation of distinct stationary states, between which transitions may or may not be allowed depending on the symmetry of the wavefunctions. The transition moment is the quantity that needs to be evaluated in order to determine whether or not a transition may occur.

We shall see in the next section that the situation is the same for the case of (harmonic) vibrational wavefunctions, for they have the same symmetry as do the particle-in-a-box wavefunctions, and their overall appearances are similar as well. Thus we see that the particle-in-a-box is a useful, simple model system to demonstrate the workings of the quantum mechanical formalism, to demonstrate the symmetry properties of wavefunctions, and to visualize the principles of orthonormality and the transition moment.

2.6 THE VIBRATIONAL SCHRÖDINGER EQUATION

The potential energy function for a molecular vibration is, of course, a much more complex function than that used in the particle-in-a-box model discussed before, in particular, since the vibration of a molecule is a $3N - 6$ dimensional problem, where N is the number of atoms in the molecule. For the discussion in this section, we restrict ourselves to a one-dimensional problem (a diatomic molecule) as we did in the case of the particle-in-a-box problem, but treat the potential energy in terms of a more sophisticated potential energy function.

Conceptually, we may visualize this function to have an energy minimum at the equilibrium bond distance (or angle) and increase relatively sharply as the atoms are pulled apart or pushed closer together. In the case of large elongations, the bond will break and thus the potential energy between the atoms will approach a constant value. Conversely, if a bond or an angle is compressed sufficiently, van der Waals forces will increase rapidly, and the potential energy curve increases very steeply. This is shown in Fig. 2.3.

This potential function is obtained by detailed quantum mechanical calculations, in which the electronic energy is computed as a function of the internuclear distance. The shape of the actual potential function has been approximated by the Morse potential:

$$V(x) = D_e[1 - e^{-a(x-x_0)}]^2 \qquad (2.6.1)$$

where D_e is the depth of the potential well (the bond dissociation energy), x is the internuclear distance, and x_0 is the equilibrium distance. The constant a is a measure of the curvature of the potential at the bottom and is given by

$$a = (k/2D_e)^{1/2} \qquad (2.6.2)$$

22 RESULTS FROM QUANTUM MECHANICS

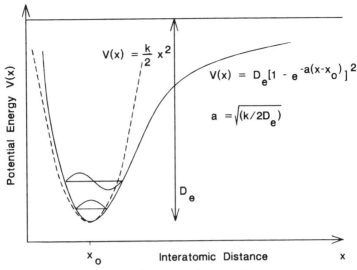

FIGURE 2.3. Approximate potential energy function for a diatomic oscillator. Solid trace, Morse potential; dashed trace, harmonic or quadratic potential.

where k is the force constant of the bond.

Solving the vibrational Schrödinger equations with the Morse expression (Eq. 2.6.1) as the potential energy term would be difficult. Thus one approximates the shape of the potential function in the vicinity of the potential energy minimum by a more simplistic function, which represents the true quantum mechanical energy function, or the Morse approximation. This is accomplished by expanding the potential energy $V(x)$ in a power series about the equilibrium distance:

$$V = v_0 + \left(\frac{dV}{dx}\right)_e dx_i + \frac{1}{2}\left(\frac{d^2V}{dx^2}\right)_e dx^2 + \cdots \qquad (2.6.3)$$

At this point, one argues that V_0 is an offset along the Y-axis and does not affect the curvature of the potential energy. The term containing the first derivative of the potential energy with respect to x is zero sine the equilibrium geometry corresponds to an energy minimum. Terms higher than the quadratic expression in Eq. 2.6.3 are ignored. Thus we approximate the potential energy $V(x)$ for small displacements ($dx \approx x$) along x by

$$V(x) = \frac{1}{2}\left(\frac{d^2V}{dx^2}\right)_e x^2 = \tfrac{1}{2}kx^2 \qquad (2.6.4)$$

where we set a harmonic force constant equal to the second derivative of the potential energy. The force constant k is totally analogous to a Hook's law force constant and has units of force/distance (*vide infra*).

The harmonic potential described by Eq. 2.6.4 is shown in Fig. 2.3 by the dotted curve next to the potential energy described by the Morse function. The parabolic approximation is acceptable for a transition from a ground state to the first excited vibrational transitions, that is, for transitions deep in the potential energy well, particularly for the vibrations of a stiff bond. Many weak bonds, such as hydrogen bonds, show large deviations from quadratic behavior and exhibit anharmonicity. This topic and the potential functions used in empirical normal coordinate calculations will be discussed in more detail in Chapter 3.

The vibrational Schrödinger equation for a two-particle system with one degree of freedom (x) is then

$$\left\{-\frac{\hbar^2}{2m}\frac{d^2}{dx^2} + \tfrac{1}{2}kx^2\right\}\psi = E\psi \tag{2.6.5}$$

The differential equation containing the second derivative of a function, and the square of this function, is known as "Hermite's" differential equation and is much more difficult to solve than the particle-in-a-box differential equation. Detailed procedures for solving this differential equation can be found in most quantum chemistry textbooks, but the mathematical procedures are tedious [Levine (I), 1970, Chapter 4]. Here, we present only the results of this derivation.

The resulting vibrational wavefunctions are of the form

$$\psi_n(x) = Ne^{-\alpha x^2/2} H_n(\sqrt{\alpha}x) \tag{2.6.6}$$

where N is a normalization factor,

$$N = (\alpha/\pi)^{1/4} \tag{2.6.7}$$

and

$$\alpha = 2\pi v m/\hbar \tag{2.6.8}$$

$H_n(\sqrt{\alpha}x)$ are the Hermite polynomials of order n in the variable $(\sqrt{\alpha}x)$, and n is the vibrational quantum number, which takes integer values, including zero. If we set $z = (\sqrt{\alpha}x)$, these polynomials can be written as

$$H_0 = 1 \qquad H_2 = 4z^2 - 2$$
$$H_1 = 2z \qquad H_3 = 8z^3 - 12z$$

and so on. Higher wavefunctions can be derived from a recursion formula:

$$zH_n(z) = nH_{n-1}(z) + \tfrac{1}{2}H_{n+1}(z) \tag{2.6.9}$$

Due to the anharmonic nature of the true potential function, the higher order harmonic wavefunctions are of little importance; however, the recursion formula 2.6.9 will be needed in the next section for the derivation of the harmonic oscillator selection rules.

The vibrational eigenvalues, that is, the eigenvalues of the vibrational Schrödinger equation (Eq. 2.6.5), are

$$E_n = (n + \tfrac{1}{2})h\nu \qquad (2.6.10)$$

These eigenvalues follow from the expansion coefficients of the trial solutions of Eq. 2.6.5, and the derivation of the two-term recursion relation of the Hermite polynomials [Levine (I), 1970, Chapter 4]. In Eq. 2.6.10, the frequency ν is in units of s^{-1}, such that the term $h\nu$ has units of energy. Vibrational spectroscopists, however, prefer to use wavenumber $\tilde{\nu}$ as a measure of photon energy, where the wavenumber is the inverse of the wavelength:

$$\tilde{\nu} = 1/\lambda = \nu/c \qquad (2.6.11)$$

Thus Eq. 2.6.10 is often written as

$$E_n = (n + \tfrac{1}{2})hc\tilde{\nu} \qquad (2.6.12)$$

For the remainder of this book, the expressions $h\nu$, $hc\tilde{\nu}$, and $\hbar\omega$ will be used to denote the energy of photons, where ω is the angular frequency, defined by

$$\omega = 2\pi\nu \qquad (2.6.13)$$

The harmonic oscillator wavefunctions and energy eigenvalues are shown in Fig. 2.4, along with the quadratic function used for the potential energy in the definition of the vibrational Schrödinger equation. Contrary to the particle in a box, where the energy levels increased with n^2, the harmonic oscillator energy levels are spaced equidistantly, but the symmetry properties of the vibrational wavefunctions are the same as those of the particle-in-a-box example. Thus one can see conceptually that the dipole selection rules should be the same for the two cases, which will be discussed in detail in the next section.

At this point, however, we present some calculations and examples in order to transmit a feeling about magnitudes of the numbers involved. For a common chemical moiety, the C—H bond, the bond energy is typically around 100 kcal/mol or 420 kJ/mol. Expressing this quantity in molecular units (by dividing by Avogadro's number), one finds that the bond dissociation energy D_e is

$$D_e(\text{C—H}) \approx 7 \cdot 10^{-19} \, \text{J} = 7 \cdot 10^{-12} \, \text{erg} \qquad (2.6.14)$$

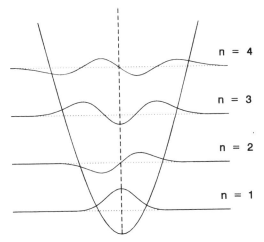

FIGURE 2.4. Vibrational wavefunctions for a harmonic oscillator. Note that the wavefunctions extend outside the potential energy function.

Here, and in most other sections of this book that involve calculations of vibrational parameters, cgs [cm gram sec] units are used, although they are not the most recently adopted IUPAC units. However, vibrational frequencies are measured in cm^{-1} and not m^{-1}, and thus the cgs system is more appropriate. Thus for the conversions from frequency to wavenumber according to Eq. 2.6.11, the value for the velocity of light used is $3 \cdot 10^{10}$ cm/sec. All remaining quantities, such as energies and forces, are expressed in erg and dyn, with

$$10^7 \text{ erg} = 1 \text{ J} \quad \text{and} \quad 10^5 \text{ dyn} = 1 \text{ N} \quad (2.6.15)$$

Also, force constants are usually expressed in cgs units, the proper units being dyn/cm, although mdyn/Å units are found quite frequently in the literature.

The C—H stretching fore constant is about $5 \cdot 10^5$ [dyn/cm] or 5 mdyn/Å. Using Eq. 2.6.2, one finds the constant "a" in the Morse potential to be

$$a \approx 2 \cdot 10^8 \text{ cm}^{-1} \quad (2.6.16)$$

The parameters a and D_e and an equilibrium bond distance of $1 \cdot 10^{-8}$ cm, were used to plot the potential curve in Figure 2.3. Finally, it is instructive to calculate the number of vibrational levels before bond dissociation occurs. The vibrational frequency of a simple stretching motion of a diatomic molecule is given by

$$v = (\tfrac{1}{2}\pi)\sqrt{(k/M_R)} \quad (2.6.17)$$

where M_R is the reduced mass,

$$M_R = m_C m_H/(m_C + m_H) \tag{2.6.18}$$

Using the force constant $k = 5 \cdot 10^5$ [dyn/cm = g/s^2], and the reduced mass of the C—H group as $1.5 \cdot 10^{-24}$ g, one finds the frequency v of a C—H stretching vibration to be 10^{15} Hz. Converting to wavenumbers, using Eq. 2.6.11, we find that $\tilde{v} \approx 3000 \, \text{cm}^{-1}$ for a C—H stretching vibration. Wavenumber is a unit directly proportional to energy; thus we can convert this result to erg, using the following conversion:

$$1 \, \text{cm}^{-1} = 2.0 \cdot 10^{-16} \, \text{erg} \tag{2.6.19}$$

Thus

$$3000 \, \text{cm}^{-1} \approx 6 \cdot 10^{-13} \, \text{erg} \tag{2.6.20}$$

We find that the spacing between (harmonic) vibrational energy levels for a C—H stretching vibration is about $6 \cdot 10^{-13}$ erg, and the depth of the potential well is $7 \cdot 10^{-12}$ erg. Thus one sees that many vibrational quanta of light are necessary to raise the energy of a bond stretching vibration such that bond dissociation occurs. However, bond breakage virtually never happens through consecutive absorption of infrared photons due to numerous factors, among them the short lifetime of vibrational states and the fast (nonradiative) dissipation of the vibrational energy into other vibrational modes. In addition, since the spacing between vibrational levels gets smaller for real molecular systems due to the anharmonicity of the potential function (*vide infra*), photons of different energy would be needed to raise the energy to the higher levels through consecutive absorption processes. On the other hand, the process of raising the vibrational energy from the ground state directly to an excited state near the dissociation limit is low because of the small transition moment. Thus we conclude that bond breakage occurs mostly through mechanisms different from direct vibrational excitation.

2.7 THE VIBRATIONAL TRANSITION MOMENT

The symmetry properties of the vibrational wavefunctions depicted in Fig. 2.4 suggest that a selection rule $\Delta n = \pm 1$ holds for the vibrational wavefunctions introduced in the previous section, similar to the case of particle-in-a-box wavefunctions. However, the mathematical derivation of the vibrational selection rule, $\Delta n = \pm 1$, is more difficult than that of the particle-in-a-box. There, the transition integrals could easily be evaluated analytically; whereas the integration of the vibrational wavefunctions is much more difficult. However, a simplification can be invoked, using the recursion properties of the Hermite polynomials alluded to in Section 2.6. A Hermite polynomial of degree n is related to that of degree $n + 1$ and $n - 1$ by the recursion formula:

$$zH_n(z) = nH_{n-1}(z) + \tfrac{1}{2}H_{n+1}(z) \tag{2.6.9}$$

The wavefunction of a vibrational state n can be written as

$$\psi_n(z) = Ke^{-z^2/2}H_n(z) \tag{2.7.1}$$

where K is a normalization constant. For a one-dimensional problem, we need to evaluate the transition moment $\langle \psi_n|\mu|\psi_m \rangle$. Writing the dipole transition operator as a multiplication along the direction of the variable z, we arrive at an expression for the transition moment given by the integral

$$\int \psi_n(z)\underline{z\psi_m(z)}\,dz \tag{2.7.2}$$

According to the recursion formula above, we can expand the underlined part of the transition moment as

$$z\psi_m(z) = K'e^{-z^2/2}H_{m-1}(z) + \tfrac{1}{2}K''e^{-z^2/2}H_{m+1}(z) \tag{2.7.3}$$

$$\psi_{m-1}(z) \qquad\qquad \psi_{m+1}(z)$$

Thus the integral in Eq. 2.7.2 can be evaluated by left-multiplying Eq. 2.7.3 by $\psi_n(z)$ and integrating the appropriate terms written below Eq. 2.7.3 to give

$$\int \psi_n(z)z\psi_m(z)\,dz \propto \int \psi_n(z)\psi_{m-1}(z)\,dz + \int \psi_n(z)\psi_n(z)\psi_{m+1}(z)\,dz \tag{2.7.4}$$

Since the vibrational wavefunctions are orthonormal, the two integrals on the right-hand side of Eq. 2.7.4 will be nonzero only if the following conditions hold:

$$\int \psi_n(z)\psi_{m-1}(z)\,dz = \delta_{n,m-1} = \begin{cases} 1 & \text{if } n = m - 1 \\ 0 & \text{if } n \neq m - 1 \end{cases} \tag{2.7.5}$$

for the first term, and

$$\int \psi_n(z)\psi_{m+1}(z)\,dz = \delta_{n,m+1} = \begin{cases} 1 & \text{if } n = m + 1 \\ 0 & \text{if } n \neq m + 1 \end{cases} \tag{2.7.6}$$

for the second term. It follows that a change in the vibrational quantum number by one unit, $\Delta n = \pm 1$, leads to a nonzero transition moment and an electrically allowed transition.

2.8 ELECTROMAGNETIC RADIATION

Atomic and molecular systems can be promoted into excited states by a number of pathways. In spectroscopy, the most common methods for excitation are those mediated through direct collisional energy transfer or via absorption of energy from electromagnetic radiation. Transitions via collisions, which do not involve radiation, are very important and common. For example, the thermal equilibration of the rotational and vibrational states of gases at moderate pressures occur through collisions. These radiationless transitions allow states to be excited that cannot be reached by direct absorption or emission of a photon, because of selection rules that prohibit the transition. Thus the collisional excitation or deactivation can lead to the population of states with long lifetimes. The deactivation of these states may then proceed via fluorescence or phosphorescence. This mechanism is also used to create population inversion for laser action.

Since the spectroscopy presented in this book deals nearly exclusively with radiation-induced transitions, the perturbation invoked in Section 2.3, which causes a transition from one state to another, is electromagnetic radiation. Thus we need to introduce the principles and equations required for the description of radiation in this section. The expressions light and electromagnetic radiation are used interchangeably in the following discussion, and the expression "light" does not necessarily imply visible light. In fact, the electromagnetic radiation referred to most frequently below is invisible infrared radiation, which is used in vibrational spectroscopy, and is the long wavelength continuation of visible radiation.

Electromagnetic radiation can exhibit different properties that are best described by two different models: a wave model and a particle model. Which of these properties are exhibited depends, to some extent, on the experiments carried out to determine the nature of light. The interaction of two light beams produces interference patterns that are analogous of those observed in the interference of wave patterns in a shallow trough filled with water. Other experiments, such as light scattering or the photoelectric effect, are best described using a particle model.

Many effects of spectroscopy can adequately be described using the particle model. In a typical absorption experiment, which occurs on-resonance, the energy of a single photon of the electromagnetic radiation must match exactly the energy difference of the stationary state energy eigenvalues of the molecule or atom. Since this book deals mostly with vibrational spectroscopy, the photon energies are on the order of $100-4000 \text{ cm}^{-1}$, corresponding to wavelengths of 0.1 mm to about 2.5 μm.

Thus a beam of monochromatic light can be viewed as a steady flow of photons, or light particles. A simple calculation shows that a photon with a wavelength of 514.5 nm (the intense green line of an Ar^+ laser) has an energy of $3.8 \cdot 10^{-19}$ [J = w·s]. Thus a laser beam of this color with a power of 1 watt carries about $2.5 \cdot 10^{18}$ photons/s. Although the light powers commonly utilized in spectrocopy are much lower, the number of photons in a beam of light still exceeds billions of photons/second. Thus our everyday picture of light ignores the enormous number of photons just as it ignores the enormous number of atoms or molecules that we encounter when we consider matter on a macroscopic, rather than atomic or molecular, scale.

The other description of light, namely, the wave model, is based on a continuous beam of light, composed of many photons. To facilitate description of the light, we may assume that all photons are in phase (coherent) and linearly polarized (as in a laser beam). In such a case, the amplitudes of the electric field of all photons add coherently, and one obtains the picture of a beam of electromagnetic radiation, which can be represented by oscillating electric and magnetic fields, as shown in Fig. 2.5. Here, the instantaneous values of the electric and magnetic fields, **E** and **H**, are given by

$$\mathbf{E} = \mathbf{e}_x E_x^0 \cos(2\pi v t - w\pi z/\lambda) = \mathbf{e}_x E_x^0 \cos(2\pi v t - 2\pi v z/c)$$
$$= \mathbf{e}_x E_x^0 \cos(\omega t - \omega z/c) \qquad (2.8.1)$$
$$\mathbf{H} = \mathbf{e}_y H_y^0 \cos(2\pi v t - 2\pi z/\lambda)$$

where the electric and magnetic fields oscillate in phase, and with the same time and space variations, but in different planes perpendicularly to the propagation direction, z. The perpendicular directions of **E** and **H** are along the unit vectors $\mathbf{e}_x$ and $\mathbf{e}_y$, which are colinear with the x and y axes, respectively. Either of the equations may be written in terms of exponential expressions in complex space (where $e^{ix} = \cos x + i \sin x$):

$$\mathbf{E} = \mathbf{e}_x E_x^0 e^{i(\omega t - \omega z/c)} \qquad (2.8.2)$$

Often, electromagnetic radiation is represented by just one quantity, the

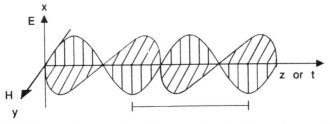

FIGURE 2.5. Time and space dependence of the electric and magnetic fields in electromagnetic radiation. The horizontal bar indicates wavelength or period.

so-called vector potential $\mathbf{A}(t, R)$, which is related to the electric and magnetic fields as follows [Michl and Thulstrup, 1986]:

$$\mathbf{E}(t, R) = -\left(\frac{1}{c}\right)\frac{\partial \mathbf{A}(t, R)}{\partial t} \tag{2.8.3}$$

and

$$\mathbf{H}(t, R) = \nabla \times \mathbf{A}(t, R) \tag{2.8.4}$$

where R is a component of the Cartesian system of axes. Thus we can write all properties of the electromagnetic wave in terms of one equation:

$$\mathbf{A}(t, R) = A^0 \cos(2\pi\nu t - 2\pi z/\lambda) \tag{2.8.5}$$

According to de Broglie's relation, the momentum of a photon is proportional to the inverse of the wavelength. Thus the second term in the parentheses in Eq. 2.8.5 is proportional to the photon's momentum. It is advantageous to define the wave vector $\mathbf{k}$ as

$$\mathbf{k} = (2\pi/\lambda)\mathbf{e}_z \tag{2.8.6}$$

where $\mathbf{e}_z$ is a unit vector in the propagation direction, z. $\mathbf{k}$ has the units of length^{-1} and is proportional to the momentum of a photon. Using the wave vector notation, Eq. 2.8.5 can be written as

$$\mathbf{A}(t, R) = \mathbf{e}_z A^0 \cos(2\pi\nu t - \mathbf{k} \cdot R) \tag{2.8.7}$$

This notation is convenient, since the direction, frequency, and momentum of the light wave are contained in it. In spectroscopy, the momentum of the interacting system must be conserved. Thus if a photon is absorbed by an atom or molecule, the photon's energy and momentum are transferred to the atom or molecule. In some of the higher order Raman effects discussed in Chapter 5, a molecule may interact with up to four photons simultaneously, but it ends up in the same state it started out, namely, the ground state. Thus the momentum vectors of the four photons that interact with the molecule must add up to zero in order to fulfill the conservation of momentum condition. The direct consequence of this condition is that the directions of the photons have to be matched carefully to conserve the momenta and allow interaction of the photons to occur.

Aside from the frequency (or wavelength) of light, its polarization state is of interest. The electromagnetic wave shown in Fig. 2.5, and described in Eqs. 2.8.1 through 2.8.5, represents only one special case, namely, that of plane, or linearly, polarized light. In this case, the electric vector describes a sinusoidal oscillation in one plane, and viewed from a point along the propagation

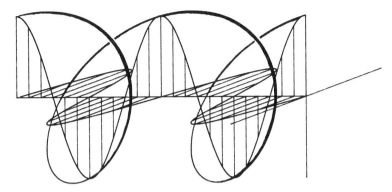

FIGURE 2.6. Superposition of two perpendicular electric fields to produce circularly polarized light. (From M. Diem, in *Magill's Survey of Science: Physical Science*, F. Magill, Ed., Salem Press, Pasadena, CA, 1992, p. 1912.)

direction, the oscillation of the electric vector would appear as a line. In the discussion of the polarization states of light, the magnetic vector is normally not discussed, since it is always perpendicular to, and in phase with, the electric vector.

Light emitted from an unordered ensemble of atoms or molecules will normally be unpolarized. In unpolarized, or randomly polarized, light, there is no preferred direction of the electric vector; that is, all orientations of the direction of the electric vector perpendicular to the propagation are equally probable. Therefore one may view randomly polarized light as a superposition of linearly polarized waves with random orientations of their electric vectors.

Circularly polarized light is another special case. In it, the electric vector traces a helical path about the propagation direction, as shown in Fig. 2.6, completing one turn for each wavelength traveled. Viewed from a point along the propagation direction, the electric vector appears to describe a clockwise or counterclockwise circular motion. Circularly polarized light can be described by a superposition of two linearly polarized rays, which are perpendicular to each other, and have a phase shift of $\pm\pi/2$ or $\pm\lambda/4$, as shown in Fig. 2.6. Spectroscopy with circularly polarized light will be introduced in Chapter 9.

2.9 EINSTEIN COEFFICIENTS OF ABSORPTION AND EMISSION

The time-dependent perturbation treatment presented in Section 2.3 resulted in an expression for the time-dependent expansion coefficients, $c_m(t)$, for the wavefunctions of a system under the influence of electromagnetic radiation. Since the square of the wavefunction is a measure of finding the system in the excited state, the square of the coefficients $c_m(t)$ yields the transition probability from one state to the other, if the transition moment is nonzero (cf. Section 2.4) and if the frequency of the incident light matches the energy difference between ground and excited state. Under these conditions, the values of the time-dependent terms in Eq. 2.3.11 need to be determined. The term for absorption (the second term in the bracket in Eq. 2.3.11) at the resonance condition, $\omega = \omega_{mn}$, becomes, according to l'Hôpital's rule,

$$c(t) \propto \lim_{\omega \to \omega_{mn}} \frac{e^{i(\omega_{mn} - \omega)t} - 1}{(\omega_{mn} - \omega)} = it \qquad (2.9.1)$$

According to Eq. 2.9.1, the transition probability will increase as the square of time, and we should see that the transition becomes more and more intense as the exposure time to the light increases. This is, of course, not in agreement with experimental observations, which demonstrate that the intensity of a transition is time independent.

There are two reasons for the rate of absorption to assume a constant value, after some time, rather than to increase with time. One of them is an argument, discussed in detail by Kauzmann [1957, Chapter 16], that the light impinging on the sample is never sufficiently monochromatic that its entire linewidth fits under the envelope of the time-dependent part of Eq. 2.3.11.

In addition, the first term in the square brackets in Eq. 2.3.11 reaches the same value as Eq. 2.9.1 if the initial state is the excited state and the final state is the ground state. Thus Eq. 2.3.11 includes both the radiation-induced absorption and radiation-induced emission. Since both absorption and emission occur simultaneously in the presence of electromagnetic radiation, the intensity of a transition will not increase indefinitely with time, but will level off at a constant intensity. This is explained in kinetic terms via the Einstein coefficients for stimulated absorption and emission, summarized below.

We define the Einstein coefficient for stimulated absorption or emission from state 0 to state k (or reverse) by

$$B_{0k} = 2\pi |\langle \psi_0 | \mu | \psi_k \rangle|^2 / 3\hbar^2 = B_{k0} \qquad (2.9.2)$$

which is a rate constant for the transition. We see that both absorption of a photon, accompanied by a transition from a lower to a higher state, as well as emission of a photon, accompanied by a transition from a higher to a lower state, are induced by the electromagnetic radiation and are equally probable. Thus in time a steady state rate of absorption and emission is reached, resulting in time-independent transition intensities.

Using Boltzmann's expression for the population N of the excited state,

$$N_k / N_0 = e^{-(E_k - E_0)/kT} \qquad (2.9.3)$$

we find that the rate of absorption of a photon is given by

$$N_0 B_{0k} \rho(\nu_{0k}) \qquad (2.9.4)$$

while the emission of a photon is determined by two parts, the stimulated and the spontaneous emission:

$$N_k(A_{k0} + \rho(v_{0k})B_{k0}) \qquad (2.9.5)$$

here, $\rho(v_{0k})$ denotes the radiation density at frequency v_{0k}. The spontaneous emission does not depend on the presence of a radiation field, as the name implies. The spontaneous emission is responsible for deactivation processes such as phosphorescence, which occur after excitation in the absence of light. The rate of such deactivation processes depends on the "allowedness" of the transition, which is very low in the case of phosphorescence.

To summarize the discussion in this section, the transition probability derived from quantum mechanical principles (Eq. 2.3.11) must be augmented by arguments involving the kinetics of activation and deactivation processes to arrive at expressions that describe the probability of a transition when a system is exposed to electromagnetic radiation for finite lengths of time.

2.10 ROTATIONAL ENERGIES OF RIGID MOLECULES

Although the majority of the material presented in this book deals directly with the principles of vibrational spectroscopy, Section 2.10 is included to provide the necessary theoretical background to explain certain effects that occur when a molecule undergoes vibrational and rotational transitions simultaneously. These effects are observed mostly in the infrared absorption spectra of gaseous molecules, and one finds that the vibrational peaks are much narrower in the gas phase than in liquid or solution phases, but also are accompanied by broad absorption features on both sides of the vibrational absorption peaks (cf. Chapter 7, Fig. 7.2). These additional absorptions may appear as broad features or as a progression of nearly equidistant, sharp lines on both sides of the vibrational peak (see Fig. 7.3), depending on the mass of the molecule and experimental parameters (such as sample pressure and instrument resolution).

The additional absorption features result from transitions between rotational energy sublevels accompanying the vibrational transition. In order to interpret these results properly, one needs to determine expressions for the rotational energy levels of molecules, and the selection rules governing rotational spectroscopy. However, rotational (or microwave) spectroscopy is an entirely different branch of spectroscopy, the detailed discussion of which has appeared in books significantly more elaborate than this present volume. Thus we shall limit ourselves to a discussion of only a few principles of rotational spectroscopy and use this information together with the vibrational discussion to investigate the rotational–vibrational fine structure of absorption bands of gaseous molecules.

To a first approximation, the energies (and Hamiltonians) of molecular rotation and molecular vibration are additive, and the total molecular

34 RESULTS FROM QUANTUM MECHANICS

wavefunctions can be written as products of electronic, vibrational, and rotational wavefunctions, as indicated before:

$$\Psi = \psi_{elec} \cdot \psi_{vib} \cdot \psi_{rot} \tag{2.2.4}$$

Referring to the short discussion of Section 2.2 (cf. Eqs. 2.2.5–2.2.8), the rotational Hamiltonian can be written as

$$H = \frac{L_a^2}{2I_a} + \frac{L_b^2}{2I_b} + \frac{L_c^2}{2I_c} \tag{2.10.1}$$

where the L_α are the components of the angular momentum operators defined in Eq. 2.2.8, and the I_α are the principal moments of inertia, defined by

$$I_\alpha = \sum_{i=1}^{N} m_i r_{\alpha i}^2 \tag{2.10.2}$$

where the sum is over all atoms N, and $r_{\alpha i}$ is the distance of particle i from the axis α. If the axes α are selected arbitrarily, the moment of inertia is obtained as a 3×3 matrix known as the inertial tensor. For one set of axes, known as the *principal axes of inertia*, the inertial tensor contains diagonal terms only, and the classical rotational energy may be expressed in terms of the components I_a, I_b, and I_c of the principal moments of inertia. The subscripts a, b, and c in Eq. 2.10.1 refer to these principal axes of inertia. Conventionally, one defines these axes such that the moment of inertia along a is smaller than or equal to the others:

$$I_a \leq I_b \leq I_c \tag{2.10.3}$$

A problem arises when one attempts to solve Eq. 2.10.1, because the three opeators L_a, L_b, and L_c do not commute with each other:

$$[L_a, L_b] = -i\hbar L_c \tag{2.10.4}$$

and, more importantly, they do not commute with the total Hamiltonian of the system. Operators that do not commute cannot be diagonalized simultaneously in the same set of eigenfunctions. Thus the rotational Hamiltonian cannot be solved in terms of the three moments of inertia if all moments of inertia are different. This case is referred to as the asymmetric top rotor, and approximations for its rotational energy in terms of symmetric top rotors (*vide infra*) are utilized. This is a complicated case and will not be treated here in any detail. The reader is referred to a number of standard texts on rotational spectroscopy [Townes and Schawlow, 1975; Wollrab, 1967]. Here, we shall concentrate on the situation where the molecular symmetry makes two or three of the components of the angular momentum equal. In these cases, the rotational Schrödinger equation can be solved explicitly by expressing the rotational energy as a function of the total angular momentum and one of the components.

2.10.1 Spherical Top Rotors

Let the *total* angular momentum operator, L^2, be defined as

$$L^2 = L_a^2 + L_b^2 + L_c^2 \tag{2.10.5}$$

L^2 commutes with each of the components:

$$[L^2, L_a^2] = 0 \tag{2.10.6}$$

and the total Hamiltonian,

$$[H, L^2] = 0 \tag{2.10.7}$$

The relationships expressed in Eqs. 2.10.6 and 2.10.7 are verified in most thorough texts on quantum chemistry [Levine (I), 1970, Section 5.3]. In a spherical top, all components of the moment of inertia are equal,

$$I_a = I_b = I_c \tag{2.10.8}$$

and the molecule does not have any preferred axes of rotation. Equation 2.10.1 takes the form

$$H = \frac{L^2}{2I_b} \tag{2.10.9}$$

By transforming L^2 into spherical polar coordinates θ and φ, a differential equation in these variables is obtained. Its solutions are the spherical harmonic functions $Y_J^K(\theta, \varphi)$, which are derived from the associated Legendre functions. The eigenvalues of the L^2 operator in the eigenspace of the spherical harmonics is

$$L^2 Y_J^K(\theta, \varphi) = J(J+1)\hbar^2 Y_J^K(\theta, \varphi) \tag{2.10.10}$$

where J and K are integer values for which certain conditions hold (*vide infra*). The differential equation for the spherical top rotor is the same as the angular part of the hydrogen atom Schrödinger equation and is discussed in more detail in most quantum chemistry texts [Levine (I), 1970, Section 5.3].

Since H and L^2 commute, and because of the relationship between the total energy operator H and L^2 expressed in Eq. 2.10.9, we may write

$$H\psi_{rot} = E\psi_{rot} \qquad (2.10.11)$$

and

$$L^2\psi_{rot} = J(J+1)\hbar^2\psi_{rot} \qquad (2.10.12)$$

from which we obtain the eigenvalues for the rotational Hamiltonian as

$$E = BJ(J+1) \qquad (2.10.13)$$

where B, the rotational constant, is given by

$$B = \hbar^2/2I_b \qquad (2.10.14)$$

At this point, it is useful to take a look at the units involved. The angular momenum, $L(=r \times p)$ has units of $[g \cdot cm^2/s]$, and the moments of inertia $[g \cdot cm^2]$. Thus the rotational energy (Eq. 2.10.1) is indeed in units of energy. Using $[erg \cdot s]$ as units for $\hbar$, we find that B in Eq. 2.10.14 is also given in units of energy $[erg]$.

2.10.2 Linear Molecules

Regardless of their symmetry, linear molecules have only one distinct moment of inertia, since we assume that the moment of inertia along the molecule is zero, and the moments about any other set of perpendicular axes are equal. Therefore the total rotational energy for a linear molecule is identical to that of the spherical top (Eq. 2.10.9), and the quantum mechanical treatment yields the same energy eigenvalues given in Eq. 2.10.14.

2.10.3 Symmetric Top Molecules

If, and only if, a molecule possesses symmetry that includes a threefold or higher axis of symmetry (cf. Chapter 4), two components of the moment of inertia are equal and different from the third one. Two cases may arise, referred to as the oblate and the prolate symmetric top. In the prolate top,

$$I_a < I_b = I_c \qquad (2.10.15)$$

and in the oblate top,

$$I_a = I_b < I_c \qquad (2.10.16)$$

The first case physically appears elongated like a football, and the second flattened like a frisbee. In both cases, the total rotational Hamiltonian, given by Eq. 2.10.1, may be simplified, using Eq. 2.10.5, to result in a Hamiltonian that contains L^2 and the distinct component of the angular momentum. For example, if $I_b = I_c$ (prolate top), Eq. 2.10.1 becomes

$$H = \frac{L^2}{2I_b} + L_a^2 \left(\frac{1}{2I_a} - \frac{1}{2I_b} \right) \tag{2.10.17}$$

Otherwise, if $I_a = I_b$ (oblate top)

$$H = \frac{L^2}{2I_b} + L_c^2 \left(\frac{1}{2I_c} - \frac{1}{2I_b} \right) \tag{2.10.18}$$

Since L^2 commutes with any of the individual components L_α^2, we can write common eigenfunctions for the following three operators, which implies that they may be diagonalized simultaneously in the same eigenfunction space:

$$H\psi_{rot} = E\psi_{rot} \tag{2.10.11}$$

$$L^2\psi_{rot} = J(J+1)\hbar^2 \psi_{rot} \tag{2.10.12}$$

and

$$L_c^2 \psi_{rot} = K^2 \hbar^2 \psi_{rot} \tag{2.10.19}$$

Here, $K^2\hbar^2$ are the eigenvalues of the L_α^2 operator, and K can have integer values from $-J, -J+1, \ldots, 0, \ldots, J-1, J$. The rotational energies for symmetric top rotors are then given by

$$E = BJ(J+1) + K^2(C - B)$$

for the case of oblate tops, and

$$E = BJ(J+1) + K^2(A - B) \tag{2.10.20}$$

in the case of prolate tops, where

$$A = \hbar^2/2I_a \quad \text{and} \quad C = \hbar^2/2I_c \tag{2.10.21}$$

an energy level diagram for spherical, linear, and symmetric top rotors is given in Fig. 2.7.

38 RESULTS FROM QUANTUM MECHANICS

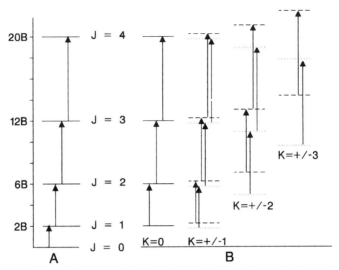

FIGURE 2.7. Rotational energy levels and transitions for (A) spherical and linear molecules and (B) prolate (dashed) and oblate (dotted) symmetric top molecules. Note that for $K = 0$, the symmetric and spherical molecules have the same energy levels.

2.10.4 Rotational Spectra and Selection Rules for Linear Molecules

An analysis of the mathematical form of the spherical harmonic functions and the recursion properties of the associated Legendre functions yields the selection rules for these molecules:

$$\Delta J = 0, \pm 1 \qquad (2.10.22)$$

This derivation is very similar to the one given for the Hermite polynomials in Section 2.7 and is discussed in detail in texts on rotational spectroscopy [Townes and Schawlow, 1975]. For linear and spherical molecules, the rotational energy is given by Eq. 2.10.13 (neglecting higher order effects such as centrifugal distortion), which contains only one variable, the rotational quantum number J. Thus the energies of rotational transitions can be evaluated, using J as the quantum number associated with the *lower* state, and $J' = J + 1$ (cf. Eq. 2.10.22) as the quantum number of the *higher* state, and Eq. 2.10.13. The transition energy is then

$$\begin{aligned}\Delta E &= BJ'(J' + 1) - BJ(J + 1) \\ &= B(J + 1)(J + 2) - BJ(J + 1) \\ &= 2B(J + 1)\end{aligned} \qquad (2.10.23)$$

These transitions are indicated as the vertical arrows in Fig. 2.7. Since the energy levels are not equally spaced, as in the case of vibrational states, the energies of the transition depend on the quantum number of the state, as indicated by Eq. 2.10.23. For example, the transition from $J = 0$ to $J = 1$ will have an energy of $2B$, and the transition from $J = 1$ to $J = 2$ has an energy of $4B$, and so on. Thus the rotational spectrum of a linear or a spherical molecule consists, to a first approximation, of equidistant lines with a spacing of $2B$.

2.10.5 Rotational Energies and Spectra of Symmetric Top Rotors

For symmetric top molecules, the following selection rules hold:

$$\Delta J = 0m \pm 1; \Delta K = 0 \qquad (2.10.24)$$

The rule that ΔK is zero implies that transitions from one to another energy ladder in Fig. 2.7 are forbidden. Since the energy correction due to the K^2 term is independent of the J value, all energies within one K-ladder are raised or lower equally. Thus the transition energies are the same, and the symmetric top selection rules (Eq. 2.10.24) lead to the same rotational spectrum as before in the case of linear and spherical molecules.

Actually, observed rotational spectra often are much more complicated: if centrifugal distortion plays a significant role, one observes each line in the rotational spectrum lowered in energy and split into $(2J + 1)$ subbands. Under these circumstances, the description of the rotational energy must be augmented beyond the rigid rotor approximation to include the centrifugal distortion explicitly. On the other hand, the rotational spectra of asymmetric top rotors are always complicated and do not allow an easy interpretation comparable to the one given so far for more symmetric species.

2.10.6 Rotational–Vibrational Spectra

We now may return to the original goal of this section, namely, the interpretation of the envelopes or of the distinct rotational bands observed in the infrared absorption spectra of gaseous molecules. As long as Eq. 2.2.4 is valid, the energies of molecular vibration and rotation are strictly additive. Thus we can write the energies for a spherical top or linear molecule as

$$E = hc\tilde{v}(v + \tfrac{1}{2}) + BJ(J + 1) \qquad (2.10.25)$$

with a similar equation, involving the term in K^2, holding for the symmetric top molecules. In Eq. 2.10.25, both energies on the right-hand side are given in units of [erg].

With the general selection rules for vibration and rotation,

$$\Delta v = \pm 1; \Delta J = 0, \pm 1 \qquad (2.10.26)$$

transitions into different rotational sublevels of a vibrational level are allowed, which can lead to the addition or subtraction of rotational quanta to or from the vibrational transition. This is shown schematically in Fig. 2.8 for the case of a spherical molecule. In the absence of any of the effects discussed in Section 2.11, one expects to observe equally spaced rotational lines about a center peak, for which the rotational quantum number does not change ($\Delta J = 0$). This peak is known as the Q branch.

To lower energies (or wavenumber) are the transitions for which $\Delta J = -1$; that is, the rotational level of the vibrational ground state is higher than the rotational level of the vibrationally excited state. These transitions are referred to as the P branch. To higher energies are the R-branch transitions, for which $\Delta J = +1$. The spacing between the rotational lines is $2B$, as discussed before. In methane, for example (cf. Chapter 7), which is a light molecule with a small moment of inertia and therefore a large rotational constant B, the spacing is on the order of $10 \, \text{cm}^{-1}$. For heavier molecules, such as chloroform, the spacing between rotational levels is much smaller and is usually not resolved into individual components. Rather, a broad envelope is observed, which is due to the superposition of the intensities of individual rotational–vibrational bands.

There are two factors influencing the intensities of these rotational–vibrational transitions. One of them is, of course, the transition moment, defined before:

$$\mu = \langle \psi_n | \mu | \psi_m \rangle \tag{2.4.1}$$

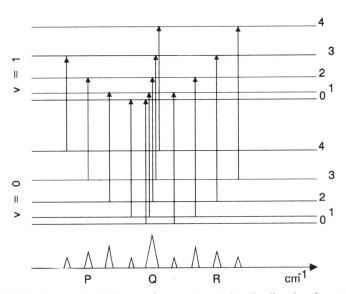

FIGURE 2.8. Energy level diagram for simple rotational–vibrational spectra.

which needs to be evaluated for the rotational wavefunctions of the molecule. These are the spherical harmonics, as discussed above, and the derivations of the particular transition moments between two rotational energy levels are a messy mathematical affair. However, the numerical value of the transition moment for a $J = 1 \to J = 2$ transition is not all that different from that of a $J = 2 \to J = 3$ transition. Thus the intensity pattern observed in the P and R branches is due to the second factor alluded to above, which takes into account the population differences between the rotational energy levels. For a given vibrational level, the ratio of the population of the rotational sublevel with quantum number J to that of the rotational ground state ($J = 0$) is given by

$$\frac{N_J}{N_0} = (2J + 1)e^{-BJ(J+1)/kT} \qquad (2.10.27)$$

where the preexponential factor is due to the $(2J + 1)$-fold degeneracy of each level. The expression on the right-hand side of Eq. 2.10.27 has a maximum for certain values of J, giving rise to the distinct maxima in the P and R branches.

Depending on the symmetry of the molecule, the Q branch may or may not be allowed in the vibrational–rotational spectrum. In heteronuclear diatomic molecules, such as HCl, the Q branch is not allowed. In such a molecule, the change in the dipole moment during the vibration occurs along the direction of largest (and only) moment of inertia; therefore the rotational–vibrational spectrum, devoid of the Q branch, is referred to as a parallel band. Incidentally, homonuclear diatomic molecules have no dipole moment, and neither the vibrational nor rotational transitions may be observed in absorption, although they are allowed in Raman scattering (cf. Chapter 5).

Linear molecules with more than two atoms can have vibrations that change the dipole moment either along the direction of the moment of inertia or perpendicularly to it. The antisymmetric stretching mode, v_3 of CO_2, observed around $2350\,cm^{-1}$ and shown in Fig. 2.9, belongs to the first case scenario and is a parallel band devoid of a Q-branch. On the other hand, the CO_2 bending mode v_2 at around $665\,cm^{-1}$ changes the dipole moment in a direction perpendicular to the largest moment of inertia. The corresponding vibrational–rotational band exhibits a Q branch and is referred to as a perpendicular band (Fig. 2.10).

Similar rules hold for symmetric top rotors as well, where parallel and perpendicular bands are observed, depending on the direction of the dipole change of a vibration moment within the framework of inertial axes. The rotational constant B can be deduced from the analysis of parallel bands, whereas the other rotational constant (C or A for oblate and prolate tops, respectively) can be obtained from the perpendicular bands.

In asymmetric top molecules, the situation is even more cumbersome. Three different band envelopes are obtained and designated A-, B-, and C-type envelopes, depending on whether the change of the dipole moment occurs

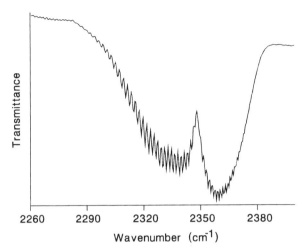

FIGURE 2.9. Example of a parallel rotational–vibrational band. Observed spectrum of the v_3 mode of CO_2 at ambient pressure in air.

along the axis of least, intermediate, or largest moment of inertia. v_3 (the antisymmetric stretching mode) of water at about $3750\,\mathrm{cm}^{-1}$ in the gas phase is a typical A-type transition with distinct P, Q, and R structure. B-type transitions are devoid of a center branch and resemble the parallel bands in symmetric and linear molecules. Detailed analyses of the rotational–vibrational spectra of many small asymmetric top rotors, such as H_2O, H_2S, H_2CO, and many more, have been summarized [Herzberg, 1945, Chapter IV]. In asymmetric top rotors with no symmetry at all, the observed vibrational–rotational spectra are all mixtures of A-, B-, and C-type envelopes. An analysis

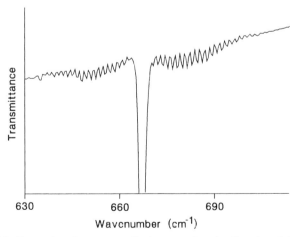

FIGURE 2.10. Example of a perpendicular rotational–vibrational band. Observed spectrum of the v_2 mode of CO_2 at ambient pressure in air.

of these envelopes can provide information that may aid in the assignment and interpretation of vibrational spectra [Diem et al., 1978].

2.11 ANHARMONICITY AND VIBRATIONAL–ROTATIONAL INTERACTION

Experimentally, one finds deviations from the models developed so far for the interpretation of molecular vibrational and rotational–vibrational spectra. In the case of vibrational spectra, one finds that not one but a propagation of vibrational transitions is observed for each vibrational coordinate. Thus, in the case of a diatomic molecule, such as HCl, absorption are observed at 2886, 5668, 8347, 10,923,... cm^{-1}, which are assigned the $v = 0 \to v = 1$, $v = 0 \to v = 2$, $v = 0 \to v = 3$, and the $v = 0 \to v = 4$ transitions. Within the harmonic oscillator approximation, these bands are forbidden, but they become partially allowed because of the nonquadratic potential function in real systems (cf. Fig. 2.3). Note that the wavenumber of the overtone is not twice that of the fundamental, but a slightly reduced value. For diatomic molecules, the progression of frequencies can be fit by an expression of the form [Levine (II), Chapter 3.2]

$$E_{\text{vib}} = (v + \tfrac{1}{2})hc\tilde{v} - (v + \tfrac{1}{2})^2 h\chi\tilde{v} \qquad (2.11.1)$$

where $\chi\tilde{v}$ is the anharmonicity contribution. The anharmonicity will almost always have the effect of lowering the vibrational energy level. Typical values for the $\chi\tilde{v}$ term range from about 2–3% of $\tilde{v}$ for single bond stretching vibrations to about 0.5% for triple bonds.

Another deviation of experimental data from the simple model proposed so far is the nonequal spacing of the rotational sublevels shown in Fig. 2.8. In fact, one observes that the spacing of the rotational lines in the P branch increases for lines further away from the Q branch, whereas the spacing of the lines in the R branch decreases. This effect is due to the vibrational–rotational interaction, which is included in the total energy expression as follows:

$$E = (v + \tfrac{1}{2})hc\tilde{v} - (v + \tfrac{1}{2})^2 h\chi\tilde{v} - (v + \tfrac{1}{2})J(J+1)h\alpha_e \qquad (2.11.2)$$

where α_e is called the vibrational–rotational coupling constant. The third term on the right-hand side of Eq. 2.11.2 implies that Eq. 2.2.4 is no longer valid; that is, that interaction between the rotation and vibration does occur. Therefore the approach of writing the total molecular energy as a sum of rotational and vibrational energies is only an approximation.

Conceptually, vibrational–rotational interaction arises because the moment of inertia is different in the vibrational excited state than in the ground state. This, in turn, is due to the fact that the anharmonicity of the vibrational potential function causes the bond length to increase in the excited vibrational state, which increases the moment of inertia and decreases the effective rotational constant.

The effect of this rotational–vibrational coupling constant is quite pronounced and can easily be observed, as demonstrated by the high-resolution spectra of methane (see Fig. 7.3). The coupling constant α_e lowers the energies of the rotational levels with increasing J and lowers the splitting of the bands in the R-branch positions by a term $-\alpha_e(J + 1)^2$. In the P branch, where the rotational quantum number decreases by one unit, the splitting between rotational lines increases by an expression proportional to $\alpha_e J^2$.

The final correction to Eq. 2.11.2 would be the inclusion of the centrifugal distortion, which takes into account the increase in bond length due to centrifugal force at high rotational quantum numbers. As discussed previously, this effect perturbs the pure rotational spectra via a term $DJ^2(J + 1)^2$.

REFERENCES

(In the following list, reference is made to a two-volume edition of I. N. Levine, *Quantum Chemistry*, Allyn & Bacon, Boston, 1970. These very comprehensive volumes have served the author as the main references in the discussion of quantum mechanical and spectroscopic principles. They are referred to in the text as Levine (I) and Levine (II), respectively.)

M. Diem, L. A. Nafie, and D. F. Burow, *J. Mol. Spectrosc.*, *71*, 446 (1978).

W. Kauzmann, *Quantum Chemistry: An Introduction*, Academic Press, New York, 1957.

I. N. Levine, *Quantum Chemistry, Volume I: Quantum Mechanics and Molecular Structure*, Allyn & Bacon, Boston, 1970.

I. N. Levine, *Quantum Chemistry, Volume II: Molecular Spectroscopy*, Allyn & Bacon, Boston, 1970.

J. Michl and E. W. Thulstrup, *Spectroscopy with Polarized Light: Solute Alignment by Photoselection in Liquid Crystals, Polymers and Membranes*, VCH Publishers, New York, 1986.

C. H. Townes and A. L. Schawlow, *Microwave Spectroscopy*, Dover Publications, New York, 1975.

J. E. Wollrab, *Rotational Spectra and Molecular Structure*, Academic Press, New York, 1967.

3

POLYATOMIC MOLECULES

Up to now, the theoretical principles of molecular vibrations and rotations have been introduced for model systems as simple as possible. In order to keep the discussion in the previous chapter clear and simple, most problems were reduced to one-dimensional examples. Consequently, many vibrational problems, among them the potential function and selection rules, were discussed for diatomic molecules. The next step toward understanding vibrational spectroscopy is the treatment of polyatomic species. First, the classical treatment of polyatomic systems, modeled by a system of point masses connected by springs, will be presented. This simple model introduces the concepts of *normal modes of vibration*, and we shall see that a completely analogous picture exists for molecules.

Next, a connection between the classical normal mode picture and the quantum mechanical models developed in Sections 2.5 and 2.6 will be worked out, and molecular vibrations of a polyatomic molecule will be discussed in terms of a simple, quantum mechanical description.

Finally, the methods for the computations of normal modes of vibration for molecules will be introduced, using the concept of transferable force constants. This subject will be accompanied by detailed examples and intermediate results, such that the reader may become familiar with the methodology and the terminology of these calculations.

3.1 THE SEPARATION OF TRANSLATIONAL AND ROTATIONAL COORDINATES

Conceptionally, one may expect that a system of N atoms has $3N$ degrees of vibrational freedom: a motion for each atom along three Cartesian coordinates, or linear combinations of them. However, vibrational spectroscopy depends, as pointed out in Section 2.1, on a restoring force to bring the atoms of a molecule back to their equilibrium position. If, for example, all atoms in a molecule move simultaneously in the x-direction, by the same amount, no bonds are being compressed or elongated. Thus this motion is not that of an internal vibrational coordinate, but that of a translation. There are three translational degrees of freedom, corresponding to a motion of all atoms along the x, y, or z axes, respectively. Although translation is quantized as well, the energy levels are far from those encountered in vibrational spectroscopy, and the three translational modes of freedom are ignored in vibrational spectroscopy. Another view of the same fact is that these modes have zero frequencies since there is no restoring force acting during the atomic displacements. This view is particularly useful, since the normal mode analyses discussed later in this chapter achieve the separation of translation and vibration exactly by means of computing zero-valued roots for the translational modes.

Similarly, one can argue that certain combinations of Cartesian displacements correspond to a rotation of the entire molecule, where there is no change in the intermolecular potential of the atoms. There are three degrees of rotational freedom of a molecule, corresponding to rotations about the three axes of inertia. Subtracting these from the remaining number of degrees of freedom, one arrives at $3N - 6$ degrees of vibrational freedom for a polyatomic, nonlinear molecule. Linear molecules have one more ($3N - 5$) degree of vibrational freedom, since they have only two moments of inertia. This is because one assumes a zero moment of inertia for a rotation about the longitudinal axis.

Mathematically, the separation of rotation and translation from the vibration of a molecule proceeds as follows: in order for the translational energy of the molecule to be zero at all times, one defines a coordinate system that translates with the molecule. In this coordinate system, the translational energy is zero by definition. The translating coordinate system is defined such that the center of mass of the molecule is at the origin of the coordinate system at all times. This leads to the condition

$$\sum_{\alpha=1}^{N} m_\alpha \xi_\alpha = 0 \qquad (3.1.1)$$

where the ξ_α are the x, y, and z coordinates of the αth atom of a molecule with N atoms. Similarly, the rotational energy can be reduced to zero by defining a coordinate system that rotates with the molecule. This requires that the angular momenta of the molecule in the rotating coordinate frame are zero, which leads to three more equations. These six equations are needed to define a coordinate system in which both translational and rotational energies are zero. Details of this derivation can be found in Wilson et al. [1955, Chapters 2 and 11].

3.2 CLASSICAL VIBRATIONS IN MASS-WEIGHTED CARTESIAN COORDINATES

To gain insight into the complex problem of molecular vibrations, one approaches the problem by a classical analogue, where the molecule is approximated by a system of masses, held in their equilibrium positions by springs obeying Hook's law to a first approximation. This system describes a number of important qualities of a vibrating molecule, such as the concept of *normal modes* of vibration, and presents the analogous mathematical formalism to solve the problem for molecular vibrations. This kind of discussion is treated in complete detail in the classic books on vibrational spectroscopy, for example, in Wilson et al. [1955, Chapter 2].

The treatment starts with Lagrange's equation of motion:

$$\frac{d}{dt}\frac{\partial T}{\partial \dot{x}_i} + \frac{\partial V}{\partial x_i} = 0 \quad (3.2.1)$$

where T and V are, as usual, the kinetic and potential energies, the x_i are any set of Cartesian displacement coordinates, and the dot denotes the derivative with respect to time. This equation is another statement of Newton's equation of motion:

$$F_i = m_i \frac{d^2}{dt^2} x_i \quad (3.2.2)$$

in terms of the kinetic and potential energies. In Eq. 3.2.2, F represents the force, which is related to the potential energy by

$$F_i = -\frac{dV}{dx_i} \quad (3.2.3)$$

The acceleration $d^2 x_i/dt^2$ can be related to the kinetic energy as follows:

$$T = \frac{1}{2}\left(\sum_i m_i \dot{x}_i^2\right) \quad (3.2.4)$$

48 POLYATOMIC MOLECULES

Thus

$$\frac{dT}{d\dot{x}_i} = m_i \dot{x}_i \tag{3.2.5}$$

Substituting Eqs. 3.2.3 and 3.2.5 into Eq. 3.2.2 yields Lagrange's equation of motion. The reason for using this equation follows from quantum mechanical considerations discussed earlier, where it is desirable to write the Hamiltonian in terms of separate expressions for kinetic and potential energies.

Next, one defines the problem in terms of mass-weighted coordinates, q_i. The reason for this is that the amplitude of a particle's oscillation depends on its mass. When mass-weighted coordinates are used, all amplitudes are properly adjusted for the different masses of the particles. In addition, the use of mass-weighted coordinates simplifies the formalism quite a bit. Let

$$q_i = \sqrt{m_i} x_i \tag{3.2.6}$$

Then the kinetic energy can be written as

$$2T = \sum_{i=1}^{3N} (\dot{q}_i)^2 \tag{3.2.7}$$

Note that Eq. 3.2.7 has only diagonal terms; that is, only terms in q with the same subscript i appear in the summation.

Next, the potential energy of the particles needs to be defined. For an atom in a molecule, the potential energy along each Cartesian coordinate is given by an expression such as the one given by Eq. 2.6.1:

$$V(x) = D_e [1 - e^{-a(x-x_0)}]^2 \tag{2.6.1}$$

When one encounters a complex expression such as Eq. 2.6.1, one attempts to simplify it by any means. For the purpose of vibrational spectroscopy, one argues that the molecule oscillates with a small amplitude of vibration about the equilibrium position, and that in the neighborhood of the equilibrium position the actual potential function can be expanded in a Taylor series:

$$V = V_0 + \sum_{i=1}^{3N} \left(\frac{\partial V}{\partial q_i}\right)_e dq_i + \frac{1}{2} \sum_{i=1}^{3N} \sum_{j=1}^{3N} \left(\frac{\partial^2 V}{\partial q_i \partial q_j}\right)_e dq_i \, dq_j + \cdots \tag{3.2.8}$$

This expansion is the multidimensional equivalent of Eq. 2.6.3 discussed before. Quite typical for physical chemists, we use only the second (quadratic) term of this expansion, arguing that the term containing the first derivative is zero, since the derivative is taken at $q = q_e$; and the first term, V_0, does not influence the curvature and is a constant. The cubic and higher expansion terms are ignored within the harmonic approximation.

Thus the potential energy is expressed as

$$2V = \sum_{i=1}^{3N} \sum_{j=1}^{3N} \left(\frac{\partial^2 V}{\partial q_i \, \partial q_j}\right)_e dq_i \, dq_j \tag{3.2.9}$$

$$= \sum_{i=1}^{3N} \sum_{j=1}^{3N} f_{ij} \, dq_i \, dq_j \tag{3.2.10}$$

with

$$f_{ij} = \left(\frac{\partial^2 V}{\partial q_i \, \partial q_j}\right)_e \tag{3.2.11}$$

The f_{ij} are known as the mass-weighted Cartesian force constants and express the change in potential energy as an atom or group is moved along the directions given by q_i and q_j. For small displacements about the equilibrium position, Eq. 3.2.10 is often written as

$$2V = \sum_{i=1}^{3N} \sum_{j=1}^{3N} f_{ij} q_i q_j \tag{3.2.12}$$

Equation 3.2.12 is Hook's law (Eq. 2.6.4) written for N particles. Taking the required derivatives and substituting the expressions for

$$\frac{d}{dt}\frac{\partial T}{\partial \dot{q}_i} \quad \text{and} \quad \frac{\partial V}{\partial q_i}$$

into Lagrange's equation of motion (Eq. 3.2.1) yield

$$\ddot{q}_i + \sum_{j=1}^{3N} f_{ij} q_j = 0 \tag{3.2.13}$$

Here, $\ddot{q}$ denotes the second derivative of q with respect to time. Equation 3.2.13 is a short form for $3N$ simultaneous differential equations, with the index i running from 1 to $3N$. Note that the double summation in Eq. 3.2.12 disappears when the derivative with respect to one of the displacements, $\partial V/\partial q_i$, is taken.

50　POLYATOMIC MOLECULES

Equation 3.2.13 is a system of $3N$ simultaneous linear differential equations:

$$\frac{d^2q_1}{dt^2} + f_{11}q_1 + f_{12}q_2 + \cdots + f_{1,3N}q_{3N} = 0$$

$$\frac{d^2q_2}{dt^2} + f_{21}q_1 + f_{22}q_2 + \cdots + f_{2,3N}q_{3N} = 0$$

$$\vdots$$

$$\frac{d^2q_{3N}}{dt^2} + f_{3N,1}q_1 + f_{3N,2}q_1 + \cdots + f_{3N,3N}q_{3N} = 0$$

In each equation, only one term in the summation $\Sigma f_{ij}q_j$ has the same index as the term containing the derivative. Thus these equations can be simplified to

$$\ddot{q}_i + f_{ii}q_i + C = 0 \tag{3.2.14}$$

where C is a constant. The solutions of these simultaneous, linear differential equations are

$$q_i = A_i \sin(\sqrt{\lambda}t + \varepsilon) \tag{3.2.15}$$

where the A_i are amplitude factors, the ε are phase angles, and λ is a frequency factor that is determined by the force constants. Following standard practice in solving linear differential equations, one takes the solution given by Eq. 3.2.15, differentiates twice with respect to time,

$$\frac{d^2q_i}{dt^2} = -\lambda q_i \tag{3.2.16}$$

and substitutes both Eqs. 3.2.15 and 3.2.16 back into Eq. 3.2.13:

$$-A_1 \lambda \sin(\sqrt{\lambda}t + \varepsilon) + f_{11}A_1 \sin(\sqrt{\lambda}t + \varepsilon) + f_{12}A_2 \sin(\sqrt{\lambda}t + \varepsilon) + \cdots = 0$$

for each of the $3N$ equations. Canceling the constant terms $\sin(\sqrt{\lambda}t + \varepsilon)$ in each equation, one obtains

$$\begin{aligned}(-A_1\lambda + f_{11}A_1) + & f_{12}A_2 & + & f_{13}A_3 & + \cdots = 0 \\ f_{21}A_1 & + (-A_2\lambda + f_{22}A_2) + & f_{23}A_3 & + \cdots = 0 \\ f_{21}A_1 & + & f_{32}A_2 & + (-A_3\lambda + f_{33}A_3) + \cdots = 0\end{aligned} \tag{3.2.17}$$

and so on, or

$$\begin{aligned}(f_{11} - \lambda)A_1 + & f_{12}A_2 & + & f_{13}A_3 & + \cdots = 0 \\ f_{21}A_1 & + (f_{22} - \lambda)A_2 + & f_{23}A_3 & + \cdots - 0 \\ f_{31}A_1 & + & f_{32}A_2 & + (f_{33} - \lambda)A_3 + \cdots = 0\end{aligned} \tag{3.2.18}$$

which yields the following system of homogeneous linear equations in the unknown amplitude coefficients A_i:

$$\sum_{i=1}^{3N} (f_{ij} - \delta_{ij}\lambda)A_i = 0 \qquad (3.2.19)$$

Thus we proceeded from a system of $3N$ simultaneous linear differential equations,

$$\ddot{q}_i + f_{ji}q_i + C = 0 \qquad (3.2.14)$$

to $3N$ simultaneous homogeneous linear equations given in Eq. 3.2.19. Homogeneous equations in x can always be written as

$$a_{11}x_1 + a_{12}x_2 + a_{13}x_3 + \cdots = 0$$
$$a_{21}x_1 + a_{22}x_2 + a_{23}x_3 + \cdots = 0$$

in contrast to inhomogeneous equations, which are of the form

$$a_{11}x_1 + a_{12}x_2 + a_{13}x_3 + \cdots = b_1$$
$$a_{21}x_1 + a_{22}x_2 + a_{23}x_3 + \cdots = b_2$$

Homogeneous equations have two kinds of solutions. One of these is the so-called trivial solution in which all coefficient a_i are zero. This condition indeed fulfills Eq. 3.2.19, but it is of no interest here, since it implies that all particles are at rest; that is, there is no vibrational motion at all.

The other solution for Eq. 3.2.19 is obtained when the determinant of the coefficients of A is zero in order for the left-hand side to be zero:

$$|f_{ij} - \delta_{ij}\lambda| = 0 \qquad (3.2.20)$$

This is called the nontrivial solution. Thus we require that

$$\begin{vmatrix} f_{11} - \lambda & f_{12} & f_{13} & f_{14} \cdots f_{1,3N} \\ f_{21} & f_{22} - \lambda & f_{23} & f_{24} \cdots f_{2,3N} \\ f_{31} & f_{32} & f_{33} - \lambda & f_{34} \cdots f_{3,3N} \\ & & \vdots & \\ f_{3N,1} & f_{3N,2} & f_{3N,3} & f_{3N,4} \cdots f_{3N,3N} - \lambda \end{vmatrix} = 0 \qquad (3.2.21)$$

This is known as the vibrational secular equation. The solution of this equation gives us the eigenvalues λ_i unambiguously, which are related to the vibrational frequencies of the system. The amplitude factors, A_i in Eq. 3.2.19, however, are

52 POLYATOMIC MOLECULES

not determined unambiguously, but they depend on the choice of a given value λ_i. This is because a set of homogeneous equations (i.e., a set of equations where the right-hand side is zero) does not have unique solutions, whereas a set of inhomogeneous equations has unique solutions and can be solved via Cramer's rules or other methods.

This is a problem inherent in all vibrational systems: the frequency of vibration and the amplitude of vibration are unrelated. This is true for classical systems, such as a pendulum or a mass m suspended from a solid bar by a spring with force constant k. For the latter example, the oscillatory frequency was introduced earlier to be

$$v = (1/2\pi)\sqrt{k/M_R} \qquad (2.6.16)$$

and is independent of the amplitude. Thus the frequency is the same for a very small and a very large oscillation, but the velocities of the moving mass are, of course, different.

Since the amplitudes are unknown, it seems at first impossible to obtain a description of the vibration in terms of the displacements of the atoms. However, in order to obtain a physical picture of the vibrational motions in terms of relative atomic displacements, one proceeds as follows. One selects one value λ_k, to fulfill Eq. 3.2.19. Let the amplitude factors be A_{ik} for this particular eigenvalue:

$$\sum_{i=1}^{3N} (f_{ij} - \delta_{ij}\lambda)A_{ik} = 0 \qquad (3.2.22)$$

Then, it is possible to normalize all other sets A_{il} (where l is another eigenvalue) with respect to A_{ik}, and one obtains relative displacement vectors for each given solution λ_i.

We define a *normal mode of vibration* to be one of $3N$ solutions of Eq. 3.2.22 where all atoms oscillate with the same frequency

$$v_k = \sqrt{\lambda_k}/2\pi \qquad (3.2.23)$$

and in-phase but with different amplitudes. This definition is one of the most important ones in vibrational spectroscopy. It implies that in a normal mode of vibration all atoms are in motion, which is required to maintain the center of mass of the molecule. If $3N$ mass-weighted Cartesian coordinates are defined, six rotational and translational modes will appear in these calculations as eigenvalues with zero frequencies. The displacement vectors will confirm that these motions are, indeed, translations and rotations.

At this point, it is convenient to introduce the concept of the *normal coordinate*. Obviously, the frequencies corresponding to the normal modes of vibration are an observable of the system. Thus it is advantageous to associate one degree of vibrational freedom, or a normal coordinate Q, with each of the $3N - 6$ normal modes of vibration.

The normal coordinates Q are related to the mass-weighted Cartesian coordinates by some transformation $\mathbf{L}$ (in matrix notation)

$$\mathbf{Q} = \mathbf{Lq} \tag{3.2.24}$$

which will be discussed below. In normal coordinate space, the kinetic and potential energies can be written as

$$2T = \sum_{k=1}^{3N-6} (\dot{Q}_k)^2 \quad \text{and} \quad 2V = \sum_{k=1}^{3N-6} \Lambda_k Q_k^2 \tag{3.2.25}$$

whereas in mass-weighted Cartesian coordinates, these expressions are

$$2T = \sum_{i=1}^{3N} (\dot{q}_i)^2 \quad \text{and} \quad 2V = \sum_{i,j=1}^{3N} f_{ij} q_i q_j \tag{3.2.26}$$

The expressions for the kinetic energy is the same in normal coordinate and mass-weighted Cartesian coordinate spaces and is given in diagonal form; that is, the expressions for T contain the squares of coordinates and no cross terms with unequal indices.

The potential energy in normal coordinates is diagonal as well, and the coefficients Λ_k are the energy eigenvalues. In q space, the potential energy is not diagonal but is the product of the force constant matrix times two Cartesian displacement vectors. Thus the computation of normal modes of a molecule involves the diagonalization of the force constant matrix: the eigenvalues of the diagonalized force constant matrix are the vibrational frequencies, and the eigenvector matrix $\mathbf{L}$ that diagonalizes the force constant matrix transforms from q space to Q space (Eq. 3.2.24). This will be elaborated on in more detail in Section 3.7.

3.3 QUANTUM MECHANICAL TREATMENT OF THE VIBRATIONS OF POLYATOMIC MOLECULES

In this section, the transition from the results of the quantum mechanical treatment of diatomic (cf. Sections 2.5–2.7) to polyatomic molecules will be made. We assume that the total wavefunction of a molecule can be separated into electronic, vibrational, rotational, and spin wavefunctions as follows:

$$\psi_{\text{total}} = \psi_{\text{elec}} \cdot \psi_{\text{vib}} \cdot \psi_{\text{rot}} \cdot \cdots \tag{3.3.1}$$

and that the corresponding energies are additive:

$$E_{\text{total}} = E_{\text{elec}} + E_{\text{vib}} + E_{\text{rot}} + \cdots \tag{3.3.2}$$

54 POLYATOMIC MOLECULES

Furthermore, one assumes that the total vibrational wavefunction can be separated into wavefunctions associated with one and only one normal coordinate Q according to

$$\psi_{\text{vib}} = \psi_1(Q_1) \cdot \psi_2(Q_2) \cdot \psi_2(Q_2) \cdots \qquad (3.3.3)$$

This separation of the vibrational wavefunction into products of wavefunctions associated with one and only one normal coordinate succeeds since the expressions for T and V are both diagonal (cf. Eq. 3.2.25) when expressed in terms of normal coordinates.

Thus the total vibrational Schrödinger equation is

$$-\frac{\hbar}{2} \sum_{k=1}^{3N-6} \frac{d^2 \psi_k}{dQ_k^2} + \frac{1}{2} \sum_{k=1}^{3N-6} \Lambda_k Q_k^2 \psi_k = E_{\text{vib}} \psi_k \qquad (3.3.4)$$

Substituting Eq. 3.3.3 into 3.3.4, one obtains $3N - 6$ separate Schrödinger equations, one for each normal coordinate Q_k:

$$-\frac{\hbar}{2} \frac{d^2 \psi_k}{dQ_k^2} \rightarrow \frac{1}{2} \Lambda_k Q_k^2 \psi_k = E_k \psi_k \qquad (3.3.5)$$

The solutions of each of the $3N - 6$ equations described by Eq. 3.3.5 yield a vibrational wavefunction similar to the one given by Eq. 2.6.6, where the variable x needs to be replaced by Q. The total vibrational energy of the molecules is given simply by the sum of the vibrational energy along each coordinate:

$$E_{\text{vib}} = \sum_{k=1}^{3N-6} (v_k + \tfrac{1}{2}) hc\tilde{v}_k \qquad (3.3.6)$$

Here, v_k and v_k are the quantum number and vibrational frequency associated with the normal coordinate k.

The selection rules ($\Delta v = 1$) discussed in Section 2.6 holed for displacements in normal coordinate space as well. Radiative (i.e., light mediated) transitions will occur within one normal coordinate only. This is shown by the vertical arrows in Fig. 3.1, which depicts the energy levels of a hypothetical molecule with three degrees of vibrational freedom, marked Q_1, Q_2, and Q_3. The relative energy levels shown are approximately those found in water. Note that for each coordinate Q_i, the vibrational ground state energy is $\tfrac{1}{2} h\tilde{v}$, and that within the harmonic approximation used so far in Eq. 3.3.5, the energy levels are equally spaced. The only allowed transitions are the vertical arrows, and all the dashed lines represent forbidden transitions. Finally, any transitions depicted in Fig. 3.1 that do not result from the vibrational ground state are "hot bands." As

TREATMENT OF THE VIBRATIONS OF POLYATOMIC MOLECULES 55

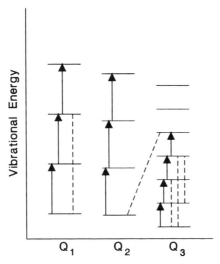

FIGURE 3.1. Vibrational energy for a hypothetical molecule with three degrees of vibrational freedom. The arrows represent allowed vibrational transitions, whereas dashed lines represent transitions that are not allowed within the harmonic oscillator approximation.

the name implies, the likelihood of transitions from an excited state into an even more excited state increases if the sample is at elevated temperature, which causes an increased thermal population of excited states. Since anharmonicity (cf. Section 2.11) lowers the energy spacing between the levels, the "hot bands" are observed shifted to lower wavenumber.

An excited vibrational coordinate can dissipate its energy nonradiatively from one to another normal mode or by collisional deactivation; however, radiative transitions will occur within one coordinate within the approximation outlined in Eq. 3.3.1.

From the discussion so far it appears that molecules vibrate only and exactly along the normal coordinates defined and discussed above. This is, of course, not so, since the thermal vibrations of the atoms are random motions, which are usually given as the thermal ellipsoids, or a contour within which one finds an atom 90% of the time. However, this random motion can be decomposed into linear combinations of the normal modes of vibration; furthermore, an absorption of a photon that excites the molecules along a given normal coordinate will increase the random motion in the direction of this coordinate.

The separation of the total vibrational wavefunction into the product of wavefunctions associated with only one normal coordinate can break down under certain circumstances. This leads to phenomena such as Fermi resonance and other mixing phenomena, such as vibrational exciton interactions. These will be discussed in the next section.

There are other deviations from the model presented so far, the most serious one being anharmonicity. This effect results from the fact that the potential function is not quadratic but is shaped similar to the Morse potential shown in Fig. 2.3. As discussed in Chapter 2, the deviation from a harmonic potential energy function is accounted for by including cubic terms in the expansion of the potential (Eq. 3.2.8) and treating these additional terms via perturbation theory. One obtains corrections to the vibrational energies as well as somewhat relaxed selection rules: the fundamental transitions with $\Delta v_k = \pm 1$ are still most predominant, but weaker overtones occur with $\Delta v_k = \pm 2, \pm 3$, and so on. The frequency of the $\Delta v_k = \pm 2$ transition does not occur at twice the frequency of the fundamental transition, but at a lower frequency. The energy levels, taking into account anharmonicity, are given for a polyatomic molecule with $3N - 6$ vibrational degrees of freedom by

$$E_{\text{vib}} = \sum_{i=1}^{3N-6} (v_i + \tfrac{1}{2})hc\tilde{v}_i - \sum_{i=1}^{3N-6}\sum_{j \geq i}^{3N-6} (v_i + \tfrac{1}{2})(v_j + \tfrac{1}{2})h\chi_{ij} \qquad (3.3.7)$$

Here, χ is the anharmonicity constant, and the effect of anharmonicity is always to lower the energy levels. This is shown in Fig. 3.2A, which depicts the potential energy function for a simple bond stretching or angle deformation coordinates in a polyatomic molecule. For the ground to first excited state transition, the anharmonic correction is usually relatively small but becomes more predominant for transitions originating from more highly excited states,

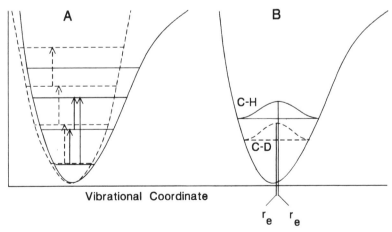

FIGURE 3.2. (A) Anharmonic vibrational potential energy and energy levels. Solid vertical arrows represent transitions allowed for an anharmonic oscillator. Dashed arrows represent the transitions allowed within the harmonic oscillator approximation. (B) Effect of the anharmonic potential energy function on the equilibrium bond length r_0 of isotopic species.

which are more populated at increased temperatures and give rise to the so-called "hot bands" (*vide supra*). The anharmonicity also makes the overtone, shown by the long vertical arrow in Fig. 3.2A, weakly allowed, whereas in the harmonic approximation, only the dashed arrows represent allowed transitions.

The anharmonic shape of the potential well shown in Fig. 3.2B has one further interesting consequence. The ground state vibrational wavefunctions of isotopic species in general, and therefore of the C—H and C—D stretching coordinates, occur in the same potential energy well, because the potential energy is assumed to be independent of the atomic masses. The ground state ($v = 0$) vibrational energy for these two vibrations differs by about $500 \, cm^{-1}$, with the C—D stretching vibration lower in the potential well than the C—H stretching vibration. Due to the asymmetry of the potential energy function, the maximum of the ground state wavefunction occurs at lower equilibrium distance for C—D than for C—H, and consequently, the average C—D bond is shorter than the C—H bond. Of course, it also requires more energy to break a C—D bond because it is lower in the potential energy well. Both these facts are responsible for the kinetic isotope effect.

There are some vibrational motions for which the potential energy functions are quite different from the one discussed so far. There are, for example, potential energy functions with two minima, as shown in Fig. 3.3. Typical molecular vibrations possessing such potential energies are low frequency ring puckering vibrations in molecules such as cyclopentene, for which there are two identical and symmetry-related minimum energy conformations. In this case, the puckering coordinate allows the molecule to flip from one conformation shown to the other. A similarly shaped potential energy well is found for the inversion mode of ammonia [Herzberg, 1945, Chapter II, 5].

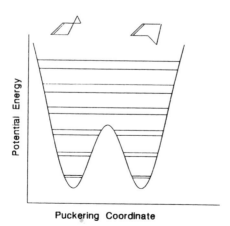

FIGURE 3.3. Double minimum potential function for ring puckering modes or inversion vibrations such as in ammonia. All energy levels are split, with alternating symmetry states.

58 POLYATOMIC MOLECULES

Internal rotations are other motions with multiple energy maxima and minima. These vibrations are hindered rotations about a single bond such as the propeller motion of the methyl group in toluene, or the methyl group in CH_3—CCl_3. At room temperature these groups do not rotate freely, but rather in a back-and-forth rotatory motion, which may have a threefold energy minimum. These potential functions can be approximated by $\cos^2$ functions with three energy minima (Fig. 3.4) [Durig and Sullivan, 1985]. At room temperature, the vibrations of internal rotation lie well below the barriers, and only at elevated temperatures does free rotation occur.

Finally, to close this discussion, a comment on the symmetry requirements of vibrational transitions of molecules more complicated than the diatomics discussed in Chapter 2 is appropriate. The details of this subject will be taken up again in Chapter 4. Thus let it suffice to state here that, for a polyatomic molecule, the symmetry of the normal coordinate must be taken into account. For absorption to occur, the transition moment of the form

$$\int \psi'(Q_k) Q_K \psi(Q_k) \, d\tau \qquad (3.3.8)$$

must be evaluated. In Eq. 3.3.8, the prime denotes the excited state wavefunction. We shall see later that certain vibrations of symmetric polyatomic molecules are forbidden because the expression 3.3.8 is zero. In vibrational spectroscopy, a symmetry forbidden transition is often truly unobservable, unlike in electronic spectroscopy, where forbidden transitions are sometimes made partially allowed by vibronic effects or magnetic interactions.

3.4 ACCIDENTAL DEGENERACY AND FERMI RESONANCE

We have assumed so far that the total molecular vibrational wavefunction is simply the product of the wavefunctions associated with the normal coordinate

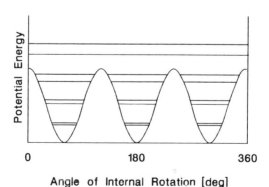

FIGURE 3.4. Potential function for internal rotation for a threefold top. All energy levels are split, with alternating symmetry states.

Q_k, and that the vibrational eigenfunctions form a vector space in which the vibrational Hamiltonian

$$H\psi_{vib} = E\psi_{vib} \tag{3.4.1}$$

yields a diagonal eigenvalue matrix:

$$H \begin{bmatrix} \psi(Q_1) \\ \psi(Q_2) \\ \psi(Q_3) \\ \vdots \\ \psi(Q_k) \end{bmatrix} = \begin{bmatrix} E(Q_1) & 0 & 0 & 0 \cdots \\ 0 & E(Q_2) & 0 & 0 \cdots \\ 0 & 0 & E(Q_3) & 0 \cdots \\ & & \vdots & \\ 0 & 0 & 0 & E(Q_k) \end{bmatrix} \begin{bmatrix} \psi(Q_1) \\ \psi(Q_2) \\ \psi(Q_3) \\ \vdots \\ \psi(Q_k) \end{bmatrix}$$

Then, the total vibrational energy is just the sum of all energies $E(Q_k)$, as discussed before (Eq. 3.3.6).

The picture discussed so far describes the vibrational states well if all energy levels are well separated and do not interact. However, under certain circumstances it is necessary to include in this simple picture the possibility that some vibrational coordinates interact. This may happen if certain symmetry requirements are fulfilled, and if the energy of the two states are similar to permit their mixing. When such mixing occurs, the corresponding vibrational modes lose their identity, and the resulting modes must be described as linear combinations of the two interacting modes, and the approximation of noninteracting vibrational coordinates (Eq. 3.3.3) breaks down.

Let us assume that in a polyatomic molecule two energy levels E_m and E_n, corresponding to the wavefunctions ψ_m and ψ_n, are accidentally degenerate or nearly degenerate, and that there is a perturbation operator H', such as anharmonic potential terms or dipolar coupling, which causes the mixing of the two states n and m:

$$E_{nm} = \langle \psi(Q_n) | H' | \psi(Q_m) \rangle \tag{3.4.2}$$

Thus the original Hamiltonian cannot be written in diagonal form, but the energy matrix will appear as follows:

$$\begin{bmatrix} E(Q_1) & 0 & 0 & 0 & 0 & 0 \cdots \\ 0 & E(Q_2) & 0 & 0 & 0 & 0 \cdots \\ & & \vdots & & & \\ 0 & 0 & 0 & E_n & E_{nm} & 0 \cdots \\ 0 & 0 & 0 & E_{nm} & E_m & 0 \cdots \\ & & \vdots & & & \\ 0 & 0 & 0 & 0 & 0 & E(Q_k) \end{bmatrix}$$

The partial determinant containing the perturbation can be solved separately:

$$\begin{vmatrix} E_n - E & E_{nm} \\ E_{nm} & E_m - E \end{vmatrix} = 0 \qquad (3.4.3)$$

From this, we obtain the energy perturbation that is added to the average of the nonperturbed energies as follows [Herzberg, 1945, Section II, 5c]:

$$E = \tfrac{1}{2}(E_n + E_m) \pm \sqrt{4|E_{nm}|^2 + \delta^2} \qquad (3.4.4)$$

where $\delta = E_n - E_m$, that is, the energy difference of the unperturbed levels. The new wavefunctions ψ^+ and ψ^- for the mixed states can be written as follows:

$$\psi^+ = (1/N)\{a\psi_m + b\psi_n\} \quad \text{and} \quad \psi^- = (1/N)\{a\psi_m - b\psi_n\} \qquad (3.4.5)$$

where a and b are the mixing coefficients, which can be expressed in terms of E_{nm} and δ according to

$$a^2 = \{\sqrt{(4|E_{nm}|^2 + \delta^2)} + \delta\}/2\sqrt{4|E_{nm}|^2 + \delta^2} \qquad (3.4.6)$$

and

$$b^2 = \{\sqrt{(4|E_{nm}|^2 + \delta^2)} - \delta\}/2\sqrt{4|E_{nm}|^2 + \delta^2} \qquad (3.4.7)$$

Thus the effect of this interaction is to increase the splitting between the energy levels and to mix them. In the mixed state it is not possible to assign one of the peaks to the state n; but it has to be assigned to the ψ^+ or ψ^- combination state. This splitting will be larger if the original energy difference is small and if the coupling energy is large. The interacting states must be of the same symmetry species (or contain the same irreducible representations, cf. Chapter 4) for the integral in Eq. 3.4.2 to be nonzero. The mixing of states is also accompanied by an equalizing effect of the vibrational intensities: the normally very weak overtone or combination band can obtain significant intensity from the fundamental with which it is in Fermi resonance.

A common example will be discussed here qualitatively to convey the order of magnitudes of the splittings and the interaction energy involved. In the C—H stretching region of a methyl group with C_{3v} symmetry (cf. Chapter 4), three peaks are observed between 2850 and 3050 cm^{-1}. The detailed assignment of these is presented in Section 7.3.2 for methyl chloride. For the time being, let it suffice to state that the highest wavenumber peak, at 3042 cm^{-1}, is the antisymmetric stretching motion of E symmetry, and the two other peaks at about 2960 and 2880 cm^{-1} are a Fermi resonance doublet of the symmetric methyl stretching vibration of A_1 symmetry and an overtone of the methyl antisymmetric deformation mode of E symmetry. This latter mode occurs typically at 1455 cm^{-1}; thus its overtone would be expected at 2910 cm^{-1}, ignoring anharmonicity.

From an analysis of overtones and combination bands, the symmetric stretching mode is expected to occur at 2930 cm^{-1}. Thus we use for the unperturbed energy eigenvalues E_n and E_m 2930 and 2910 cm^{-1}, respectively, and we set

$$\delta = E_n - E_m = 20 \text{ cm}^{-1} \qquad (3.4.8)$$

and

$$\tfrac{1}{2}(E_n + E_m) = 2920 \text{ cm}^{-1} \qquad (3.4.9)$$

Thus Eq. 3.4.4 yields

$$2960 = 2920 + \sqrt{4|E_{nm}|^2 + \delta^2} \quad \text{and} \quad 2880 = 2920 - \sqrt{4|E_{nm}|^2 + \delta^2}$$

from which we conclude that $(4|E_{nm}|^2 + \delta^2) = 1600$.

Using $\delta^2 = 400$, we get a value for E_{nm} of about 17 cm^{-1}, and for the coefficients a^2 and b^2 values of 0.75 and 0.25, respectively. Thus the wavefunction of the higher component of the Fermi doublet could be written as

$$\psi_{2960} = 0.87\psi(v_s^{CH_3}) + 0.5\psi(2\delta_{as}^{CH_3}) \qquad (3.4.10)$$

with an analogous expression holding for the lower component. We shall demonstrate in Chapter 4 that the overtone of the antisymmetric deformation ($2\delta_{as}^{CH_3}$) contains the totally symmetric irreducible representation, which allows the interaction of the two energy levels.

The example above demonstrates the most common case of Fermi resonance, where one of the states is a combination band or an overtone. Overtones of nondegenerate fundamentals always contain the totally symmetric representation of a group, and those of degenerate states often do. Thus overtones and combination bands often interact with totally symmetric vibrations. Many more examples exist in the vibrational spectra of simple molecules: in CCl_4, to be discussed in more detail in Chapter 7, the antisymmetric C—Cl stretching vibration at ~ 785 cm^{-1} (T_1) interacts strongly with the combination mode $A_1 + T_2$, which is estimated to occur at 770 cm^{-1}. In the last example of CCl_4, the fundamental and the combination are equally strong in the Raman spectra. Another common example is that of CO_2, which will be discussed in Chapter 7 as well.

3.5 GROUP FREQUENCIES

One of the most useful aspects of vibrational spectroscopy is the fact that a given group or bond in a molecule will produce spectral features that are characteristic of this group or bond and are therefore referred to as "group frequencies." Similar group-specific signatures are not uncommon in some other forms of spectroscopy: in NMR spectroscopy, methyl groups are usually observed at fairly reproducible chemical shifts, and in UV/visible absorption spectroscopy, a carbonyl group exhibits a transition of medium intensity at about 280 nm. Thus these techniques can be used as qualitative analytical tools.

Microwave (rotational) spectroscopy, on the other hand, samples a quantity that is typical of the entire molecule, and not certain portions of it: in rotational spectroscopy, the (principal) moments of inertia are observed. Even a simple substitution of a methyl for a deuteriomethyl group will alter the center of gravity, thereby shifting the principal axes of inertia, the moments of inertia, and the rotational spectra. This spectral change cannot be predicted a *priori* with simple rules but requires detailed computations.

Thus infrared (and Raman) spectroscopies offer the enormous advantage that they permit qualitative identification of compounds by examining their group vibrations. Furthermore, the group frequencies allow for an initial vibrational assignment, which often determines the quality of the fit of a force field (*vide infra*).

Let us discuss these concepts further using concrete examples. A single C—Cl "group," or bond, will produce a strong infrared absorption (and a strong Raman emission) around $750 \, \text{cm}^{-1}$, which is more or less independent of the rest of the molecule. This vibration is a group frequency typical of a C—Cl group. However, if a molecule possesses two C—Cl groups in close proximity (e.g., in a CCl_2 group), a symmetric and an antisymmetric Cl—C—Cl stretching mode are observed, in addition to a Cl—C—Cl bending vibration, all of which are again characteristic group frequencies of a CCl_2 group.

Introductory organic chemistry textbooks, in the discussion of infrared spectroscopy, typically list several dozen characteristic vibrational frequencies, for example, of methyl groups, methylene groups, methine hydrogens, C—H groups in an aromatic or vinylic environment, carbonyl groups, aromatic rings, and the vibrations of many heteroatomic groups. A listing of such group frequencies is presented in Appendix I. For a more complete compilation of group frequencies and their use, see, for example, Colthup et al. [1990] and Dollish et al. [1974].

Although the discussion of molecular vibrations in terms of group vibrations may appear somewhat crude, particularly after the discussions of normal modes of vibration in this chapter, it is amazing how well the group frequency approach holds, and how frequently it can be used for the identification of compounds. However, for the detailed understanding of vibrational spectroscopy, one has to keep in mind that a normal mode of vibration involves the entire molecule, but that the major contribution to a mode may involve just one group, which gives rise to the "group frequency."

Obviously, the concept of group frequencies is responsible for the introduction of internal coordinates in the following discussions in this chapter. The argument here is, of course, the following: if a stretching vibration of a group, such as a C—Cl bond, is more or less independent of the rest of the molecule, its vibrational frequency can be approximated by

$$v = (1/2\pi)\sqrt{k/M_R} \qquad (2.6.16)$$

where M_R, the reduced mass, is given by

$$M_R = \frac{m_C m_{Cl}}{m_C + m_{Cl}} \qquad (3.5.1)$$

and k is the force constant (or a combination of several force constants) of the vibrational motion under consideration. Consequently, we introduce an internal coordinate for this particular vibrational mode, and we find experimentally that this coordinate does not mix much with other vibrational coordinates (except another C—Cl stretching motion, with which it is degenerate). In either case, the force constant k associated with the C—Cl stretching motion is independent of the environment; the combination of two C—Cl stretching motions in a Cl—C—Cl group into a symmetric and antisymmetric vibration occurs automatically during the computations, and the proper splitting between the two vibrations is reproduced computationally by fitting appropriate stretch–stretch interaction constants.

Thus group frequencies are of importance to chemists in general for the identification and qualitative interpretation of compounds. However, they are also extremely important to vibrational spectroscopists involved in the calculation of normal modes of vibrations and molecular force fields. In the perturbation of a force field to reproduce the observed vibrational frequencies, it often happens that the calculated frequencies fit the observed ones very nicely. A closer inspection of the PED (*vide infra*), however, reveals that the calculated vibration is not in agreement with the vibrational assignment, based on group frequencies, isotopic data, and so on. Thus a vibrational assignment, based on the observed spectra, has to be used as a criterion for the quality of the force constants. This step, unfortunately, has been omitted in a number of vibrational calculations, resulting in improper force fields.

3.6 ISOTOPIC SUBSTITUTION AND THE TELLER–REDLICH PRODUCT RULE

One of the most powerful techniques for the interpretation of vibrational spectra is the incorporation of isotopic species in a molecule. Assuming that the potential energy functions experienced by isotopomers (i.e., molecules that are identical except for the incorporation of an isotopic atom) are identically the same, isotopic substitution can be used to identify vibrations, provide information on the mixing of internal coordinates in normal modes of vibration, and provide information on the shape of the potential energy curve.

The identification of group frequencies, as discussed in the previous section, is greatly aided by isotopic data. Aside from the straightforward case, namely, that of an isolated C—H stretching mode such as in chloroform, which shifts by nearly a factor of 1.4 toward lower wavenumber (cf. Section 7.3), deuterium substitution is essential for the assignment of the methyl rocking modes, for example, in complicated molecules. These modes, which occur between 900 and 1100 cm^{-1} (cf. Section 7.3) mix with C—C stretching coordinates and are sometimes hard to identify. However, their frequencies are lowered by about 200 cm^{-1} upon deuteriation and thus can be recognized unambiguously. ^{13}C, ^{15}N, and ^{18}O also provide isotopic labels that are used frequently to aid in vibrational assignments. The magnitude of the expected shift can be estimated qualitatively, as in the case of deuterium substitution, by using Eq. 2.6.16. The maximum possible shift one could expect can be estimated by setting the mass of the rest of the molecule to infinity. In that case, the effect of ^{18}O substitution on the carbonyl stretching frequency in a heavy molecule should be about

$$\sqrt{18/16} = 1.06$$

which amounts to a nearly 100 cm^{-1} shift. In practice, much smaller shifts are observed, since the carbonyl carbon atom has a large amplitude in this vibration as well, and the ratio of reduced masses (Eq. 3.5.1) should be used.

Using the square root of the ratio of reduced masses (1.025) predicts a maximal frequency shift for a double bonded carbon/oxygen oscillator of 40 cm^{-1}. If the observed shift is significantly less than this number, one may conclude that the vibrational motion is not pure, but mixed with those of other groups in the molecule.

Without synthesis of special isotopomers (such as D, ^{13}C, or ^{18}C containing species), the most readily observed isotopic effect is that of ^{35}Cl/^{37}Cl, since naturally occurring chlorine consists of these isotopic species in a natural abundance of about 75%:25%. Thus many modes involving significant chlorine motion exhibit a splitting of 2–3 cm^{-1}, with a 3:1 intensity ratio, due to the chlorine isotopes (cf. Chapter 7). The natural abundance of ^{13}C is sufficiently high to permit observation of the vibrational shifts due to this isotope. In benzene, for example, the strong ring breathing mode at 992 cm^{-1} shows a weak low frequency satellite peak in the Raman spectrum, due to ^{13}CC$_5$H$_6$.

Quantitatively, the frequencies of an isotopically substituted molecule can be related to those of the parent molecule via the Teller–Redlich product rule [Herzberg, 1945, Section II, 6]. This rule relates the ratio of the products of all frequencies of isotopomers to the masses and moments of inertia and can be written for the case of an asymmetric molecule as follows:

$$\prod_{i=1}^{3N-6} \frac{v'_i}{v_i} = \left[\frac{I'_x I'_y I'_z}{I_x I_y I_z}\right]^{1/2} \cdot \left[\frac{M'}{M}\right]^{3/2} \cdot \prod_{k=1}^{3N} \left[\frac{m_k}{m'_k}\right]^{1/2} \quad (3.6.1)$$

Here, the I_α denote the principal moments of inertia, M the molecular mass, m

the masses of the atoms (*vide infra*), and v the (harmonic) vibrational frequencies. In Eq. 3.6.1, the primed quantities refer to the isotopically substituted species. For molecules with symmetry, Eq. 3.6.1 can be used separately for vibrations with different symmetry species. In this case, Eq. 3.6.1 can be simplified somewhat. If, for example, vibrations are considered which transform like T_z and R_z (cf. Chapter 4), then the translational and rotational terms in Eq. 3.6.1 can be modified as follows:

$$\left[\frac{I'_z}{I_z}\right]^{1/2} \cdot \left[\frac{M'}{M}\right]^{1/2}$$

and only the vibrations belonging to this particular representation need to be included in the product on the left-hand side of Eq. 3.6.1. Details of the derivation of the Teller–Redlich product rule and further applications may be found in the literature [Wilson et al., 1955, Section 8.5; Herzberg, 1945, Section II, 6].

3.7 NORMAL MODE CALCULATIONS

3.7.1 Aim of Normal Mode Analysis

The aim of normal coordinate computations is the calculation of the vibrational frequencies (eigenvalues) and atomic displacements for each normal mode of vibration from structural parameters, atomic masses, and a force field or, alternatively, the determination of the force field from observed vibrational frequencies. A normal mode, as defined before, is a motion in which all atoms vibrate at the same frequency and phase but different amplitudes. These amplitudes cannot be determined unambiguously, however, the normal mode calculations yield a *relative* amplitude of the motion of each atom during a given normal mode.

The discussion presented in the following two sections introduces the theory and background for the computation of vibrational frequencies and normal modes of vibration. These calculations follow a format developed by Schachtschneider [1964], who wrote the first set of modern programs for the purpose of normal coordinate calculations in FORTRAN IV in the early 1960s. These programs were adapted and reworked for operation on personal computers by this author in the late 1980s, with particular emphasis on the computations for asymmetric molecules and the use of these programs in the classroom environment. Other programs for the computation of normal modes exist [e.g., see Quantum Mechanical Program Exchange QeMP 067], but more for historical reason than any other ones, the nomenclature and methodology of Schachtschneider's original work are followed here.

3.7.2 Internal Coordinates

In Section 3.2, the principles underlying the computations of normal modes of vibrations were introduced. For the calculations, one assumes that the atoms in a molecule oscillate about their equilibrium position on a harmonic potential surface, and that there exist $3N - 6$ degrees of vibrational freedom for nonlinear molecules, which are referred to as the normal modes of vibration and are associated with the $3N - 6$ fundamental vibrational transitions accessible via infrared and Raman spectroscopies.

Thus the principles of the computations appear at first glance straightforward: one needs only to set up the potential energy matrix in Cartesian or mass-weighted Cartesian coordinates (cf. Eq. 3.2.26) and diagonalize this matrix to obtain the energy (vibrational) eigenvalues. The transformation from the Cartesian to normal coordinate space is obtained from the eigenvector matrix. However, the difficulty lies in setting up the potential energy matrix: in order to calculate $3N - 6$ vibrational frequencies, one needs to define a $3N \times 3N$ potential energy matrix, which requires $3N(3N + 1)/2$ force constant, which are the second derivatives of the potential energy with respect to two coordinate components. Thus, for a simple molecule such as water, there are six Cartesian (mass-weighted) force constants and only three vibrational frequencies. Since these force constants are not known *a priori*, the attempt to determine six force constants to reproduce three frequencies leads to a problem that is indeterminate. Furthermore, if in another computation the molecule is oriented differently in the coordinate system, all force constants may be different; that is, there is no transferability of force constants between one and the next molecule if the potential energy is defined in Cartesian coordinates. Thus one finds that expressing the potential energy in any *space fixed* coordinates (i.e., a displacement coordinate system that is colinear with external laboratory coordinates) is chemically meaningless. Rather, one attempts to define the potential energy in terms of coordinates that are chemically relevant and orientation independent (i.e., they are "attached" to the molecule). By choosing these coordinates appropriately, they assume "chemical meaning": as a chemist, one prefers to think in terms of the stiffness of a bond, or the ease with which an angle can be deformed, and not in terms of the Cartesian components of these forces along arbitrary axes. Thus one introduces a coordinae system in which the potential energy is expressed in terms of chemically significant groupings.

These coordinates are known as "internal coordinates" and are given the symbol R. Force constants associated with the internal coordinates (bond stretching, angle deformation, wagging and torsional motions) will be designated by script f_{ij}. The diagonal terms f_{ii} of the internal coordinate force constant matrix $\mathscr{F}_{ij}$ are intuitively understandable as the force it takes to stretch a given bond, or squeeze a bond angle, independent of the orientation of the bond in the molecule. These force constants are—to some degree— transferable between molecules.

NORMAL MODE CALCULATIONS 67

Internal coordinates are defined as follows. Assume that all atoms in a molecule are at their equilibrium coordinates X, Y, and Z, and that unit Cartesian displacement vectors **x**, **y**, and **z** are attached to all atoms. To avoid confusion, we shall use capital X, Y, and Z to denote the *space fixed equilibrium coordinates* of the atoms, and lowercase x, y, and z to denote *Cartesian displacement coordinates*.

An internal coordinate is defined as the displacement of two or more atoms along unit vectors $\mathbf{e}_i$, which are linear combinations of the Cartesian displacement coordinates, such that motions along $\mathbf{e}_i$ produce a maximal change in a given internal coordinate R [Wilson et al., 1955, Section 4-1]. For a simple bond stretching motion between atom i and j, the internal coordinate is defined as

$$R = e_i - e_j \tag{3.7.1}$$

where the e's are along the bond direction between atom i and j, and e_i and e_j are entered in the definition of R with opposte signs (cf. Section 3.9). For an angle deformation and wagging or torsional motions, the expressions for the definition of the internal coordinates are more cumbersome and are derived in detail in Wilson et al. [1955, Section 4-1].

The transformation between internal and Cartesian displacement coordinates is given by the **B** matrix, defined in Eq. 3.7.2, which provides a description of an internal coordinate in terms of Cartesian displacement coordinates:

$$\mathbf{R} = \mathbf{Bx} \tag{3.7.2}$$

In this equation, and the followig discussion, both **R** and **x** are column vectors. The B matrix is rectangular, since there are $3N$ Cartesian displacement vectors and about $3N - 6$ internal coordinates. The number of internal coordinates one defines may exceed $3N - 6$, if redundant coordinates are defined. This will be discussed later. A detailed example of how to set up these transformations and an example of the form of the matrices involved are given in Section 3.9.

3.7.3 Kinetic and Potential Energies in Internal Coordinates

Next, the potential energy needs to be expressed in internal coordinates. For this, we expand the potential energy of the molecule in a Taylor series about the equilibrium positions in terms of the internal coordinates R, in complete analogy to the expansion presented in Eqs. 3.2.8 to 3.2.11. Furthermore, the arguments about the V_0 term, as well as the terms cubic in R, can be applied to the expansion in internal coordinates as well, and one obtains the potential energy in terms of the internal coordinates R as follows:

$$2V = \mathbf{R}^\mathrm{T}\mathscr{F}\mathbf{R} = \sum \mathscr{F}_{ij} R_i R_j \tag{3.7.3}$$

Here, $\mathbf{R}^T$ is a row vector, $\mathbf{R}$ a column vector, as before, and $\mathscr{F}$ a square matrix composed of the second derivatives of the potential energy with respect to the internal coordinates:

$$\mathscr{F}_{ij} = \frac{\partial^2 V}{\partial R_i \partial R_j} \tag{3.7.4}$$

At this point, the kinetic energy may be transformed into internal coordinate space (cf. Eq. 3.7.8) as well:

$$2T = \dot{\mathbf{x}}^T \mathbf{M} \dot{\mathbf{x}} = \dot{\mathbf{R}}^T \mathbf{B}^T \mathbf{M} \mathbf{B} \dot{\mathbf{R}} \tag{3.7.5}$$
$$= \dot{\mathbf{R}}^T \mathbf{G}^{-1} \dot{\mathbf{R}}$$

with

$$\mathbf{G}^{-1} = \mathbf{B}^T \mathbf{M} \mathbf{B} \tag{3.7.6}$$

Equation 3.7.6 contains the definition of the G matrix, which has historical significance because the "GF matrix method" was used in the precomputer era in the computation of normal modes of vibration. G matrix elements were tabulated for most common groups, which permitted the convenient computation of the kinetic energy part of the molecular energy. In order to solve the secular equation in internal coordinate space, the determinant

$$GF - E\lambda = 0 \tag{3.7.7}$$

needs to be evaluated. This method was devised originally to simplify the (mostly manual) computations of normal coordinates 40 years ago. Even the original programs written by Schachtschneider used this method; however, this method is relatively slow compared to more direct methods, to be described below.

Instead of transforming the kinetic energy to internal coordinates, the potential energy, defined in internal coordinates, is transformed back to mass-weighted Cartesian coordinate space, and the remainder of the computations is carried out in Cartesian coordinates. This procedure saves considerable computation time in the numerical matrix diagonalization, since the GF matrix product in internal coordinate space is a nonsymmetric matrix. Diagonalization of a nonsymmetric matrix is much more time consuming than that of a symmetric matrix.

Thus we need to concern ourselves with the problem of setting up the potential energy in terms of internal coordinates, since they are chemically significant and offer some transferability in the force constants. Subsequently, the potential energy is transformed back into Cartesian displacement coordinates, and the diagonalization is carried out in Cartesian space. In the next pages, these transformations will be presented.

3.7.4 Transformation Between Cartesian, Internal, and Normal Coordinates

In this section, the detailed procedure for the computations in Cartesian coordinates will be presented. Since the molecules studied in the author's laboratory are usually devoid of symmetry, no presymmetrization of the potential energy via the use of symmetry coordinates (cf. Sections 3.75 and 4.5) is presented, and the calculations involve the transformation between Cartesian, internal, and normal coordinates only. We shall continue to utilize the symbol f_{ij} for force constants in mass-weighted Cartesian space, and $\mathbf{F}$ for the corresponding matrix, and f_{ij} and $\mathscr{F}$ for the same quantities in internal coordinate space. Boldface type always implies a matrix (or vector).

In terms of Cartesian and mass-weighted Cartesian displacement coordinates, the kinetic energy is

$$2T = \sum m_i(\dot{x}_i)^2 \quad \text{or} \quad 2T = \dot{\mathbf{x}}^T \mathbf{M} \dot{\mathbf{x}} \qquad (3.7.8)$$

and

$$2T = \sum (\dot{q}_i)^2 \quad \text{or} \quad 2\mathbf{T} = \dot{\mathbf{q}}^T \dot{\mathbf{q}} \qquad (3.7.9)$$

Here, $\mathbf{M}$ is a diagonal matrix of the masses 1 through N, and the summation extends from $i = 1$ to $3N$. $\dot{\mathbf{x}}^T$ and $\dot{\mathbf{x}}$ are row and column vectors of the x_i, respectively, such that the matrix product $\dot{\mathbf{x}}^T \mathbf{M} \dot{\mathbf{x}}$ contains entries only along the main diagonal. Thus it is evident that the kinetic energy is in a diagonal form at this point. Care has to be taken when the M matrix is defined: each of the N masses appears three times along the diagonal, since each atom has three degrees of freedom.

Since the kinetic energy is diagonal, we transform the potential energy into the same coordinate system by using Eq. 3.7.2 and its transpose:

$$\mathbf{R} = \mathbf{B}\mathbf{x} \qquad (3.7.2)$$

$$\mathbf{R}^T = \mathbf{x}^T \mathbf{B}^T \qquad (3.7.10)$$

Thus we obtain for V

$$2V = \mathbf{R}^T \mathscr{F} \mathbf{R} \qquad (3.7.3)$$

$$= \mathbf{x}^T \mathbf{B}^T \mathscr{F} \mathbf{B}\mathbf{x} \qquad (3.7.11)$$

Rewriting the definition of mass-weighted Cartesian coordinates (Eq. 3.2.6) as

$$\mathbf{q} = \mathbf{M}^{1/2}\mathbf{x} \quad \text{or} \quad \mathbf{x} = \mathbf{M}^{-1/2}\mathbf{q} \qquad (3.7.12)$$

we obtain for the potential energy

$$2V = \mathbf{q}^T \mathbf{M}^{-1/2} \mathbf{B}^T \mathscr{F} \mathbf{B} \mathbf{M}^{-1/2} \mathbf{q} \qquad (3.7.13)$$

Writing Eq. 3.2.10 in matrix notation,

$$2V = \mathbf{q}^T \mathbf{F} \mathbf{q} \tag{3.7.14}$$

and comparing this expression with Eq. 3.7.13, we find that

$$\mathbf{F} = \mathbf{M}^{-1/2} \mathbf{B}^T \mathscr{F} \mathbf{B} \mathbf{M}^{-1/2} \tag{3.7.15}$$

Since $\mathbf{M}^{-1/2}$ is diagonal, it need not be transposed for the first matrix multiplication in Eqs. 3.7.13 and 3.7.15. Thus Eq. 3.7.15 transforms the force constant matrix $\mathscr{F}_{ij}$ from internal coordinate space to Cartesian coordinate and, subsequently, to mass-weighted Cartesian coordinates. The eigenvalues Λ of the matrix H,

$$\mathbf{H} = \mathbf{M}^{-1/2} \mathbf{B}^T \mathscr{F} \mathbf{B} \mathbf{M}^{-1/2} \tag{3.7.16}$$

are the vibrational frequencies (after proper unit conversion), which are obtained by numeric diagonalization of H. This latter process is accomplished by finding a linear transformation matrix C such that

$$\mathbf{C}^{-1} \mathbf{H} \mathbf{C} = \Lambda \tag{3.7.17}$$

where Λ is a diagonal matrix of the eigenvalues, and the matrix C that diagonalizes H is known as the eigenvector matrix C. This matrix defines the transformation S between Cartesian and normal coordinates:

$$\mathbf{x} = \mathbf{S}\mathbf{Q} \tag{3.7.18}$$

$$= \mathbf{M}^{-1/2} \mathbf{C} \mathbf{Q} \tag{3.7.19}$$

The transformation from internal coordinates to normal coordinates is given by

$$\mathbf{R} = \mathscr{L} \mathbf{Q} \tag{3.7.20}$$

where

$$\mathscr{L} = \mathbf{B}\mathbf{S} = \mathbf{B}\mathbf{M}^{-1/2}\mathbf{C} \tag{3.7.21}$$

These matrix manipulations are very common in linear algebra to solve sets of homogeneous equations, and the procedures are described well in many texts on mathematical physics [e.g., see Butkov, 1968].

Along with the vibrational frequencies, two more pieces of information may be obtained from the normal mode calculations. The first is the potential energy distribution (PED), which is a listing of the normalized contribution of each of the internal coordinate force constants to a given frequency. The PED is important since it gives the only indication whether or not a computed eigenvalue (vibrational frequency) conforms to the vibrational assignment.

Since the vibrational eigenvectors are obtained via a transformation from Cartesian or internal coordinate space into normal coordinate space,

$$\Lambda = \mathscr{L}^T \mathscr{F} \mathscr{L} \qquad (3.7.22)$$

the eigenvalue of each normal mode, λ_i, is given by

$$\lambda_i = \sum_{j,k} \mathscr{L}_{ji} \mathscr{L}_{ki} \mathscr{F}_{jk} \qquad (3.7.23)$$

where the summation in Eq. 3.7.23 is over all rows and columns of the matrix product. The PED, that is, the fractional contribution of the kth force constant to a vibrational eigenvalue λ_i, is then given by

$$\mathscr{L}_{ji} \mathscr{L}_{ki} \mathscr{F}_{jk} / \lambda_i \qquad (3.7.24)$$

Equation 3.7.24 expresses the PED in terms of force constants in internal coordinate force space, rather than the Cartesian components. This is desirable, since inspection of the PED is supposed to yield the contribution of a given internal coordinate to the normal mode. The contributions of off-diagonal force constants to the PED may be negative, and the sum of all contributions is unity [Schachtschneider, 1964].

The final pieces of information available from normal mode calculations are the atomic displacement vectors in Cartesian displacement coordinates. These displacements are actually contained in the columns of the S matrix, which need to be decomposed into triads of displacement coordinates x for each of the N atoms. As pointed out before, the actual amplitudes of vibration are not known. Thus the displacements are relative and normalized to unit displacement lengths. However, the relative motions of all atoms with respect to each other for a given normal mode of vibration are accurate.

A final word about the methods to define $\mathscr{F}_{ij}$ in internal coordinates is appropriate. Since a given force constant may occur more than once in a potential energy matrix, it is advantageous to define the $\mathscr{F}$ matrix as a product

$$\mathscr{F}_{ij} = \sum_k Z_{ijk} \Phi_k \quad \text{or} \quad \mathscr{F} = \mathbf{Z}\boldsymbol{\Phi} \qquad (3.7.25)$$

where Φ is a one-dimensional vector of force constants, and Z is a matrix that determines where a force constant Φ_i will occur in the matrix $\mathscr{F}$. Again, the example in Section 3.9 explains more details of these procedures.

3.7.5 Symmetry Coordinates

It is sometimes advantageous to group together certain internal coordinates and define a new set of coordinates known as symmetry coordinates. This procedure was used extensively in the precomputer era, when the reduction of matrix size was enormously important. Indeed, the use of symmetry coordinates can reduce the size of matrices quite significantly by breaking an overall energy matrix into smaller submatrices, the eigenvalues of which are independent. However, the time savings offered by this presymmetrization is relatively small with the high-speed computers available these days.

However, symmetry coordinates are important for a variety of reasons other than the time factor and therefore should be discussed for completeness sake. However, since the concept of molecular symmetry will not be introduced until Chapter 4, the discussion of symmetry coordinates is postponed until the end of the chapter on group theory and symmetry.

3.8 FORCE FIELDS

The successful computation of vibrational frequencies and atomic displacement vectors from force constants depends on the use of a correct force field. Unfortunately, the determination of force constants from vibrational frequencies, which is the reverse problem, is difficult since, with the exception of diatomic molecules, there are always more (harmonic) force constants to be determined than there are vibrational frequencies. When anharmonic (cubic) force constants are to be determined, the problem is even worse. Thus the evaluation of molecular force constants from vibrational data is always indeterminate, and methods have to be devised to reduce the inherent ambiguity.

The first method to overcome the problem of indeterminacy is that of transferring force constants from other molecules. Since moieties such as a methyl group exhibit similar vibrational frequencies in different molecules, it appears appropriate to transfer the force constant of the methyl group from one molecule to another. This method works reasonably well for the diagonal force constants.

Another method is that of using as many isotopomers as possible. One assumes that an isotopic species vibrates in the same force field as the parent molecule. The additional vibrational frequencies available from the isotopomer(s) allow a better determination of more force constants, particularly off-diagonal values. These off-diagonal force constants often are responsible for complicated mixing pattern of internal coordinates, which are manifested in sometimes unpredictable shifts in vibrational frequencies upon isotopic substitutions. This will be discussed in Chapter 7.

Finally, force fields that use fewer force constants by collecting a number of off-diagonal force constants into one force constant have been used, and force fields have been calculated quantum mechanically. The first of these approaches is known as the Urey–Bradley method. The computation of force constants from *ab initio* quantum mechanical methods has just recently become practical and may offer the best source for numerical values. In the following sections, some of the choices for force fields will be introduced.

3.8.1 The Generalized Valence Force Field (GVFF)

When force constants were defined in internal coordinate space (Eq. 3.7.3), it was implied that these force constants are GVFF force constants. In this approach, the diagonal force constants f_{ii} are intuitively understandable as the forces required to stretch or deform a chemical bond or bond angle. Thus their units are in force/length [dyn/cm or mdyn/Å] or force/radian (see the comments on units of force constants in Section 2.6) and they can be associated with stiffness of a bond.

The off-diagonal elements

$$f_{ij} = \frac{\partial^2 V}{\partial R_i \partial R_j} \qquad (3.8.1)$$

give a measure of how the potential energy of coordinate *i* changes after coordinate *j* has been extended or compressed. These off-diagonal force constants are often not well known, and they do not transfer all that well between molecules.

The enormous number of force constants needed in the GVFF approach makes this force field less useful: listed below are the distinct force constants needed for a group as simple as a methyl group attached to an another C atom:

Two diagonal stretching force constants (f_{C-H} and f_{C-C})
Two diagonal bending constants (f_{H-C-H} and f_{C-C-H})
Two stretch–stretch interaction constants ($f_{C-H/C-H}$ and $f_{C-H/C-C}$)
Six stretch–bend interaction constants ($f_{C-H/C-C-H'}$, $f_{C-H/C-C-H}$, $f_{C-H/H-C-H}$, $f_{C-H/H'-C-H}$, $f_{C-C/C-C-H}$, $f_{C-C/H-C-H}$)
Four bend–bend interactions ($f_{H-C-H/H-C-H}$, $f_{C-C-H/C-C-H}$, $f_{C-C-H/H-C-H}$, and $f_{C-C-H/H'-C-H}$)

where the prime denotes whether or not the same internal coordinate is involved in the interaction. Thus a total of 16 distinct force constants are required, which contrasts with only six observable vibrational frequencies. The number of observed frequencies is lower than the expected $3N - 6 \, (=9)$ modes because of the high symmetry of the methyl group, which causes some vibrational coordinates to be degenerate (cf. Chapter 4). Thus it is clear that some of the methods discussed above to reduce ambiguity need to be employed to derive a reliable force field.

Yet, the GVFF is useful to calculate vibrational frequencies and displacement vectors, in particular when a set of isotopic molecules is available. For example, GVFF calculations for the molecule CH_3Cl, which exhibits six vibrational frequencies and requires 16 force constants, can be made significantly more reliable if vibrational data of CD_3Cl are available, in particular, when the ^{35}Cl and ^{37}Cl isotopic splitting can be observed as well. The refinement of the force constants in such a case proceeds most logically via the definition of the F matrix as shown before, using Eq. 3.7.25, in particular when a set of trial force constants can be transferred from the literature. In the refinement, one often attempts to maintain these force constants close to the literature values for related molecules. A number of commonly used force constants are given in the Appendix II.

When defining the potential energy around a tetrahedrally substituted atom, such as the central C atom in CH_3Cl, one is faced with the problem that there are six internal coordinates around the central atom, as shown in Fig. 3.5. These internal coordinates are not linearly independent, since all six angles cannot increase simultaneously. There are a number of methods to counteract the problem. One may just only define five internal coordinates and leave one angle deformation out arbitrarily. This bears the disadvantage that it is often not obvious which coordinate to omit.

Another method is to define six internal angle deformation coordinates R_1 to R_6, which are later converted to five linearly independent "symmetry"

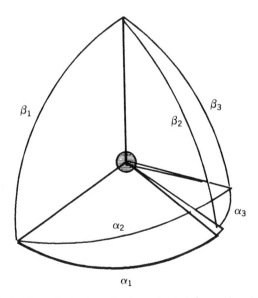

FIGURE 3.5. Definition of six linearly dependent deformation internal coordinates about a tetrahedrally substituted C atom.

(orthogonal) coordinates S_1 to S_6 as follows:

$$S_1 = \alpha_1 + \alpha_2 + \alpha_3 - \beta_1 - \beta_2 - \beta_3 \text{ (symmetric deformation)}$$
$$S_2 = 2\alpha_1 - \alpha_2 - \alpha_3 \text{ (antisymmetric deformation)}$$
$$S_3 = \alpha_2 - \alpha_3 \text{ (antisymmetric deformation)}$$
$$S_4 = 2\beta_1 - \beta_2 - \beta_3 \text{ (rocking mode)}$$
$$S_5 = \beta_2 - \beta_3 \text{ (rocking mode)}$$
$$S_6 = \alpha_1 + \alpha_2 + \alpha_3 + \beta_1 + \beta_2 + \beta_3 \equiv 0$$

where S_6 is zero, thereby reducing the number of coordinates by one. The final method is to define all six deformation coordinates around the central atom and not worry about the extra coordinate, which will yield a zero root since it corresponds to a molecular motion with no amplitude.

Although this method seems preferable, because it is simplest, it bears the disadvantage that as soon as redundant coordinates are defined, one of the original assumptions utilized in the simplification of Eq. 3.2.8,

$$V = V_0 + \sum_{i=1}^{3N} \left(\frac{\partial V}{\partial q_i}\right)_e dq_i + \frac{1}{2} \sum_{i=1}^{3N} \sum_{j=1}^{3N} \left(\frac{\partial^2 V}{\partial q_i \, \partial q_j}\right)_e dq_i \, dq_j + \cdots \quad (3.2.8)$$

namely, that the linear terms (dV/dq) are zero, no longer holds. Thus, for mathematical rigor, one should include linear force constants. This is naturally inconsistent with the principles of trying to make the calculations less ambiguous and indeterminate. Thus these linear constants are commonly ignored, and we shall see later that they are generally relatively small.

3.8.2 Urey–Bradley Force Field (UBFF)

Another method to reduce the number of force constants necessary to describe the potential energy of the molecules was proposed by Urey and Bradley [1931]. In this approach, which is particularly suitable for molecules with low or no symmetry, the stretch–stretch and stretch–bend interaction force constants are lumped together into "nonbonded interaction" force constants.

However, these nonbonded force constants introduce redundancy; that is, for a triad of atoms A, B, and C, there are diagonal A—B and B—C stretching force constants, and an A—B—C deformation force constant. In addition, a nonbonded A····C interaction is defined. Since the motion that elongates the distance A—C may be described by combination of the A—B and the B—C stretching coordinates as well as the A—B—C deformation coordinate, it is clear that the A····C coordinate is redundant. The effect of redundancy is that the second term in Eq. 3.2.8 is no longer zero; that is, linear force constants (first derivative force constants) need to be introduced. These were shown to be small, and they can be expressed as functions of the harmonic nonbonded force constants.

In terms of the Urey–Bradley force field, the potential energy is written as [Urey and Bradley, 1931]

$$2V = \sum [K_i \Delta r_i^2 + 2K'_i \Delta r_i r_i] + \sum [H_i \Delta \alpha_i^2 + 2H'_i \Delta \alpha_i]$$
$$+ \sum [F_{ij} \Delta q_{ij}^2 + 2F'_{ij} q_{ij} \Delta q_{ij}] \qquad (3.8.2)$$

where K' and K are linear and quadratic bond stretching force constants, H' and H are linear and quadratic angle deformation force constants, and F' and F are linear and quadratic nonbonded interaction constants, respectively. In most applications, one assumes that the linear force constants can be approximated by

$$X' = -0.1X \qquad (3.8.3)$$

where X denotes F, K, or H.

Schachtschneider [1964] defined an elegant method to incorporate UBFF calculations into the sequence of programs for the computation of normal modes of vibration, written in the mid-1960s in his research group. For these calculations, internal coordinates for UBFF calculations are defined identically as in the case of computations in internal coordinate space. Bond stretching coordinates are associated with a bond stretching force constant, as before; however, the numeric value of these is usually lower than in the GVFF approach, since nonbonded interactions affect stretching coordinates as well (*vide infra*).

Angle deformation coordinates are treated in the Urey–Bradley approach as follows. For a simple triad of atoms A···C···B, there is a diagonal A—C—B angle deformation force constant, which is numerically lower than a corresponding GVFF force constant. In addition, quadratic and linear nonbonded interaction constants A···B are defined, as discussed earlier. Their contribution to the vibrational potential energy enters the calculations via the Z matrix: Schachtschneider's program UBZM (Urey–Bradley Z-Matrix) evaluates the contribution of the Urey–Bradley (UB) force constants in internal coordinate space and transforms the UB force constants to contributions *added* to the diagonal terms of the potential energy matrix. Thus the same set of programs may be used for computations utilizing GV and UB force fields. In the case of GVFF computations, the Z matrix is defined manually, indicating the position in the $\mathscr{F}$ matrix of each force constant; whereas in the UBFF approach, the (much more complicated) Z matrix is computed by the program UBZM.

The UB approach, at this stage, does not appear to simplify the vibrational computations significantly. However, a little reflection shows that even for a simple molecule, such as a planar, triangular molecule such as a carbonate ion, all GVFF stretch–stretch, stretch–bend, and bend–bend interaction constants can be expressed, in the UB approach, by one single $O \cdots O$ nonbonded interaction constant.

For tetrahedrally substituted moieties, such as the center C atom in methyl chloride (cf. Section 3.8.1), the reduction of force constants is particularly impressive: in the GVFF, 16 force constants were required, whereas in the standard approach to UBFF calculations, only seven values are needed:

Two diagonal stretching force constants (K_{C-H} and K_{C-Cl})
Two diagonal bending constants (H_{H-C-H} and H_{C-C-H})
Two nonbonded interaction constants ($F_{C \cdots H}$ and $F_{H \cdots H}$)
One "intermolecular tension" constant ρ

Schachtschneider defined UB interactions for basically three different configurations: a simple, bend triad of atoms $A \cdots C \cdots B$, tetrahedrally substituted moieties, and cis-interactions extending over one additional bond, such as the dotted interactions between A and D in a structure

$$\begin{array}{c} B\text{---}C \\ / \quad \backslash \\ A \cdots\cdots D \end{array}$$

A number of Urey–Bradley force constants, refined for various small molecules, are listed in Appendix II.

3.8.3 *Ab Initio* Force Fields

These force fields are derived using *ab initio* quantum mechanical methods to calculate the second derivative of the molecular energy with respect to the Cartesian displacement coordinates. Great strides have been made over the past decade in optimizing the calculations to predict the total molecular energy, as well as the second derivatives (or curvature) of the energy with respect to the Cartesian coordinates.

The best methodology available to date appears to be via geometry optimization and subsequent analytical second differentiation of the restricted Hartree–Fock energy with the GAUSSIAN 86 program using the internal or a modified 6-31G* basis set. These calculations produce force constants that are larger than those derived from vibrational spectroscopy, and consequently, normal mode calculations using *ab initio* force constants produce vibrational frequencies that are about 10–12% too high. Yet, since all force constants, including diagonal and off-diagonal values, seem uniformly high by about the same percentage, scaling of the entire force field by a factor of about 0.88–0.9 produces adequate frequencies.

Although an arbitrary scaling of the quantum mechanically derived force constants by a factor does not seem like a very sophisticated and quantitative method, it appears that the future of normal mode calculations, especially for small molecules, belongs to *ab initio* methods, since the ambiguity of fitting off-diagonal force constants in the GVFF approach is even larger. With the availability of high-performance small computers, many of these calculations will be possible outside the mainframe computer environment, and improved computational techniques may make much more accurate force constants readily available. Finally, since vibrational intensity calculations, based on *ab initio* quantum mechanical calculations, have become practical (cf. Section 3.10), it appears likely that the entire vibrational information—frequencies and intensities—will be available computationally in the near future.

3.9 EXAMPLE FOR THE COMPUTATION OF NORMAL MODES OF VIBRATION

An example will be presented here, and carried through the ensuing discussion, to illuminate the concepts introduced. The formalism of the calculations, the sequence of programs, and the format of the data are taken from the original work by Schachtschneider [1964]. In the discussion that follows, the reader should not perceive the data presented as state-of-the-art knowledge, but rather as a guide for the computations. This applies to geometric data and observed frequencies, as well as force constants.

Consider the water molecule, with an equilibrium geometry and an orientation presented in Fig. 3.6. Its Cartesian coordinate system is defined as shown in Fig. 3.6. The Cartesian *equilibrium coordinates* (in the ensuing jargon the X matrix, or XMAT) is given in Table 3.1. These data were obtained from the bond lengths and angle using the program CART [Barlow and Diem, 1991].

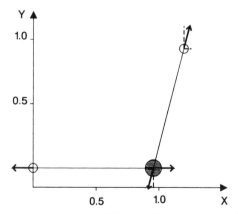

FIGURE 3.6. Definition of the equilibrium Cartesian coordinates and Cartesian displacement components of internal coordinates R_1 and R_2 for water. All dimensions are in angstroms (Å).

EXAMPLE FOR THE COMPUTATION OF NORMAL MODES OF VIBRATION 79

TABLE 3.1. Cartesian Equilibrium Coordinates for Water[a]

Coordinate	H_1	O	H_2
X	0.000	0.960	1.200
Y	0.000	0.000	0.929
Z	0.000	0.000	0.000

[a] Bond distances = 0.96 Å; bond angle = 104.5°. See also Fig. 3.6.

For a simple molecule, such as water, this matrix can easily be established using a hand calculator. For complicated molecules, there exist a variety of programs to compute the X matrix from assumed or correct bond angles and distances. For the computational programs discussed below, the X matrix is stored in the following format:

$$\begin{array}{ccc} 1 & 2 & 0.960 \\ 1 & 3 & 1.200 \\ 2 & 3 & 0.929 \end{array}$$

where the first index denotes X, Y, or Z coordinates (for 1, 2, and 3, respectively), the second index the atom number, and the following value the actual coordinate. Zero values are not stored.

For water, the internal coordinates R_1, R_2, and R_3 may be defined as the O—H_1 and O—H_2 bond stretching and the H—O—H angle deformation coordinates. In terms of the *Cartesian displacement coordinates*, the B-matrix elements, obtained from program BMAT [Barlow and Diem, 1991], are

$$\mathbf{R} = \mathbf{Bx} \tag{3.7.2}$$

$$\begin{bmatrix} R_1 \\ R_2 \\ R_3 \end{bmatrix} = \begin{bmatrix} -1.00 & 0.00 & 0.00 & 1.00 & 0.00 & 0.00 & 0.00 & 0.00 & 0.00 \\ 0.00 & 0.00 & 0.00 & -0.25 & -0.97 & 0.00 & 0.25 & 0.97 & 0.00 \\ 0.00 & -1.04 & 0.00 & -1.01 & 1.30 & 0.00 & 1.01 & -0.26 & 0.00 \end{bmatrix} \begin{bmatrix} x_1 \\ x_2 \\ x_3 \\ x_4 \\ x_5 \\ x_6 \\ x_7 \\ x_8 \\ x_9 \end{bmatrix}$$

80　POLYATOMIC MOLECULES

Note that the Cartesian displacements (the **x** vector) are written as x_1 through x_9, corresponding to the x, y and z displacement coordinates of atoms 1, 2, and 3. For the two stretching modes R_1 and R_2, the Cartesian displacement coordinates that contribute are shown in Fig. 3.6. It is easily seen that R_1 has only two contributions, $-x_{H_1}(-x_1)$ and $x_O(x_4)$. Similarly, the figure shows that four components of the **x** vectors are needed to describe R_2.

For computations, the **B** matrix may be stored in a format similar to the **X** matrix before. The first index denotes the number of the internal coordinate, the second number the component of the x vector, followed by the actual matrix element.

1	1	−1.000		1	4	1.000		2	4	−0.250		2	7	0.250
2	5	−0.968		2	8	0.968		3	7	1.008		3	4	−1.008
3	2	−1.042		3	8	−0.261		3	5	1.302				

The **B** and **X** matrices are the two computed quantities that depend on the geometry of the molecule. Next, the potential energy is defined in internal coordinate (or in UB) space, and then converted to Cartesian coordinate space via the transform described in Eq. 3.7.15 for speedy diagonalization.

The definition of the $\mathscr{F}$ matrix proceeds via Eq. 3.7.25:

$$\mathscr{F}_{ij} = \sum Z_{ijk} \Phi_k \quad \text{or} \quad \mathscr{F} = \mathbf{Z}\mathbf{\Phi} \qquad (3.7.25)$$

where $\mathbf{\Phi}$ is a vector of force constants, and the **Z** matrix determines where each force constant appears in the $\mathscr{F}$ matrix. Thus the three indices of the **Z** matrix denote the row (i) and column (j) position of the force constant Φ_k in the $\mathscr{F}$ matrix.

In the water example, the two O—H stretching coordinates R_1 and R_2 have one force constant associated with them, namely, the O—H stretching force constant f_{O-H}, which is numerically the same for both coordinates. This is, of course, due to the fact that internal coordinates are chosen such that chemically equivalent bonds have the same force constants. In order to facilitate the perturbation of the force field to best reproduce the observed frequencies, it is advantageous to input the value of the O—H stretching force constant in the above example only once. The **Z** matrix is used to place the appropriate value of f_{O-H} into the appropriate positions of the $\mathscr{F}$ matrix.

For the above example of two equivalent O—H stretches associated with one force constant f_{O-H}, and the bending coordinate associated with the deformation constant f_{H-O-H}, the **Z** matrix is

1	1	1	1.000	for R_1
2	2	1	1.000	for R_2
3	3	2	1.000	for R_3

EXAMPLE FOR THE COMPUTATION OF NORMAL MODES OF VIBRATION 81

where it is assumed that f_{O-H} is the first force constant in the vector $\mathbf{\Phi}$, and f_{H-O-H} is the second force constant in the vector. The coefficients 1.000 are used in the case of the GVFF and specify that the unscaled value of the force constant be used in the $\mathscr{F}$ matrix.

Using only these two diagonal force constants is tantamount to using the valence force field (VFF), which was abandoned in favor of the GVFF, because it generally gives much better results. Using numerical values for

$$f_{O-H} = 7.7 \text{ mdyn/Å}$$

$$f_{H-O-H} = 0.65 \text{ mdyn/radian}$$

and masses of 1.008 and 16.000 for H and O, respectively, the data summarized in Table 3.2 are obtained. A number of comments about Table 3.2 are in order. This output was obtained from program NOCO [Barlow and Diem, 1991]. The observed frequencies are input into the computations along with an assignment of the vibration. Thus the input file contains a line

$$3756 \quad v_{as}(O-H)$$

indicating that the vibration at 3756 cm^{-1} is the antisymmetric O—H stretching vibration. This assignment must be in agreement with the PED, which is listed in the line(s) following the observed and calculated frequencies for each mode, enumerating all force constants contributing more than 10% (or a fraction of 0.1) to the vibration. In Table 3.2, each mode has only one force constant listed in the PED, indicating 100% contribution of the force constant listed to the normal mode.

TABLE 3.2. Calculated Frequencies for Water Using a Valence Force Field

Mode	Observed Frequency [cm^{-1}]	Calculated Frequency [cm^{-1}]	Difference [cm^{-1}]	Percent Error
1	3756.0	3740.9	15.1	0.402
2	3652.0	3687.2	−35.2	−0.964
3	1595.0	1597.9	−2.9	−0.182

Average error = 17.7 cm^{-1} or 0.52%
Potential energy distribution:
obs: 3756.0, calc: 3740.9, assignment: $v_{as}(O-H)$
  1.00 Force constant #1 $f(O-H)$
obs: 3652.0, calc: 3687.2, assignment: $v_s(O-H)$
  1.00 Force constant #1 $f(O-H)$
obs: 1595.0, calc: 1597.9, assignment: $\delta(H-O-H)$
  1.00 Force constant #2 $f(H-O-H)$

Since the splitting of the symmetric and antisymmetric stretching modes is underestimated in the computations, it is necessary to input a stretch–stretch interaction constant, $f_{\text{O-H/O-H}}$. Thus the Z matrix becomes

$$
\begin{array}{cccc}
1 & 1 & 1 & 1.000 \\
2 & 2 & 1 & 1.000 \\
3 & 3 & 2 & 1.000 \\
1 & 2 & 3 & 1.000
\end{array}
$$

indicating that force constant 3 causes interaction between internal coordinates 1 and 2. After a minor readjustment of the force constants to the values indicated below,

$$f_{\text{O-H}} = 7.65$$
$$f_{\text{H-O-H}} = 0.648$$
$$f_{\text{O-H/O-H}} = -0.10$$

new vibrational frequencies and a new PED are obtained, which are listed in Table 3.3. Again, a number of features are interesting about these results. Obviously, the fit between observed and calculated frequencies has improved drastically, as compared to the results shown in Table 3.2. Second, the contributions in the PED have changed. The antisymmetric stretching mode at 3753 cm^{-1} now has a fractional contribution of $f_{\text{O-H}}$ of 0.99, and presumably of 0.01 of $f_{\text{O-H/O-H}}$. However, since only contributions above 10% are listed, this latter contribution is not shown. The contribution of 101% of $f_{\text{O-H}}$ to the symmetric stretching mode at 3651 cm^{-1} is due to the fact that the contribution of $f_{\text{O-H/O-H}}$ is negative for this mode.

An inspection of the results in Table 3.3 suggests that the force constants are very well chosen. However, this is not quite the case. Including in the calculations the water isotopomer D_2O, one sees that the agreement there is much poorer, as shown by the discrepancies between observed and calculated frequencies. This error is unacceptable and a force field has to be derived that fits both molecules. Thus the remaining off-diagonal force constant needs to be included, which is the stretch–bend interaction constant, $f_{\text{O-H/H-O-H}}$, which appears twice in the potential energy matrix:

$$
\begin{array}{cccc}
1 & 1 & 1 & 1.000 \\
2 & 2 & 1 & 1.000 \\
3 & 3 & 2 & 1.000 \\
1 & 2 & 3 & 1.000 \\
1 & 3 & 4 & 1.000 \\
2 & 3 & 4 & 1.000
\end{array}
$$

EXAMPLE FOR THE COMPUTATION OF NORMAL MODES OF VIBRATION   83

TABLE 3.3. Calculated Frequencies for H_2O and D_2O, Using a GVFF

	H_2O			
Mode	Observed Frequency [cm^{-1}]	Calculated Frequency [cm^{-1}]	Difference [cm^{-1}]	Percent Error
1	3756.0	3753.0	3.0	0.079
2	3652.0	3651.2	0.8	0.023
3	1595.0	1595.4	−0.4	−0.027

Average error = 1.4 cm^{-1} or 0.04%
Potential energy distribution:
obs: 3756.0, calc: 3753.0, assignment: v_{as}(O—H)
  0.99 Force constant #1 f(O—H)
obs: 3652.0, calc: 3651.2, assignment: v_s(O—H)
  1.01 Force constant #1 f(O—H)
obs: 1595.0, calc: 1595.4, assignment: δ(H—O—H)
  1.00 Force constant #2 f(H—O—H)

	D_2O			
Mode	Observed Frequency [cm^{-1}]	Calculated Frequency [cm^{-1}]	Difference [cm^{-1}]	Percent Error
1	2784.0	2749.0	35.0	1.258
2	2666.0	2642.1	23.9	0.897
3	1179.0	1162.4	16.6	1.428

Average error = 25.2 cm^{-1} or 1.19%

indicating interaction of the stretching coordinate R_1 with the deformation and the stretching coordinate R_2 with the deformation. An attempt to fit all six fundamentals properly with these four force constants is by no means a trivial task, as anyone with access to the normal coordinate programs can verify.

It should be noted that there exist some discrepancies in the way internal coordinates are scaled. Many researchers have multiplied the angle deformation coordinates by a bond length in order to preserve the same units as those obtained for stretching coordinates. Thus the bond angle deformation constants reported by different workers may disagree by ±15%.

As mentioned in the discussion in Section 3.8, an alternative approach to the GVFF is the Urey–Bradley force field, for which only three force constants are needed to define the entire potential energy matrix. These force constants are the diagonal H—O—H deformation and the O—H stretching force

84 POLYATOMIC MOLECULES

constants, and the nonbonded H⋯H interaction. The Z matrix is calculated using on "GEM" configuration [Schachtschneider, 1964], which describes the nonbonded interaction energy terms for a triad of atoms A⋯B⋯C. Using Schachtschneider's program UBZM, slightly modified by Barlow and Diem, the Z-matrix elements for the water molecule are calculated to be

```
1  1  1   1.000      2  2  1   1.000      3  3  2   1.000
1  1  3   0.588      2  2  3   0.588      3  3  3   0.403
1  2  3   0.663      2  3  3   0.418      1  3  3   0.418
```

The first three entries give the contributions of the diagonal force constants to the potential energy in complete analogy to the case of a GVFF force field. The next three terms indicate that all three internal coordinates change the (nonbonded) distance between the H atoms; thus force constant #3, the nonbonded interaction constant $f_{H\cdots H}$, contributes to the potential energy of all three internal coordinates in addition to the appropriate diagonal force constants. The last three terms represent the stretch–stretch and stretch–bend interactions; the term

$$1 \quad 2 \quad 3 \quad 0.663$$

is the contribution of the nonbonded interaction constant expressed as a stretch–stretch interaction constant in GVFF, and the terms

$$2 \quad 3 \quad 3 \quad 0.418 \qquad 1 \quad 3 \quad 3 \quad 0.418$$

represent the corresponding stretch–bend interactions. All matrix elements discussed in this section are listed to a maximum of four significant figures, although it is necessary in the computations to carry them to the limit of the compiler to reduce round-off errors.

Using numerical values for the UB force constants,

$$1 \quad 7.450 \quad f(O\text{—}H)$$
$$2 \quad 0.420 \quad f(H\text{—}O\text{—}H)$$
$$3 \quad 0.600 \quad k(H\cdots H)$$

the results listed in Table 3.4 are obtained. These force constants represent trial values, and no effort was made to refine their values. Nevertheless, the fit between observed and calculated frequencies is reasonable for trial values of force constants and can be improved by adjusting their values. This fit was achieved using one less force constant than in the case of the GVFF. Although this reduction in the number of force constants may seem irrelevant, for larger molecules this reduction can make the difference between a realistic and

TABLE 3.4. Calculated Frequencies for H_2O and D_2O using UBFF

	H_2O			
Mode	Observed Frequency [cm^{-1}]	Calculated Frequency [cm^{-1}]	Difference [cm^{-1}]	Percent Error
1	3756.0	3797.5	−41.5	−1.105
2	3652.0	3668.6	−16.6	−0.453
3	1595.0	1596.7	−1.7	−0.109

Average error = 19.9 cm^{-1} or 0.56%
Potential energy distribution:
obs: 3756.0, calc: 3797.5, assignment: v_{as}(O—H)
  0.91 Force constant #1 f(O—H)
obs: 3652.0, calc: 3668.6, assignment: v_s(O—H)
  1.01 Force constant #1 f(O—H)
obs: 1595.0, calc: 1596.7, assignment: δ(H—O—H)
  0.65 Force constant #2 f(H—O—H)
  0.33 Force constant #3 k(H···H)

	D_2O			
Mode	Observed Frequency [cm^{-1}]	Calculated Frequency [cm^{-1}]	Difference [cm^{-1}]	Percent Error
1	2784.0	2736.5	47.5	1.708
2	2666.0	2687.1	−21.1	−0.792
3	1179.0	1168.3	10.7	0.910

Average error = 26.5 cm^{-1} or 1.14%
Potential energy distribution:
obs: 2784.0, calc: 2736.5, assignment: v_{as}(O—H)
  0.91 Force constant #1 f(O—H)
obs: 2666.0, calc: 2687.1, assignment: v_s(O—H)
  1.01 Force constant #1 f(O—H)
obs: 1179.0, calc: 1168.3, assignment: δ(H—O—H)
  0.65 Force constant #2 f(H—O—H)
  0.33 Force constant #3 k(H···H)

unreliable force field. All the force constants listed in this example, and the GVFF calculations before, are unrefined and should be construed as starting values for a detailed force field analysis.

In the case of the deformation modes, it can be seen that two force constants

86 POLYATOMIC MOLECULES

contribute to the PED; thus it appears that the deformation mode is affected most by the nonbonded interaction.

Finally, the atomic displacement vectors can be obtained from these calculations. For H_2O, they are listed in Table 3.5. The reason that there is no apparent symmetry in these displacement vectors is due to the choice of the coordinate system in which the Cartesian equilibrium coordinates of water were defined (see Fig. 3.6).

3.10 VIBRATIONAL INTENSITIES: ABSORPTION

Most of Chapter 3 has dealt with the problem of calculating the normal vibrations of a molecule, and the question of the vibrational intensity has been neglected so far. The omission of the discussion of vibrational intensities is unfortunately a common feature of books on vibrational spectroscopy; the reason for this omission is, of course, the difficulties one encounters when one attempts to compute the infrared absorption strength of molecular vibrational transitions. Incidentally, recent progress in the field of computing infrared absorption intensities must be credited, to a large extent, to the researchers in

TABLE 3.5. Atomic Displacement Vectors for H_2O

Atom	Mode 1: $v_{as}(O-H)$		
	ΔX	ΔY	ΔZ
H_1	0.6893	0.0450	0.0000
O	−0.0298	0.0385	0.0000
H_2	−0.2162	0.6561	0.0000

Atom	Mode 2: $v_s(O-H)$		
	ΔX	ΔY	ΔZ
H_1	−0.6781	0.0000	0.0000
O	−0.0534	−0.0414	0.0000
H_2	0.1698	0.6565	0.0000

Atom	Mode 3: $\delta(H-O-H)$		
	ΔX	ΔY	ΔZ
H_1	0.0061	0.6777	0.0000
O	−0.0416	−0.0538	0.0000
H_2	−0.6546	0.1756	0.0000

the field of infrared circular dichroism (cf. Chapter 9), who had to develop the theory of infrared intensities along with the theory in infrared CD.

The integrated infrared absorption intensity or dipole strength for the jth normal vibration is given by

$$D_{01}^j = \int (\varepsilon/v) dv = |\mu_{01}|^2 \tag{3.10.1}$$

$$= \frac{\hbar}{4\pi v_j} \left|\frac{\partial \mu}{\partial Q_j}\right|^2 \tag{3.10.2}$$

where v_j is the frequency of the jth normal vibration associated with normal coordinate Q_j. The difficulty lies, of course, in deriving a method that predicts the dipole derivatives from any available experimental or theoretical procedures.

Early attempts to quantify infrared intensities came from the argument that the observed absorption intensity for a diatomic molecule is given by only one term $(\partial \mu / \partial Q)$. If the internuclear distance and the permanent dipole moment are known, a numerical value for the dipole change can be obtained. These dipole changes were subsequently translated into static partial charges attached to, and vibrating with, the atoms. Originally, it was hoped that "fixed partial charge" parameters could be obtained and transferred between atoms in different molecules to compute infrared intensities.

In this "fixed partial charges" approach, as generalized to polyatomic molecules, the dipole change may be expressed as

$$\left(\frac{\partial \mu}{\partial Q_j}\right) = \sum_n \xi_n \left(\frac{\partial \mathbf{R}_n}{\partial Q_j}\right) \tag{3.10.3}$$

where ξ is the partial, or shielded, nuclear charge of nucleus n, and $\mathbf{R}$ is its equilibrium position. The term $(\partial \mathbf{R}_n / \partial Q_j)$ is the trajectory of the nth nucleus in the jth normal mode [Freedman and Nafie, 1987].

Unfortunately, this method is accurate in predicting vibrational intensities for very selected vibrational modes only. For example, CH stretching intensities may be predicted reasonably well from Eq. 3.10.3, since the very nonpolar character of the C—H bond does not change appreciably upon bond elongation.

A more sophisticated approach, known as the "polar tensor" model, improves the predictions somewhat. In the polar tensor approach, the dipole change is expressed as

$$\left(\frac{\partial \mu}{\partial Q_j}\right) = \sum_n \left(\frac{\partial \mu}{\partial \mathbf{R}_n}\right) \cdot \left(\frac{\partial \mathbf{R}_n}{\partial Q_j}\right) \tag{3.10.4}$$

Here, the terms $(\partial \mu / \partial \mathbf{R}_n)$ are the derivatives of the total molecular dipole moment with respect to the Cartesian displacements of nucleus n. This derivative matrix is known as the polar tensor. For symmetric molecules, Newton and Person [1976] were successful in reducing this molecular polar tensor to an atomic polar tensor format. It was found that these experimentally derived atomic polar tensors were transferable to some degree from one molecule to another. However, even this model was far from a generally applicable approach to predict infrared intensities.

The reason for the shortcomings of the earlier models lies in the fact that the "partial charges" of the atoms vary enormously during vibrations. In fact, they vary so much that vibronic methods need to be invoked to predict the dipole change during a vibration properly. The charge redistributions, or "charge flows," are so significant that apparently only *ab initio* quantum mechanical calculations are able to predict the dipole changes properly.

These quantum mechanical methods have only been developed and refined in the past decade or so to the point that relatively reliable infrared intensities can be computed. In the earlier quantum mechanical models, the "localized molecular orbital" (LMO), by Nafie and co-workers [Nafie and Polavarapu, 1981], the effect of bare nuclear charges oscillating was offset by the motion of the centroids of the localized molecular orbital:

$$\left(\frac{\partial \mathbf{\mu}}{\partial Q_j}\right) = \sum_n Z_n e \left(\frac{\partial \mathbf{R}_n}{\partial Q_j}\right) - \sum_k e \left(\frac{\partial \mathbf{r}_k}{\partial Q_j}\right) \qquad (3.10.5)$$

where $Z_n e$ is the complete nuclear charge of nucleus n, and $(\partial \mathbf{r}_k / \partial Q_j)$ is the displacement of the orbital centroids during a normal vibration. This method requires the computation (preferably at the *ab initio* level) of molecular orbitals and their decomposition into localized orbitals for the atomic equilibrium positions, and the atomic positions at various positions along the normal coordinate.

REFERENCES

A. Barlow and M. Diem, *J. Chem. Ed.*, 68, 35 (1991).

E. Butkov, *Mathematical Physics*, Addison-Wesley, Reading, MA, 1968.

N. B. Colthup, L. H. Daly, and S. E. Wiberly, *Introduction to Infrared and Raman Spectroscopy*, Academic Press, New York, 1990.

F. R. Dollish, W. G. Fateley, and F. F. Bentley, *Characteristic Raman Frequencies of Organic Compounds*, Wiley, New York, 1974.

J. R. Durig and J. F. Sullivan, in *Chemical, Biological and Industrial Applications of Infrared Spectroscopy*, J. R. Durig, Ed., Wiley, New York, 1985.

T. B. Freedman and L. A. Nafie, in *Stereochemical Applications of Vibrational Optical Activity*, E. L. Eliel and S. H. Wilen, Eds., Wiley, New York, 1987.

G. Herzberg, *Molecular Spectra and Molecular Structure. II. Infrared and Raman Spectra of Polyatomic Molecules*, Van Nostrand Reinhold, New York, 1945.

L. A. Nafie and P. L. Polavarapu, *J. Chem. Phys.*, *75*, 2935 (1981).

J. H. Newton and W. B. Person, *J. Chem. Phys.*, *64*, 30306 (1976).

Quantum Mechanical Program Exchange, University of Indiana, Bloomington, IN, Program BMAT (QeMP 067).

J. H. Schachtschneider, *Vibrational Analysis of Polyatomic Molecules, Volumes* and VI, Shell Development Company Report, Emeryville, CA, 1964.

H. C. Urey and C. A. Bradley, *Phys. Rev.*, *38*, 1969 (1931).

E. B. Wilson, J. C. Decius, and P. C. Cross, *Molecular Vibrations: The Theory of Infrared and Raman Vibrational Spectra*, McGraw-Hill, New York, 1955.

4

SYMMETRY OF MOLECULAR VIBRATIONS

It was pointed out before in Chapter 2 that inspection of the wavefunctions and the transition operator sometimes can lead to a decision whether or not a transition moment is zero. In the case of the particle-in-a-box wavefunctions, for example, one can easily see that multiplication and integration of two wavefunctions with consecutive quantum numbers will produce a zero net area (cf. Fig. 2.2); that is, the wavefunctions are orthogonal. Multiplying this product of two consecutive wavefunctions by x changes the symmetry properties of the product, since the function $y = x$ is odd. Thus the integral, over the triple product, and the transition moment are nonzero. Therefore a mere inspection of the symmetry (reflection) properties of the particle-in-a-box or the harmonic oscillator wavefunctions can determine whether or not a transition is allowed.

In polyatomic molecules, however, the argument of whether or not a transition is allowed becomes more complicated. This is because an N-atomic molecule has $3N - 6$ degrees of vibrational freedom, and the total vibrational wavefunctions of ground and excited states are products of the wavefunctions of each vibrational coordinate, as discussed in Section 3.3. Thus the symmetry of the normal modes of vibration and their associated wavefunctions need to be evaluated. One finds that each normal mode of vibration can be described by a symmetry representation, and the transformation properties of such a representation will determine whether or not a given normal mode of vibration is symmetry allowed.

This is, of course, the subject of group theory. This topic is much too broad to be treated in detail in just one chapter; thus the reader is referred to specialized textbooks [Cotton, 1963; Ferraro and Ziomek, 1969] for the derivations and more examples. However, the basic concepts of group theory, as applied to vibrational spectroscopy, will be presented, and a connection

SYMMETRY OPERATIONS AND GROUPS

between this relatively unfamiliar branch of mathematics to vector spaces and linear algebra will be made. The aim of this chapter is to present sufficient information for the interpretation of vibrational spectra and the derivation of symmetry-based selection rules. Hopefully, enough of the basic treatment of molecular symmetry is contained in this chapter that no other textbook is required for a course in vibrational spectroscopy.

4.1 SYMMETRY OPERATIONS AND GROUPS

Molecules are classified into symmetry-related categories, named symmetry groups, according to the number and nature of symmetry operations that can be carried out on a molecule. Symmetry operations are procedures applied to a molecule such that after the operation has been carried out, the molecule appears in an indistinguishable form. A simple example of a symmetry operation is shown in Fig. 4.1. When a water molecule is rotated by 180° about the axis shown, the molecule appears identical to and indistinguishable from its original state aside from the numbering of the hydrogen atoms. We call this particular symmetry operation a proper, twofold rotation about a symmetry axis, and refer to it as a C_2 symmetry operation.

Investigating the symmetry properties of a number of arbitrary molecules, or even macroscopic objects such as snow crystals, a cube, a flower, certain letters, and many other items which we intuitively associate with being "symmetric," one finds that there exist only a few distinct and independent symmetry operations. These can be combined and repeated to create other symmetry operations. For the discussion of the symmetry properties of small molecules for molecular spectroscopy, there are five symmetry operations that are of interest: the identity element, proper axes of rotation, reflection by mirror plane, center of inversion, and rotation–reflection axis (or improper axes of rotation). These operations will be discussed in turn.

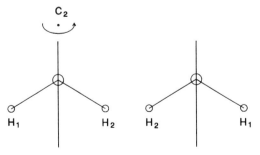

FIGURE 4.1. Definition of a proper rotation about the Z-axis by 180°. After the rotation, the molecule appears unchanged, except atoms 1 and 2 have changed positions.

1. *The Identity Element*, Designated (E). This operation leaves the molecule unchanged, and its inclusion in this discussion is necessary for two reasons. The first of these is purely mathematical and will be discussed below. The other reason can be viewed intuitively as follows: we define a symmetry operation as a procedure that will return the molecule into a state that is indistinguishable from the state before the operation. Thus we need to include in our formalism the possibility that the molecules *actually was left unchanged*, that is, the molecule was not rotated or operated on at all.

2. *Proper Axes of Rotation* (C_n). This operation was introduced above for a twofold rotation. In general, we designate such an axis as a C_n operation, where n describes how often the operation has to be repeated until a complete rotation by 360° has occurred. Thus C_n indicates a rotation about the given axis by $360/n$ degrees.

3. *Reflection by Mirror Plane* (σ). Reflections by two different mirror planes are shown in Fig. 4.2. The mirror plane can be perpendicular to the major axis of symmetry. In this case, the mirror plane is designated σ_h (Fig. 4.2B). If the mirror plane contains the axis, it is referred to as a σ_v plane (Fig. 4.2A).

4. *Center of Inversion* (i). When pairs of atoms are at positions for which the coordinates are identically the same except that all signs are reversed, then these atoms are related by a center of inversion. Planar ethene (C_2H_4) is a typical example (Fig. 4.3): the hydrogen atoms 1 and 4, for example, have equal and opposite coordinates if the coordinate origin lies at the center of the C=C bond. In this case, the center of inversion lies at the origin.

5. *Rotation–Reflection Axis or Improper Axes of Rotation* (S_n). This operation consists of rotation about an axis by $360°/n$, followed by a reflection in a plane perpendicular to the axis. The hydrogen atoms in ethane, for example,

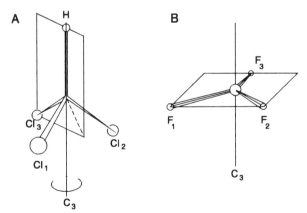

FIGURE 4.2. Definition of symmetry planes (or reflection operation). (A) One of three σ_v planes of chloroform. The reflection operation will exchange atoms 1 and 2. The plane contains the major axis (the C_3 axis). (B) The horizontal plane σ_h in BF_3, which is perpendicular to the major axis.

SYMMETRY OPERATIONS AND GROUPS   93

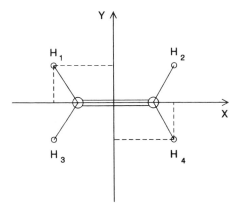

FIGURE 4.3. Definition of a center of inversion. Planar C_2H_4 has a center of inversion at the coordinate origin.

are related by an S_6 operation (Fig. 4.4): rotation of the molecule about the direction of the C—C bond by 60° brings the top three hydrogen atoms into a position opposite to the original position of the lower three hydrogen atoms. Reflection of the three upper hydrogens by a plane bisecting the C—C bond brings the upper three hydrogen atoms into the positions originally occupied by the lower three hydrogen atoms.

It is a matter of practice to identify all the symmetry elements of a molecule. Benzene, with a perfect hexagonal structure, has the following symmetry elements: one C_6 axis, two C_3 axes coincident with the C_6 axis, one C_2 axis coincident with the C_2 axis, three C_2 axes that bisect opposing sides, three C_2 axes that contain opposing atoms, a center of inversion, a horizontal symmetry

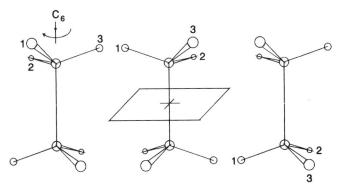

FIGURE 4.4. Definition of a sixfold rotation–reflection operation in C_2H_6. The operation can be broken down into a C_6 rotation, followed by a reflection by σ_h. After the operation, the molecule appears unchanged, except that all atoms have changed positions.

plane σ_h, three σ_v planes that contain opposing atoms, three σ_d planes that bisect opposing bonds, two S_6 axes coincident with the C_6 axes, and two S_3 axes. When one hydrogen atom is substituted by deuterium, or one carbon by a ^{13}C atom, the symmetry is lowered enormously to just contain one each of the following elements: E, C_2, σ_v, and σ'_v.

Certain molecules, such as water, benzene-d_1, and formaldehyde, H_2CO (cf. Chapter 7), possess the same symmetry elements, namely, E, C_2, σ_v, and σ'_v. (The prime denotes that there are two mutually perpendicular planes that are both parallel to the major axis. These symmetry elements are also denoted as $\sigma_v(yz)$ and $\sigma_v(xz)$, respectively.) The occurrence of the same symmetry elements in different molecules suggests that the grouping of the four symmetry elements above is, indeed, special, and it can be shown that the four symmetry elements form a group. A group in the mathematical sense is a collection of elements that may be connected according to certain rules. These rules are listed below for the case of symmetry groups but can similarly be written for other mathematical groups:

1. Successive applications of any two or more operations of the group produce another symmetry operation of the group. For example, successive applications of the C_2 operation brings the molecule back to its original orientation. Thus in the symmetry group of water and formaldehyde,

$$C_2 C_2 = E$$

2. There exists an identity element E such that for any operation A

$$AE = A$$

3. The associative law holds. Let A, B, and C be symmetry operations in a group. Then the order in which they are applied to a molecules does not matter:

$$A(BC) = (AB)C$$

4. Every operation must have an inverse. The symmetry operator A will have an inverse operation, A^{-1}, such that $AA^{-1} = E$.

Another example, aside from symmetry groups, is presented here to demonstrate that the concept of groups is common in mathematics. The integer numbers, for example, form a mathematical group for the addition operation. The four conditions above hold for all integers:

1. Addition of integers yields integers only.
2. The number zero is the identity element, since addition of zero leaves any integer unchanged.

3. The order of addition is immaterial.
4. There exists an inverse, namely, the negative of each number, such that zero is obtained by adding a number and its inverse.

For the study of molecular symmetry, we need to concern ourselves with about 40 different symmetry groups, of which only about a dozen or so are common in molecules. The groups are given special designations depending on the symmetry elements they contain, and the particular group with the symmetry elements discussed above is known as C_{2v}:

$$C_{2v}: E \quad C_2 \quad \sigma_v \quad \sigma_v'$$

In this group, there are four elements, and each one is in a separate class, which implies that none of the symmetry operations occurs more than once. Furthermore, we define the order of a group—and designate it as h—as the sum of all the symmetry operations. Thus the order of C_{2v} is $h = 4$.

Ammonia, NH_3 belongs to a symmetry group known as C_{3v}, which contains the following elements:

$$C_{3v}: E \quad 2C_3 \quad 3\sigma_v$$

The two threefold axes are clockwise and counterclockwise rotation by 120°, and the three symmetry planes are all parallel to the C_3 axes and contain one of the N—H bonds each. Thus in C_{3v} there are six symmetry elements in three classes, and some elements occur more than once in this group. The order of the group is $h = 6$. The order of a group and the number of classes will be used in the next section in the derivation of the irreducible representations (*vide infra*). Some of the common symmetry groups are listed in Appendix III. For the classification of molecules into these symmetry groups, the reader is referred to introductory texts on physical chemistry.

4.2 GROUP REPRESENTATIONS

Mathematically, we may describe the effect of a symmetry operator on the coordinates of a single atom in Cartesian coordinates by a 3×3 matrix. For example, an identity operation is represented by the matrix

$$\begin{bmatrix} 1 & 0 & 0 \\ 0 & 1 & 0 \\ 0 & 0 & 1 \end{bmatrix} = E \qquad (4.2.1)$$

indicating that the x, y, and z Cartesian coordinates transform into themselves. For an n-fold (proper) rotation about the z-axis by $\Theta = 360°/n$, the transform-

96 SYMMETRY OF MOLECULAR VIBRATIONS

ation matrix is

$$\begin{bmatrix} \cos\Theta & -\sin\Theta & 0 \\ \sin\Theta & \cos\Theta & 0 \\ 0 & 0 & 1 \end{bmatrix} = C_n \qquad (4.2.2)$$

The product of two matrices for two consecutive operations must be the same as the matrix representing the product operation, which is another member of the group. There are countless combinations of symmetry elements possible, since we need not restrict ourselves to binary combinations only.

The matrices specific for symmetry operations are known as "representations" of a group. A representation is defined as a set of matrices, representing single operations of a group, which can be combined in the same way the operations of a group can be combined. Thus any combination of the following set of matrices describe one (of many) representation for an atom in the C_{2v} symmetry group:

$$\begin{array}{cccc} E & C_2 & \sigma_{xz} & \sigma_{yz} \\ \begin{bmatrix} 1 & 0 & 0 \\ 0 & 1 & 0 \\ 0 & 0 & 1 \end{bmatrix} & \begin{bmatrix} -1 & 0 & 0 \\ 0 & -1 & 0 \\ 0 & 0 & 1 \end{bmatrix} & \begin{bmatrix} 1 & 0 & 0 \\ 0 & -1 & 0 \\ 0 & 0 & 1 \end{bmatrix} & \begin{bmatrix} -1 & 0 & 0 \\ 0 & 1 & 0 \\ 0 & 0 & 1 \end{bmatrix} \end{array}$$

Furthermore, in a molecule, one can attach a Cartesian coordinate system to each atom and determine the transformation matrices for each atom's coordinates under the symmetry operations of a group. The traces of these transformation matrices also form a representation, characteristic of the group and the molecule.

However, there is a way to represent a group in such a manner that the representation is of the simplest possible form. This can be visualized by taking general points, and studying their transformation properties under the symmetry elements of the group. Under C_{2v}, a point along the z axis (the symmetry axes) would be symmetric ($+1$) under all symmetry operations of the group and thus have a representation of $\{1 \ 1 \ 1 \ 1\}$ under the four operations E, C_2, σ_v, and σ_v'. A point on the x-axis, however, would transform antisymmetrically (-1) under C_2 and σ_v' and thus would have a representation $\{1 \ -1 \ 1 \ -1\}$. It can be shown (*vide infra*) that these representations are the equivalent of unit vectors in a space whose dimension is given by the number of symmetry elements in a group. Thus C_{2v} has four of these unit vectors, which are known as the *irreducible representations* of a group. The symmetry operations and irreducible representations of a number of common symmetry groups are discussed below and listed in Appendix III.

The derivation of these four *irreducible representations* is quite complicated and follows directly from the *orthogonality theorem of group theory* [e.g., see

Cotton, 1963]. Some of the consequences of this theorem are:

1. The number of irreducible representations in a group is equal to the number of symmetry classes. Thus C_{2v} will have four, and C_{3v} will have three irreducible representations.
2. If we define a number l_i to be the dimensions of the ith irreducible representations, then

$$\sum l_i^2 = h \qquad (4.2.3)$$

This implies for C_{2v} that

$$l_1^2 + l_2^2 + l_3^2 + l_4^2 = 4 \qquad (4.2.4)$$

and thus $l_1 = l_2 = l_3 = l_4 = 1$. Thus C_{2v} has four one-dimensional representations, denoted Γ_1 through Γ_4. Similarly, for C_{3v}, we have according to Eq. 4.2.3

$$l_1^2 + l_2^2 + l_3^2 = 6 \qquad (4.2.5)$$

which yields values for l_1, l_2, and l_3 of 1,1, and 2. Thus C_{3v} will have two one-dimensional irreducible representations and one two-dimensional irreducible representation.

At this point, one defines the entries, known as the characters X, of all the orthonormal irreducible representations Γ. One of them was already discussed for C_{2v}, namely, that of a point along the Z-axis, which is

$$\Gamma_1: 1 \quad 1 \quad 1 \quad 1$$

Since the other three irreducible representations must be orthogonal to Γ_1, one can show easily that these other irreducible representations are

	E	C_2	σ_v	σ_v'
Γ_1	1	1	1	1
Γ_2	1	1	-1	-1
Γ_3	1	-1	1	-1
Γ_4	1	-1	-1	1

Here, orthogonality is interpreted in the same fashion as in n-dimensional vector spaces: denoting the entries for each Γ as X_i, one can show easily that the inner product of two irreducible representations, defined as

$$\sum_\alpha X_\alpha(\Gamma_i) X_\alpha(\Gamma_j) = 0 \qquad (4.2.6)$$

98 SYMMETRY OF MOLECULAR VIBRATIONS

is zero for any two irreducible representations, where the summation is from 1 to 4. For example, Γ_2 and Γ_4 are orthogonal since

$$(1)(1) + (1)(-1) + (-1)(-1) + (-1)(1) = 0$$

Similarly, the irreducible representations for C_{3v} are obtained from the first one, which is totally symmetric, and again represents the transformation of a point along the Z-axis for the operations E, $2C_2$, $3\sigma_v$:

$$\Gamma_1: 1 \quad 1 \quad 1$$

Using orthogonality, one finds that the three irreducible representations are

	E	$2C_2$	$3\sigma_v$
Γ_1	1	1	1
Γ_2	1	1	-1
Γ_3	2	-1	0

In order to prove that these irreducible representations are orthogonal, one has to multiply the characters by a number g, which indicates how often an operation occurs in a class (the number of elements in each class). Thus Eq. 4.2.6 needs to be modified as follows:

$$\sum_\alpha g_\alpha X_\alpha(\Gamma_i) X_\alpha(\Gamma_j) = 0 \tag{4.2.7}$$

to indicate orthogonality, and one can show readily that the three irreducible representations are, indeed, all orthogonal:

$$\Gamma_1 \otimes \Gamma_2 = (1)(1)(1) + (2)(1)(1) \quad + (3)(1)(-1) = 0$$
$$\Gamma_1 \otimes \Gamma_3 = (1)(1)(2) + (2)(1)(-1) + (3)(1)(0) \quad = 0$$
$$\Gamma_2 \otimes \Gamma_3 = (1)(1)(2) + (2)(1)(-1) + (3)(-1)(0) = 0$$

The tables shown above for C_{2v} and C_{3v} are known as *character tables*. Normally, the irreducible representations Γ are given special designations: A denotes representations that are symmetric ($+1$) with respect to the major axis of symmetry, B are those antisymmetric (-1) with respect to this axis. E denotes doubly degenerate representations with a character of 2, and T triply degenerate representations. Subscripts g and u (which do not occur in the two character tables shown here) denote symmetric (gerade) and antisymmetric (ungerade) with respect to a center of inversion. There is always a totally symmetric representation in the group, which is always written as the first row in the character table. Its designation is always A, and its exact nomenclature varies from group to group and may be A_1, A_{1g}, A_g, and so on. Thus these two character tables are usually written in the form

C_{2v}	E	C_2	σ_v	σ'_v		
A_1	1	1	1	1	T_z	
A_2	1	1	-1	-1		R_z
B_1	1	-1	1	-1	T_x	R_y
B_2	1	-1	-1	1	T_y	R_x

C_{3v}	E	$2C_2$	$3\sigma_v$		
A_1	1	1	1	T_z	
A_2	1	1	-1		R_z
E	2	-1	0	(T_x, T_y)	(R_x, R_y)

The meaning of the symbols T_x, T_y, R_x, and so on will be discussed in the next section.

Any general representation in a group can be decomposed into irreducible representations just as a vector in three-dimensional space can be decomposed into Cartesian unit vectors. We used above the concept of the inner product to discuss orthogonality. Similarly, we shall use the principle of projecting a vector onto the direction of a unit vector to determine its component along this direction. We shall return to the example of water, to demonstrate these principles. This molecule, along with a set of Cartesian coordinates for each of the atoms, is shown in Fig. 4.5. Applying one symmetry operation, for example, the C_2 rotation, relates the old (before the operation) coordinates x_i to the new (after the operation) coordinates (x'_i) as follows:

$$\begin{bmatrix} x'_1 \\ y'_1 \\ z'_1 \\ x'_2 \\ y'_2 \\ z'_2 \\ x'_3 \\ y'_3 \\ z'_3 \end{bmatrix} = \begin{bmatrix} 0 & 0 & 0 & 0 & 0 & 0 & -1 & 0 & 0 \\ 0 & 0 & 0 & 0 & 0 & 0 & 0 & -1 & 0 \\ 0 & 0 & 0 & 0 & 0 & 0 & 0 & 0 & 1 \\ 0 & 0 & 0 & -1 & 0 & 0 & 0 & 0 & 0 \\ 0 & 0 & 0 & 0 & -1 & 0 & 0 & 0 & 0 \\ 0 & 0 & 0 & 0 & 0 & 1 & 0 & 0 & 0 \\ -1 & 0 & 0 & 0 & 0 & 0 & 0 & 0 & 0 \\ 0 & -1 & 0 & 0 & 0 & 0 & 0 & 0 & 0 \\ 0 & 0 & 1 & 0 & 0 & 0 & 0 & 0 & 0 \end{bmatrix} \begin{bmatrix} x_1 \\ y_1 \\ z_1 \\ x_2 \\ y_2 \\ z_2 \\ x_3 \\ y_3 \\ z_3 \end{bmatrix}$$

The trace of this transformation matrix is -1. Similarly, the trace of the transformation matrix under the identity operation is 9, since each Cartesian

100 SYMMETRY OF MOLECULAR VIBRATIONS

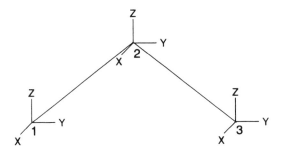

FIGURE 4.5. Atomic Cartesian displacement coordinates for water.

coordinate is transformed into itself. For σ_v and σ'_v, the traces are 3 and 1, respectively, which can be discovered by a little thought. Thus the representation of the coordinate systems attached to the atoms of water under the symmetry operations of C_{2v} is

$$\{9 \quad -1 \quad 3 \quad 1\}$$

which is referred to as a *reducible representation*. This reducible representation can be decomposed into its contribution of irreducible representations by "projecting" it onto the irreducible representations, which act as the unit vector for the group, as will be demonstrated in the next section.

4.3 REPRESENTATIONS OF MOLECULAR VIBRATIONS

In Fig. 4.5, we attached a Cartesian coordinate system to each of the atoms in a molecule. However, these coordinates could also be interpreted to be components of the Cartesian displacement vectors discussed in Section 3.2. The transformation matrix of these $3N$ Cartesian displacement vectors also forms a reducible representation under the symmetry operations of a group, and we have written the reducible representation for these displacement vectors of water as $\Gamma_C\{9 \quad -1 \quad 3 \quad 1\}$, where the subscript C denotes the representation of the Cartesian displacements. This vector can be decomposed into the four unit vectors $A_1, A_2, B_1,$ and B_2 of the C_{2v} symmetry group via the decomposition formula

$$a_i = (1/h)\sum g_\alpha X_\alpha(\Gamma_i) X_\alpha(\Gamma) \qquad (4.3.1)$$

On the left-hand side of Eq. 4.3.1, a_i is the number that defines how often the ith irreducible representation is contained in a given reducible representation. The right-hand side is again an inner product of two vectors as discussed before in Eq. 4.2.7. There, we took the inner product of two unit vectors, which was zero because of orthogonality. Here, we calculate the projection of a vector, Γ_C, onto the unit vectors, and a_i is just this projection.

In Eq. 4.3.1, $X_\alpha(\Gamma)$ are the characters of the representation to be reduced, and $X_\alpha(\Gamma_i)$ and g_α are the characters of a given symmetry representation and the number of elements in each symmetry class, respectively, as defined before.

Applying the reduction formula to the above reducible representation

$$\Gamma_C \ \{9 \quad -1 \quad 3 \quad 1\}$$

yields the following decomposition in terms of irreducible representations under C_{2v} symmetry:

$$a(A_1) = \tfrac{1}{4}[1\cdot 1\cdot 9 + 1\cdot 1\cdot(-1) + 1\cdot 1\cdot 3 + 1\cdot 1\cdot 1] = 3$$

$$a(A_2) = \tfrac{1}{4}[1\cdot 1\cdot 9 + 1\cdot 1\cdot(-1) + 1\cdot(-1)\cdot 3 + 1\cdot(-1)\cdot 1] = 1$$

$$a(B_1) = \tfrac{1}{4}[1\cdot 1\cdot 9 + 1\cdot(-1)\cdot(-1) + 1\cdot 1\cdot 3 + 1\cdot(-1)\cdot 1] = 3$$

$$a(B_2) = \tfrac{1}{4}[1\cdot 1\cdot 9 + 1\cdot(-1)\cdot(-1) + 1\cdot(-1)\cdot 3 + 1\cdot 1\cdot 1] = 2$$

Thus the reducible representation $\{9 \ -1 \ \ 3 \ \ 1\}$ can be decomposed into

$$3A_1 + A_2 + 3B_1 + 2B_2$$

This decomposition resulted in nine degrees of freedom, since three Cartesian degrees of freedom were assigned to each atom (cf. Fig. 4.5). Since there are only $3N - 6$, or three, degrees of vibrational freedom for the water molecule, the remaining six are three translational and three rotational degrees of freedom, which were included since $3N$ Cartesian displacement coordinates were used to define the reducible representation. The character table for C_{2v} above contains the irreducible representations for translation and rotation. These are marked as T_α ($\alpha = x$, y, or z) and R_α. Thus the translation of the water molecule in Fig. 4.5 along the positive Z-axis transforms as a A_1 representation.

Subtracting the irreducible representations for translation and rotation ($1A_1, 1A_2, 2B_1$, and $2B_2$), we end up with $2A_1 + B_1$ as the symmetry species for the three vibrations in water. The normal mode calculations discussed in the previous chapter readily identify these three vibrations, as shown in Fig. 4.6.

This procedure is generally applicable for any molecule. For ammonia with C_{3v} symmetry, for example, the reducible representation of the Cartesian displacement coodinates is $\Gamma_C\{12 \ \ 0 \ \ 2\}$. This representation can be derived easily using Eq. 4.2.2 for the rotation matrix applied to each of the Cartesian coordinate sets. The trace of Eq. 4.2.2, with $\Theta = 120°$, is zero, and the trace for the σ_v operation is $+1$. This trace enters into the overall transformation matrix only for the atoms that to not change position during the reflection operation.

Applying Eq. 4.3.1 to the reducible representation above yields the following

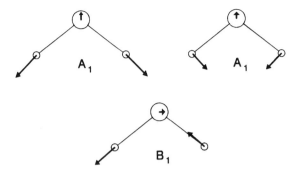

FIGURE 4.6. Normal modes of vibration for water. The displacement vectors are exaggerated with respect to the bond distances but are approximately drawn to size relative to each other.

components:

$$a(A_1) = \tfrac{1}{6}[1 \cdot 1 \cdot 12 + 2 \cdot 1 \cdot 0 + 3 \cdot 1 \cdot 2] = 3$$

$$a(A_2) = \tfrac{1}{6}[1 \cdot 1 \cdot 12 + 2 \cdot 1 \cdot 0 + 3 \cdot (-1) \cdot 2] = 1$$

$$a(E) = \tfrac{1}{6}[1 \cdot 2 \cdot 12 + 2 \cdot (-1)0 + 3 \cdot 0 \cdot 2] = 4$$

or $3A_1 + 1A_2 + 4E$, which indeed accounts for 12 degrees of freedom. The symmetry species of the *vibrational* degrees of freedom is obtained after subtracting A_1 and E for thre translational, and A_2 and E for three rotational degrees of freedom. The remaining $2A_1$ and $2E$ species are the symmetric and antisymmetric N—H stretching modes of A_1 and E symmetry, respectively, as well as the symmetric and antisymmetric H—N—H deformation modes.

Thus it is obvious that irreducible representations for each normal coordinate of a molecule can be found, and from it the number of vibrations for each symmetry species of the group. Similarly, we can use the procedure outlined above to determine whether or not an overtone or combination band contains a given irreducible representation. In the discussion of Fermi resonance of methyl vibrations (Section 3.4), we mentioned that under C_{3v} symmetry, the overtone of the degenerate (E) antisymmetric deformation mode interacts with the symmetric methyl stretching mode of A symmetry. Here, we demonstrate that the overtone $2\delta^{as}(CH_3)$ can mix with $v^s(CH_3)$. The representation of the overtone, $\Gamma_E \otimes \Gamma_E = \{2 \ -1 \ \ 0\} \otimes \{2 \ -1 \ \ 0\} = \{4 \ \ 1 \ \ 0\}$. Using Eq. 4.3.1, we find that A_1 is, indeed, contained in $\{4 \ \ 1 \ \ 0\}$, since $a(A_1) = 1$. Thus the overtone and the fundamental contain the same irreducible representation, and the two vibrations can interact.

Next, we need to establish whether or not a vibration of a given symmetry species will produce an observable infrared absorption. This is accomplished by considering how the dipole operator and the normal coordinates Q_k transform in a given symmetry group. For this, one needs to determine whether or not the dipole transition moment for a normal coordinate, which is $\int \psi'(Q_k) Q_k \psi(Q_k) \, d\tau$ (cf. Section 3.3), is zero. This will be discussed in the next section.

4.4 SELECTION RULES FOR NORMAL MODES OF VIBRATION

The energy eigenvalues of a system in a stationary state is given by the time-independent Schrödinger equation,

$$H\psi = E\psi \tag{4.4.1}$$

which, for our purposes, is the vibrational Schrödinger equation. When a symmetry operation is applied to a molecule described by this wave equation, the position and displacement vectors of the atoms may change; however, this exchange of coordinates does not affect the energy eigenvalues. Thus the energy is *invariant* under the symmetry operations. This statement is equivalent to saying that the symmetry operation R commutes with the Hamiltonian:

$$HR = RH \tag{4.4.2}$$

Furthermore, the eigenfunctions of the Hamiltonian provide the bases for the irreducible representations of a group, since the orthonormality condition and the equation above require (for a one-dimensional representation) that

$$R\psi = \pm 1\psi \tag{4.4.3}$$

Thus the transition moment must be nonzero in the space of the eigenfunctions (which are the normal coordinates)

$$\boldsymbol{\mu} = \int \psi'(Q_k) \boldsymbol{\mu}(Q_k) \, d\tau \neq 0 \tag{4.4.4}$$

in order for the transition to be allowed. Here, Q_k is the normal coordinate along which the transition occurs, and $\psi(Q_k)$ and $\psi'(Q_k)$ are the ground and excited state wavefunctions associated with this coordinate, respectively, and $\boldsymbol{\mu}$ is the electric dipole operator, with the components

$$\boldsymbol{\mu} = \mathbf{i}\mu_x + \mathbf{j}\mu_y + \mathbf{k}\mu_z \tag{4.4.5}$$

where μ_x, μ_y, and μ_z are the Cartesian components of $\boldsymbol{\mu}$. Thus we may write Eq. 4.4.4 as

$$\int \psi'(Q_k) \mu_x \psi(Q_k) \, d\tau + \int \psi'(Q_k) \mu_y \psi(Q_k) \, d\tau + \int \psi'(Q_k) \mu_z \psi(Q_k) \, d\tau \tag{4.4.6}$$

This expression implies that for a transition to occur, the integral over the excited state wavefunction, the ground state wavefunction, and the dipole operator must be nonzero for either one of the components of μ.

The integration of any product function will be zero unless the product is invariant (+1) under the symmetry operations of the group. This is the equivalent of the earlier arguments presented in Chapter 2 that the integrand of the particle-in-a-box or harmonic oscillator wavefunction transition moments must be even to be nonzero. Thus if we consider the integral

$$\int f_1 f_2 f_3 \, d\tau \tag{4.4.7}$$

we find that it is nonzero if and only if the product $f_1 f_2 f_3$ is invariant under all symmetry operations (is even). This condition is fulfilled if the product $f_1 f_2 f_3$ contains the totally symmetric representation of the group. For vibrational wavefunctions, this may be demonstrated by investigating the symmetry properties of the wavefunctions.

For each vibrational coordinate Q_k, we may write the associated wavefunction as (cf. Eq. 2.6.6)

$$\psi^v = N e^{-\gamma Q^2/2} H^{(v)}(\gamma^{1/2} Q) \tag{4.4.8}$$

where

$$H^{(0)}(\gamma^{1/2} Q) = 1$$
$$H^{(1)}(\gamma^{1/2} Q) = 2\gamma^{1/2} Q$$
$$H^{(2)}(\gamma^{1/2} Q) = 4\gamma Q^2 - 2 \quad \text{and so on}$$

Thus the ground state wavefunction, $\psi^0 = N e^{-\gamma Q^2/2}$, is always just a Gaussian function and therefore totally symmetric. The first excited state wavefunction, $\psi^1 = N 2\gamma^{1/2} Q e^{-\gamma Q^2/2}$ will always have the symmetry of Q itself. Thus the ground state $\psi(Q_k)$ in the integral in Eq. 4.4.6 above contains the symmetric representation. Therefore $\psi'(Q_k) \cdot \mu_\alpha (\alpha = x, y, \text{ or } z)$ must contain the totally symmetric representation for the entire integral in Eq. 4.4.6 to be nonzero. This is the case if the excited state representation contains the representation of the components of x, y, or z.

Again, an example may serve to illustrate this point. Consider a molecule of C_{2h} symmetry, such as planar N_2F_2 with the following structure:

for which the character table is

C_{2h}	E	C_2	i	σ	
A_g	1	1	1	1	
B_g	1	−1	1	−1	
A_u	1	1	−1	−1	T_z
B_u	1	−1	−1	1	T_x, T_y

In this molecule, the symmetric N—F stretching mode, where both F atoms move out in phase, belongs to the totally symmetric representation, A_g. If we designate this coordinate as Q_1, we find that

$$\Gamma[\psi(Q_1)] = A_g$$

The *excited state* vibrational wavefunction for this coordinate, $\psi'(Q_1)$, has the symmetry of Q_1 (which is A_g), which does not contain a component of x, y, or z. The transformation properties of each irreducible representation are obtained, of course, from the character table above, which shows that the T_x, T_y, and T_z components occur only in the *ungerade* representations.

Thus the product of the representations of the excited and ground state wavefunctions,

$$\Gamma[\psi'(Q_1)] \otimes \Gamma[\psi(Q_1)]$$

does not contain any of the dipole components, and therefore the transition is not allowed in absorption.

For the antisymmetric F—N stretching mode, one of the F atoms moves toward the N it is bonded to, whereas the F atom moves away from the N atom. If one sketches these displacement vectors, one can easily convince oneself that this motion belongs to the B_u representation (antisymmetric with respect to the C_2 axis and to the center of inversion, symmetric with respect to the σ_h plane). The ground state representation is still totally symmetric (since it is a ground state):

$$\Gamma[\psi(Q_2)] = A_g$$

but the representation of the excited state is B_u:

$$\Gamma[\psi'(Q_2)] = B_u$$

Therefore the product

$$\Gamma[\psi'(Q_2)] \otimes \Gamma[\psi(Q_2)]$$

does contain one component of the Cartesian displacement vector, x, y, and the transition is allowed.

4.5 SYMMETRY COORDINATES

In Section 3.7.5, we have pointed out that it is sometimes advantageous to group together a number of internal coordinates to form a new set of coordinates such that the new coordinates have the same symmetry properties as do the normal coordinates. This principle will be elaborated upon next, using a few common examples.

In the discussion of normal mode calculations of water, we previously defined three internal coordinates, the two O—H stretching coordinates R_1 and R_2, and the deformation coordinate R_3. In the method outlined in Section 3.9, we proceeded to transform the problem from internal to Cartesian coordinates, and to determine the vibrational frequencies by diagonalization of a 9×9 matrix. Of course, one could have solved this problem in internal coordinate space as well, particularly since the problem reduces to diagonalizing a 3×3 matrix (*vide infra*). However, using symmetry coordinates, this problem can be simplified even further.

Using the three internal coordinates R_1 to R_3, one can write the potential energy in the GVFF approach in terms of the following matrix:

$$\begin{bmatrix} f_s & f_{ss} & f_{bs} \\ f_{ss} & f_s & f_{bs} \\ f_{bs} & f_{bs} & f_b \end{bmatrix} \quad (4.5.1)$$

where f_s and f_b denote the diagonal O—H stretching and H—O—H deformation force constants, and f_{ss} and f_{bs} the stretch–stretch and stretch–bend interaction constants.

Since we have found in Section 4.3 that the vibrations of water fall into two symmetry species, $2A_1$ and B_1, we now define symmetry coordinates S, which have the same transformation properties as the normal modes:

$$\begin{aligned} S_1 &= 1R_1 + 1R_2 + 0R_3 \\ S_2 &= 0R_1 + 0R_2 + 1R_3 \\ S_3 &= 1R_1 - 1R_2 + 0R_3 \end{aligned} \quad (4.5.2)$$

where S_1 and S_2 transform as A_1, and S_3 as B_1. Equation 4.5.2 is often written in matrix form:

$$\mathbf{S} = \mathbf{U} \cdot \mathbf{R} \quad (4.5.3)$$

U is normally defined to be an orthonormal transformation, thus the coefficients in Eq. 4.5.2 are normalized:

$$\begin{bmatrix} (1/\sqrt{2}) & (1/\sqrt{2}) & 0 \\ 0 & 0 & 1 \\ (1/\sqrt{2}) & (-1/\sqrt{2}) & 0 \end{bmatrix} = \mathbf{U} \qquad (4.5.4)$$

Transforming the potential energy from internal coordinates to symmetry coordinates via the typical similarity transformation formalism yields

$$\mathbf{U}\mathscr{F}\mathbf{U}^\mathrm{T} = \begin{bmatrix} (f_s + f_{ss}) & (f_{bs}\sqrt{2}) & 0 \\ (f_{bs}\sqrt{2}) & f_b & 0 \\ 0 & 0 & (f_s - f_{ss}) \end{bmatrix} \qquad (4.5.5)$$

Thus the symmetrization of the potential energy matrix has factorized it into two symmetry blocks, which do not interact. In a tetrahedral molecule, for example, this can be very helpful since the four observable vibrations (cf. Chapter 7) fall into four different symmetry species.

The C—H stretching frequencies in a methyl group can similarly be described in symmetry coordinates. We define three C—H stretching internal coordinates R_1, R_2, and R_3, and a corresponding potential energy matrix (ignoring interaction between the methyl stretching and deformation modes) given by Eq. 4.5.6, where f_s is the diagonal stretching constant and f_{ss} the stretch–stretch interaction:

$$\begin{bmatrix} f_s & f_{ss} & f_{ss} \\ f_{ss} & f_s & f_{ss} \\ f_{ss} & f_{ss} & f_s \end{bmatrix} = \mathscr{F} \qquad (4.5.6)$$

Defining the symmetry coordinates according to the symmetry species of the three C—H stretching modes under C_{3v} symmetry (A_1 and E) as

$$\begin{aligned} S_1 &= 1R_1 + 1R_2 + 1R_3 \\ S_2 &= 2R_1 - 1R_2 - 1R_3 \\ S_3 &= 0R_1 + 1R_2 - 1R_3 \end{aligned} \qquad (4.5.7)$$

we obtain a normalized U matrix as follows:

$$\begin{bmatrix} (1/\sqrt{3}) & (1/\sqrt{3}) & (1/\sqrt{3}) \\ (2/\sqrt{6}) & (-1/\sqrt{6}) & (-1/\sqrt{6}) \\ 0 & (1/\sqrt{2}) & (-1/\sqrt{2}) \end{bmatrix} = \mathbf{U} \qquad (4.5.8)$$

The symmetrized potential energy for the methyl stretching modes appears as

$$\mathbf{U}\mathscr{F}\mathbf{U}^T \begin{bmatrix} (f_s + 2f_{ss}) & 0 & 0 \\ 0 & (f_s - f_{ss}) & 0 \\ 0 & 0 & (f_s - f_{ss}) \end{bmatrix} \quad (4.5.9)$$

Since there are two observed eigenvalues, the symmetric stretching mode due to the force constants $(f_s + 2f_{ss})$ and the degenerate antisymmetric stretching mode due to $(f_s - f_{ss})$, unique values can be obtained for both f_s and f_{ss}. This is true, as pointed out above, if one ignores the coupling of the A_1 with other A_1 vibrations allowed under C_{3v}, such as the symmetric deformation mode.

For the analysis of highly symmetric molecules, symmetry coordinates are used frequently, which simplify the vibrational problem significantly. For the discussion of the allowed vibrations in benzene (cf. Section 7.6), for example, we shall use symmetry coordinates to set up the problem. In this way, one obtains the normal modes of vibration expressed in terms of the symmetry coordinates, which helps to visualize the displacements occurring in the normal modes of vibration.

REFERENCES

F. A. Cotton, *Chemical Applications of Group Theory*, Wiley-Interscience, New York, 1963.

J. R. Ferraro and J. S. Ziomek, *Introductory Group Theory*, Plenum Press, New York, 1969.

5

INTRODUCTION TO RAMAN SPECTROSCOPY

5.1 GENERAL CONSIDERATIONS

In vibrational spectroscopy, there are several distinct and different methods to observe transitions between vibrational states. In the discussions of Chapters 2, 3, and 4, we have assumed an electric dipole operator-mediated transition, which may be observed (for molecular vibrations) via infrared absorption spectroscopy. There is, however, another commonly utilized mechanism for vibrational excitation, namely, that of inelastic light scattering, in which an incident photon with an energy much larger than the vibrational quantum loses part of its energy to the molecular vibrational excitation, and the remaining energy is scattered as a photon with reduced frequency. These processes are depicted schematically in Fig. 5.1.

In absorption spectroscopy, a beam of light is passed through the sample, and its attenuation is monitored as a function of wavelength or wavenumber of the light. This process is described by the Lambert–Beer law,

$$A(\tilde{v}) = \varepsilon(\tilde{v})Cl = \log[I_0(\tilde{v})/I(\tilde{v})] \tag{5.1.1}$$

Here, C is the concentration of the sample [in mol/L], l is the path length [in cm], and $\varepsilon(\tilde{v})$ is the molar extension coefficient [in L/mol·cm]. $I(\tilde{v})$ and $I_0(\tilde{v})$ are the light intensity emerging from, and incident on, the sample, respectively, as a function of wavenumber. The attentuation of the light, expressed in units of transmission, absorption, or ε, is plotted versus wavenumber to produce the infrared spectrum. This is shown schematically for a hypothetical system with three vibrational coordinates 1, 2, and 3 in Fig. 5.1 by the heavy lines marked $hc\tilde{v}_1$, $hc\tilde{v}_2$, and $hc\tilde{v}_3$. These lines represent a direct excitation from the ground

110 INTRODUCION TO RAMAN SPECTROSCOPY

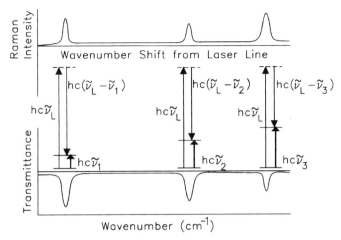

FIGURE 5.1. Schematic diagram for an infrared absorption process (solid upward arrows) and Raman scattering (downward arrows) for a hypothetical molecule with three vibrational coordinates. The infrared absorptions occur at $hc\tilde{v}_1$, $hc\tilde{v}_2$, and $hc\tilde{v}_3$ cm^{-1}. The Raman (Stokes) emissions occur at $hc(\tilde{v}_L - \tilde{v}_1)$, $hc(\tilde{v}_L - \tilde{v}_2)$, and $hc(\tilde{v}_L - \tilde{v}_3)$.

to an excited state, with the annihilation of a photon and the transfer of the photon's energy to the molecule.

Raman spectroscopy, on the other hand, is a scattering or emission phenomenon. A stream of photons, typically from a laser, impinges on the sample. The photons that have been scattered by the sample and have lost energy to the sample's vibrational levels contain the desired spectral information, whereas the transmitted laser beam, which has lost some intensity due to the scattering processes, is discarded. The light scattering is such a weak process that the loss of laser beam intensity is virtually immeasurable, unless one deals with very high laser power and secondary effects take place (cf. Section 5.8). The Raman scattered intensity is given by

$$I \propto v^4 I_0 N f(\alpha^2) \tag{5.1.2}$$

where I_0 is the intensity of the exciting light (typically a laser), N is the number of scattering molecules in a given state, v is the frequency [in s^{-1}] of the exciting light, and α is known as the polarizability of the sample. The exact functionality of α will be discussed below. The Raman scattering process can be depicted schematically as shown in Fig. 5.1: the vibrational coordinates are, of course, the same ones involved in the infrared absorption process; however, the excitation mechanism is totally different in Raman spectroscopy. Scattering phenomena are two-photon processes, where an incident photon $hc\tilde{v}_L$ is momentarily absorbed into a virtual state, indicated by the dashed energy level. A new photon is created and scattered from this virtual level. The final state

GENERAL CONSIDERATIONS   111

can be either the vibrational ground state (Rayleigh scattering) or the vibrationally excited state (Stokes–Raman scattering). The scattering process also can originate from the vibrationally excited state and proceed, via the virtual state, to the vibrational ground state. This process is known as anti-Stokes–Raman scattering. Thus the scattered photon may have the energy $hc\tilde{v}_L$, $hc(\tilde{v}_L - \tilde{v}_m)$, or $hc(\tilde{v}_L + \tilde{v}_m)$, where $\tilde{v}_m$ is a molecular vibrational transition frequency, expressed in cm^{-1}. These three processes are depicted schematically in Fig. 5.2.

Although the Raman scattered photon has lost a vibrational quantum to, or gained a vibrational quantum from, the molecular vibrational energy level, it is often still a visible photon: if, for example, a visible laser color is used for excitation, the laser photons have wavenumbers between 16,000 and 25,000 cm^{-1} (corresponding to wavelengths of 625 to 400 nm); even after losing a 1000 cm^{-1} vibrational photon, the scattered photon is still visible.

In Eq. 5.1.2, N was introduced as the number of particles involved in the scattering process. For normal Stokes scattering, which usually originates from the vibrational ground state, N would therefore indicate the number of molecules in the vibrational ground state. Anti-Stokes–Raman scattering, on the other hand, originates from a vibrationally excited state and returns the molecule to the vibrational ground state. Thus N for the anti-Stokes case would be the number of molecules originally in the vibrationally excited state. This number depends on the absolute temperature of the sample and on the energy difference between the ground and excited states for a given vibrational coordinate, according to Boltzmann's law:

$$N_k/N_0 = e^{-(E_k - E_0)/kT} \quad (2.9.3)$$

Consider a low frequency vibrational state with $\tilde{v} \approx 200$ cm^{-1}. Using the previously discussed conversion that room temperature corresponds to about

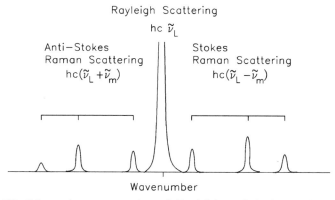

FIGURE 5.2. Schematic representation of Rayleigh and Stokes and anti-Stokes–Raman scattering.

208 cm^{-1}, the exponent in Eq. 2.9.3 will be about (negative) unity, and the population of the first excited state of this coordinate is about 38% and, consequently, the anti-Stokes intensities will be 38% of the Stokes intensity. For a lower energy vibration (e.g., 100 cm^{-1}) at room temperature, the anti-Stokes intensity is nearly 62% of the Stokes intensity. Over 500 cm^{-1}, the anti-Stokes intensities drop off rapidly. The relative intensities of Stokes– and anti-Stokes–Raman scattering can be used as a convenient probe for the temperature of the scattering molecules and allows one to determine whether or not interactions of the molecules with the laser beam may have caused local heating.

The description of the intermediate, or virtual, state that gives rise to the Raman effect is difficult and not intuitively obvious. It is not a stationary state, since a colorless liquid does not have a stationary electronic state in the visible part of the spectrum, yet Raman scattering is observed from colorless liquids. As indicated before, the original interaction of the light with the matter is an off-resonance effect, mediated by the polarizability. Various approaches will be presented in the subsequent sections to present an understanding of the origins of the Raman effect, the virtual state, and the polarizability, starting with the simplest picture, namely, that of a classical interference (beat frequency) in Section 5.2.

The spontaneous Raman effect (as opposed to the coherent effects discussed in Section 5.8) occurs off-resonance, which accounts for the fact that Raman spectroscopy is extremely feeble, despite all the technical advances to make it a practical technique. It experienced a rebirth in the 1970s with the advent of commercial lasers, which rendered this technique suddenly as applicable as infrared spectroscopy. Over the following two decades, a number of new Raman techniques have evolved, such as resonance Raman (RR), time-resolved RR (TRRR), higher order Raman effects such as hyper-Raman spectroscopy, coherent anti-Stokes–Raman spectroscopy (CARS), inverse Raman and stimulated Raman spectroscopies, and recently FT–Raman spectroscopy with excitation in the infrared region. These will be treated in turn, starting with the simplest one, the classical Raman effect. Its most basic description is based on a picture of beat frequencies between the frequency of the light and the molecular vibrational frequencies.

5.2 CLASSICAL DESCRIPTION OF THE RAMAN EFFECT

In the most basic description of Raman spectroscopy, the effect of the oscillating electric field on the molecule is described classically and macroscopically. We assume that electromagnetic radiation with an electric field **E** impinges on a transparent material and will induce a dipole moment μ, since the electromagnetic field will exert a force on the charged particles of the molecule (cf. Eq. 2.3.5). This induced dipole moment is given by

$$\mathbf{\mu} = \alpha \mathbf{E} \quad (5.2.1)$$

where α is the polarizability tensor. A tensor is a physical quantity, written as a matrix, which mathematically connects two vectors. Thus a tensor relating two vectorial quantities can change the direction of cause and effect: we shall see shortly that an electric field in the X-direction, for example, can induce a dipole component in the Y-direction. Furthermore, the rules of vector and matrix multiplication apply.

The polarizability itself can be visualized as the macroscopic response of the medium to the perturbing electromagnetic field: the least tightly bound outer electrons in a molecule or atom will respond most easily to the force exerted by the electromagnetic field by following the oscillation of the light, thereby producing a dynamic (or induced) dipole moment at the frequency of the oscillation of the light given by

$$\mathbf{E} = E_0 \cos \omega t \quad (5.2.2)$$

Thus the induced dipole moment will also oscillate at ω:

$$\mathbf{\mu} = \alpha E_0 \cos \omega t \quad (5.2.3)$$

The polarizability is a molecular quantity, to be discussed in more detail in Section 5.3, which describes the ease with which electrons can be moved by the electromagnetic radiation, relative to the nuclear framework. α is an off-resonance response of matter to an electromagnetic perturbation and is related to the dielectric constant ε and to the refractive index (cf. Section 5.3).

Since the molecule undergoes constant vibrational motion along its $3N - 6$ molecular vibrational coordinates Q, the polarizability α itself is modulated at the vibrational frequencies of the molecule:

$$Q = Q_0^m \cos \omega_m t \quad (5.2.4)$$

where ω_m is the frequency of the mth normal coordinate. The polarizability can be expanded in a Taylor series in terms of the normal coordinates Q_m about the equilibrium position

$$\alpha = \alpha_0 + (\partial \alpha / \partial Q_m) Q_m + \cdots \quad (5.2.5)$$

just as the dipole moment was expanded in terms of the normal coordinates, and one obtains from Eqs. 5.2.4 and 5.2.5

$$\alpha = \alpha_0 + (\partial \alpha / \partial Q_m) Q_0^m \cos \omega_m t \quad (5.2.6)$$

The induced dipole moment (Eq. 5.2.3) can therefore be written as

$$\mu = \alpha_0 E_0 \cos \omega t + (\partial\alpha/\partial Q)Q_0^m E_0 \cos \omega_m t \cos \omega t \qquad (5.2.7)$$

Using the trigonometric identity $(\cos \alpha)(\cos \beta) = \tfrac{1}{2}[\cos(\alpha + \beta) + \cos(\alpha - \beta)]$, one obtains

$$\mu = \alpha_0 E_0 \cos \omega t + \tfrac{1}{2}(\partial\alpha/\partial Q)Q_0^m E_0[\cos(\omega - \omega_m)t + \cos(\omega + \omega_m)t] \qquad (5.2.8)$$

Thus the induced dipole moment oscillates at the frequencies $\omega - \omega_m$, $\omega + \omega_m$, and ω, where m can have all values between 1 and the number of normal frequencies in the molecule, $1 < m < 3N - 6$. Since an oscillating electric dipole will emit radiation at its vibrational frequency, the light emitted from the induced dipole will have the same frequency components, which are referred to as Stokes, anti-Stokes, and Rayleigh scattering, and which were introduced in the previous section.

5.3 THE POLARIZABILITY TENSOR

The polarizability tensor that mediates Raman transitions needs to be investigated next. So far, we have introduced the polarizability as a proportionality between induced dipole moment and electric field. This classical view of polarizability is valuable and can be used to derive a number of off-resonance effects that occur in the interaction of light and matter. Furthermore, its derivative with respect to the normal coordinates gives rise to Raman scattered intensities. Its detailed classical derivation can be found in Kauzmann's book, *Quantum Chemistry* [Kauzmann, 1957, Chapter 16].

In short, this derivation proceeds as follows. We write the induced dipole moment, μ_{ind}, as the expectation value of the operator $\boldsymbol{\mu}$:

$$\mu_{\text{ind}} = \int \Psi^* \boldsymbol{\mu} \Psi \, d\tau \qquad (5.3.1)$$

We employ the same expansion of the time-dependent wavefunctions

$$\Psi(x, t) = \sum c_k \psi_k(x) \phi(t) \qquad (2.3.8)$$

in terms of time-dependent coefficients c_k, given by Eq. 2.3.8, and the stationary state wavefunctions $\psi_k(x)$. Using Eq. 5.2.2,

$$\mathbf{E} = E_0 \cos \omega t \qquad (5.2.2)$$

we find that for states 0 (ground state) and any excited state m, the induced dipole moment can be written as

$$\mu_{ind} = \mu_{00} + \frac{2E_0}{\hbar} \cos\omega t \sum_m \frac{\omega_{0m}|\langle\psi_0|\mu|\psi_m\rangle|^2}{\omega_{0m}^2 - \omega^2}$$

$$= \frac{2E_0}{\hbar} \sum_m \frac{|\langle\psi_0|\mu|\psi_m\rangle|^2}{\omega_{0m}^2 - \omega^2} \omega_{0m} \cos\omega_{0m}^t \quad (5.3.2)$$

The first term in Eq. 5.3.2 describes the permanent dipole moment, the second term a component of the induced dipole that oscillates at the same frequency as the light, and the third term an induced dipole that oscillates at the transition frequency ω_{0m} and is clearly not in phase with the frequency of the light inducing the dipole moment. The second term is called the polarizability, and a comparison between Eqs. 5.2.3

$$\mu = \alpha E_0 \cos\omega t \quad (5.2.3)$$

and 5.3.2 yields

$$\alpha = \frac{2}{\hbar} \sum_m \frac{\omega_{0m}}{\omega_{0m}^2 - \omega^2} |\langle\psi_0|\mu|\psi_m\rangle|^2 \quad (5.3.3)$$

This result implies that any medium exposed to electromagnetic radiation will undergo some, albeit small, change, even when the frequency of the radiation is far from a transition energy. This interaction is determined by a sum of all the dipole transition moments, each one weighted by an energy term in the denominator. This weighting term gets very large if the frequency of the light ω approaches the energy difference between two real stationary states, that is, if $\omega_{0m}^2 = \omega^2$. The contribution of a given transition moment is small when the exciting light is far from the energy of a transition. All possible excited states must be summed up to yield the polarizability, which thus can be viewed as an attempt of the system to undergo a transition, but the energy of the photon is insufficient, or two large, to actually permit the transition.

The polarizability α is related to the refractive index of a sample as follows:

$$\alpha = \frac{3}{4\pi N} \frac{n^2 - 1}{n^2 + 2} \quad (5.3.4)$$

Thus we find that the off-resonance spectroscopic properties, such as the dielectric constant or the refractive index, are related to the transition moments via the polarizability. The results of this statement are that the proximity of an electronic transition will increase the refractive index. Thus, in the visible spectral region, acetone has a higher refractive index than water, and CsI has a higher refractive index than LiF.

The polarizability, as discussed in this section, determines the macroscopic optical properties (cf. Eq. 5.3.4) but is also responsible for Rayleigh and Raman scattering. In these effects, the polarizability creates a virtual state for the incident light to interact with matter, although there is no stationary state nearby into which the photons can be absorbed. The polarizability allows this virtual state to absorb the photon momentarily. Reemission then creates the scattered photon. This description, which involves the polarizability as a means to create a virtual state for a photon of any wavelength, leads to the familiar picture of Rayleigh and Raman scattering from an intermediate, virtual state (cf. Fig. 5.1).

In the vibrational Raman effect the scattering normally occurs between different vibrational states of the same electronic state; however, the intermediate (virtual) state may belong to any electronic state.

Thus, for Raman scattering, one writes the polarizability as

$$\alpha_{\alpha\beta} = \frac{2}{\hbar} \sum_m \frac{\omega'}{\omega'^2 - \omega^2} \langle \psi_{ev}|\mu_\alpha|\psi_{e'v'}\rangle \langle \psi_{e'v'}|\mu_\beta|\psi_{ev'}\rangle \qquad (5.3.5)$$

where e and v denote electronic and vibrational ground states, and the primes excited states. ω' denotes the frequency of the transition from ev to $e'v'$, and the subscripts α and β denote the appropriate Cartesian components of the polarizability tensor. As pointed out before, the polarizability is given by a summation over all transitions into (real) intermediate states, weighted by an energy denominator.

Equation 5.3.5 shows clearly the vibronic nature of Raman intensities. One also realizes that Raman spectroscopy is a two-photon process, since there are two electronic transition moment expressions contained in the definition of the polarizability. Thus the selection rules for the Raman effect are different from those of infrared absorption spectroscopy. Because there are two transition moments contained in the definition of one component of the polarizability tensor, the Raman effect depends on the components $x \cdot x$, $x \cdot y$, $x \cdot z$ of the Cartesian displacement components, which are given in the character tables. This aspect will be discussed in Section 5.4.

The Raman intensity of a given vibrational mode is given by the appropriate derivative of the polarizability tensor elements with respect to the normal coordinate:

$$I \propto (\partial \alpha / \partial Q_k) \qquad (5.3.6)$$

The description of the polarizability tensor elements presented so far is applicable not only to the standard Raman effect, but describes qualitatively the resonance enhancement exhibited by the Raman effect when the exciting wavelength is close to a molecular absorption. In this case, the denominator in one term in the sum in Eq. 5.3.5 becomes very small, and the corresponding polarizability term becomes very large. A more detailed description, using the Raman scattering tensor, will be introduced in Section 5.6.

Resonance Raman (RR) intensities may be enhanced by up to five orders of magnitude over standard Raman scattering intensities. In addition to enhancing the generally weak Raman effect, RR exhibits the advantage that excitation into an electronic chromophore may selectively enhance the vibrational modes of the chromophore, thus allowing sections of a molecule to be probed separately. Finally, in the case of a molecule with more than one chromophore, the excitation wavelength of the laser may be scanned, and the Raman intensities as a function of exciting wavelengths are monitored. Such an "excitation profile" contains the electronic absorption spectrum as observed through the scattered intensities. These aspects will be discussed in Section 5.6.

5.4 RAMAN SELECTION RULES

Since the Raman effect is a form of two-photon spectroscopy, different selection rules apply from those for standard one-photon absorption or emission. Basically, this can be visualized from the fact that the polarizability expression, Eq. 5.3.5, contains two electric dipole terms. Since each one of them transforms as the components μ_x, μ_y, or μ_z, the product of the transition moments transforms as the binary combinations $x \cdot x$, $x \cdot y$, and so on. Thus a vibration is Raman active if the irreducible representation of the mode contains one of the Cartesian polarizability components, which are listed in the character tables as the binary products of the axes. For the C_{2v} point group, for example, the binary products transform as the following irreducible representations:

C_{2v}	E	C_2	σ_v	σ_v'		
A_1	1	1	1	1	T_z	x^2, y^2, z^2
A_2	1	1	-1	-1	R_z	xy
B_1	1	-1	1	-1	T_x, R_y	xz
B_2	1	-1	-1	1	T_y, R_x	yz

One interesting aspect of these selection rules is that for highly symmetric point groups, the components x, y, and z, and their binary combinations x^2, xy, and so on, often transform as different representations of a group. In all centrosymmetric groups, for example, there is a direct exclusion rule, which states that no representation that transforms like a component x, y, or z can also transform like a binary combination. This is demonstrated below for the C_{2h} point group:

C_{2h}	E	C_2	i	σ		
A_g	1	1	1	1		$\alpha_{xx}, \alpha_{yy}, \alpha_{zz}, \alpha_{xy}$
B_g	1	-1	1	-1		α_{yz}, α_{xz}
A_u	1	1	-1	-1	T_z	
B_u	1	-1	1	-1	T_x, T_y	

Consequently, in any centrosymmetric group, Raman allowed vibrations are infrared absorption forbidden, and vice versa. In CS_2, for example, the symmetric stretching mode (656 cm^{-1}) is Raman active (in fact, one of the strongest known Raman modes), whereas it is infrared forbidden, since it does not change the dipole moment. The other vibrational degrees of freedom, the antisymmetric stretching mode (1523 cm^{-1}) and the (degenerate) deformation mode (397 cm^{-1}), both change the molecular dipole moment and are therefore infrared allowed and Raman forbidden.

Even aside from these strict exclusions, one finds that very polar vibrations, such as a C—F stretching vibration, are very strong in infrared absorption and very weak in Raman scattering. On the other hand, bonds that involve many highly polarizable (movable) electrons, such as a S—H or C=S stretching vibration, are much more easily detected in Raman than in infrared spectroscopy. This observation can be explained in terms of a qualitative argument, which states that a strong dipole moment localizes the electrons and makes it harder for the perturbing electromagnetic radiation to move the electrons. Thus Raman and infrared absorption spectroscopies are complementary techniques, and for small, highly symmetric molecules, it is essential that both Raman and infrared data are available. On the other hand, the symmetry and shape of small molecules can often be determined on the bases of vibrational spectra alone: for example, a linear or bend structure of a triatomic molecule XY_2 can be distinguished by the mutual exclusion of the symmetric stretching modes in infrared and Raman spectra for the linear case. These aspects will be discussed in more detail in Chapter 7.

5.5 POLARIZATION OF RAMAN SCATTERING

In infrared absorption spectroscopy, a polarization response of the absorption intensities toward linearly polarized light is observed only for oriented, nonisotropic samples. Raman scattering, on the other hand, is completely or partially polarized, depending on the symmetry of the vibration responsible for the scattering even for isotropic and nonoriented samples. This polarization property is usually expressed in terms of the depolarization ratio, which is a very useful property of Raman bands, since it allows the differentiation of whether or not Raman active vibrations belong to the totally symmetric representation of the group. Moreover, even for molecules with low symmetry, the depolarization ratios are often very helpful for the assignment of spectral bands: they will be used frequently to distinguish between symmetric and antisymmetric vibrations of a symmetric part a molecule, such as a methyl group.

The derivation of depolarization ratios can be carried out using the classical description of the scattering phenomenon and averaging the scattering over a large ensemble of scattering species. For this purpose, we consider the polarizability again strictly as a proportionality constant between the exciting field and the induced moment (cf. Eq. 5.2.1):

$$\mathbf{\mu} = \alpha \mathbf{E} \tag{5.2.1}$$

where $\mathbf{\mu}$ and $\mathbf{E}$ are vectors, which are not necessarily collinear, and the polarizability is a tensor, which is written in matrix form. In terms of the Cartesian vector components, we may write Eq. 5.2.1 as

$$\begin{bmatrix} \mu_X \\ \mu_Y \\ \mu_Z \end{bmatrix} = \begin{bmatrix} \alpha_{XX} & \alpha_{XY} & \alpha_{XZ} \\ \alpha_{YX} & \alpha_{YY} & \alpha_{YZ} \\ \alpha_{ZX} & \alpha_{ZY} & \alpha_{ZZ} \end{bmatrix} \begin{bmatrix} E_X \\ E_Y \\ E_Z \end{bmatrix} \tag{5.5.1}$$

Here, capital subscripts X, Y, and Z denote space fixed (laboratory) coordinates, and the polarizability tensor elements in Eq. 5.5.1 can be construed to be either the static polarizability or the polarizability derivatives $(\partial \alpha / \partial Q)$. In the first case, one deals with the dipole moment induced only at the laser frequency, and the corresponding emission process is Rayleigh scattering. In the second case, the induced dipole moment is modulated as discussed in Section 5.2 and leads to Raman scattering. Since the formalism presented below for averaging the tensor elements over all orientations is the same for the elements or their derivatives with respect to the normal coordinates, we shall utilize the tensor elements themselves in the following derivation. A more detailed discussion of the Raman intensities is presented in Section 5.5.3.

A nonrotating molecule, located at the origin of a coordinate system (X, Y, Z), is exposed to light from a laser beam, which travels in the positive Z direction and which is polarized in the XZ-plane, as shown in Fig. 5.3. For this laser beam, $E_Y = E_Z = 0$, and only $E_X \neq 0$. In this special case of linearly polarized incident light,

$$\begin{aligned} \mu_X &= \alpha_{XX} E_X \\ \mu_Y &= \alpha_{XY} E_X \\ \mu_Z &= \alpha_{XZ} E_X \end{aligned} \tag{5.5.2}$$

Equation 5.5.2 indicates, as pointed out before, that a dipole moment is induced in all three Cartesian directions, even though the electric field oscillates along the X-direction only. This is due to the fact that the polarizability is a tensor, which can relate vectors of cause and effect that are not collinear. A permanent as well as an induced oscillating dipole will radiate energy according to

$$I = \frac{16\pi^4 \nu^4}{3c^3} \mu_0^2 \tag{5.5.3}$$

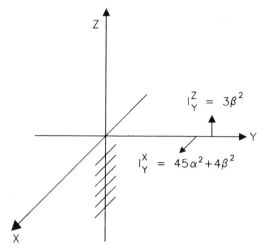

FIGURE 5.3. Definition of Raman scattering geometry for 90° (right angle) scattering. Scattered Raman intensities are shown for incident light polarized along the X-direction.

where I is average radiated intensity, v is the frequency of radiated light, and

$$\mu(t) = \mu_0 \cos \omega t \tag{5.5.4}$$

The light will be radiated into all space (4π steradians). The light intensity scattered in the Y-direction within the solid angle $\sin \theta \, \partial \theta \, \partial \phi$ is given by

$$I_T = (2\pi^3 v^4/c^3)(\mu_X^2 + \mu_Z^2)$$
$$= (2\pi^3 v^4/c^3)(\alpha_{XX}^2 + \alpha_{XZ}^2)E_X^2 \tag{5.5.5}$$

The difference in the factor $16\pi^4 v^4/3c^3$ in Eq. 5.5.3 and $2\pi^3 v^4/c^3$ in Eq. 5.5.5 is due to the fact that now only a limited cone of light is sampled. Furthermore, one can analyze the scattered light through a polarizer transmitting the intensities scattered in the Y direction, and polarized along Z and X (I^Z or I^X) separately:

$$I_Y^X = (2\pi^3 v^4/c^3)(\alpha_{XX}^2)E_X^2 \tag{5.5.6}$$
$$I_Y^Z = (2\pi^3 v^4/c^3)(\alpha_{XZ}^2)E_X^2 \tag{5.5.7}$$

In Raman spectroscopy of single crystalline compounds, oriented properly with respect to the laser propagation and polarization direction, these components of the polarizability tensor derivatives can be observed separately. However, for rapidly tumbling molecules in the liquid and gaseous phases, as well as for randomly oriented molecules in the solid phase, it is necessary to transform from the space fixed (laboratory) Cartesian coordinate system X, Y, and Z to a molecule fixed coordinate system and express the polarizability tensor in this new coordinate system. Such a coordinate transform is a simple matrix multiplication process, known in linear algebra as a similarity transformation. If this transformation is carried out properly, the polarizability tensor (or the derivatives of the polarizability tensor with respect to the normal coordinates) can be transformed into a coordinate system in which it can be written as a diagonal matrix:

$$\begin{bmatrix} \alpha_{XX} & \alpha_{XY} & \alpha_{XZ} \\ \alpha_{YX} & \alpha_{YY} & \alpha_{YZ} \\ \alpha_{ZX} & \alpha_{ZY} & \alpha_{ZZ} \end{bmatrix} \rightarrow \begin{bmatrix} \alpha_1 & 0 & 0 \\ 0 & \alpha_2 & 0 \\ 0 & 0 & \alpha_3 \end{bmatrix} \qquad (5.5.8)$$

The system of axes in which the polarizability is diagonal is called the principal axes of polarizability.

For the following discussion, we shall use these conventions. We refer to the polarizability elements in the space fixed Cartesian coordinate system as α_F, and to the polarizability tensor in the molecule fixed coordinates as α_g.

A similarity transformation involves the multiplication of the property to be transformed (here, the polarizability) by the matrix S, and its transpose S^T, as follows:

$$\alpha_F = S\alpha_g S^T \qquad (5.5.9)$$

where S transforms from the molecule fixed (g) to the space fixed coordinates (F) as follows:

$$\mathbf{F} = \mathbf{Sg} \qquad (5.5.10)$$

Before carrying out this coordinate transformation, and the subsequent averaging over all possible orientations, we need to discuss a number of tensor properties that are maintained under a linear transformation. In other words, these properties are independent of the coordinate system the tensor is expressed in and are therefore known as the *tensor invariants*. There are three of these invariants:

1. The first invariant (trace, or tensor isotropic part) is

$$\alpha_{XX} + \alpha_{YY} + \alpha_{ZZ} = \alpha_{xx} + \alpha_{xx} + \alpha_{xx} \qquad (5.5.11)$$

2. The second invariant (or first anisotropy) is the sum of all minors:

$$\begin{vmatrix} \alpha_{XX} & \alpha_{XY} \\ \alpha_{YX} & \alpha_{YY} \end{vmatrix} + \begin{vmatrix} \alpha_{XX} & \alpha_{XZ} \\ \alpha_{ZX} & \alpha_{ZZ} \end{vmatrix} + \begin{vmatrix} \alpha_{YY} & \alpha_{YZ} \\ \alpha_{ZY} & \alpha_{ZZ} \end{vmatrix} = \text{constant} \qquad (5.5.12)$$

When the determinantes in Eq. 5.5.12 are expanded and added up, one obtains for the first anisotropy (with $\alpha_{YX} = \alpha_{XY}$)

$$\alpha_{XX}\alpha_{YY} + \alpha_{XX}\alpha_{ZZ} + \alpha_{YY}\alpha_{ZZ} - \alpha_{XY}^2 - \alpha_{XZ}^2 - \alpha_{YZ}^2 \qquad (5.5.13)$$

3. The third invariant (or second anisotropy) is

$$\det|\alpha_{XX'}| = \det|\alpha_{xx'}| \qquad (5.5.14)$$

In Eqs. 5.5.11 to 5.5.14, capital and lowercase indices denote coordinate systems related by a linear transformation, for example, space or molecule fixed Cartesian coordinates. In Eq. 5.5.14, XX' and xx' stand for all permutations of the coordinate components.

The transformation referred to in Eq. 5.5.9 involves rotation of the coordinate axes by the angle θ between the X, Y, Z coordinates and the principal axes 1, 2, 3. For this case, Eq. 5.5.10 can be written as

$$\alpha_{FF'} = \sum_3 \alpha_i \Phi_{Fi} \Phi_{Fi'} \qquad (5.5.15)$$

where F and F' are any combination of the components X, Y, or Z and the elements Φ_{Fi} are the directional cosine matrix elements between the two coordinate systems. Thus the two elements of the space fixed polarizability tensor required in Eqs. 5.5.6 and 5.5.7 may be expressed as

$$\alpha_{XX} = \alpha_1 \Phi_{X1}^2 + \alpha_2 \Phi_{X2}^2 + \alpha_3 \Phi_{X3}^2 \qquad (5.5.16)$$

and

$$\alpha_{XZ} = \alpha_1 \Phi_{X1}\Phi_{Z1} + \alpha_2 \Phi_{X2}\Phi_{Z2} + \alpha_3 \Phi_{X3}\Phi_{Z3} \qquad (5.5.17)$$

α_{XX}^2 then will contain terms in α_1^2, $\alpha_1\alpha_2$, etc., and Φ_{X1}^4, $\Phi_{X1}^2\Phi_{X2}^2$,.... The directional cosine matrix elements Φ_{Fi} now need to be averaged over all orientations to account for the random tumbling motion, or the random orientation of the molecules. This integration is given in detail elsewhere [Wilson et al., 1955, Chapter 3.6 and Appendix IV].

These values, averaged over all possible orientations, are

$$\overline{\alpha_{XX}^2} = \tfrac{1}{5}(\alpha_1 + \alpha_2 + \alpha_3) + \tfrac{2}{15}(\alpha_1\alpha_2 + \alpha_1\alpha_3 + \alpha_2\alpha_3) \qquad (5.5.18)$$

and a similar expression holds for the other averaged term, $\overline{\alpha_{XZ}^2}$.

Rewriting the constants in Eq. 5.5.6 as

$$k = (16\pi^4 v^4 I_0 N/3c^3) \qquad (5.5.19)$$

we may write the two components of the scattered intensities, I_Y^X and I_Y^Z,

$$I_Y^X = k\left[3\sum_{i=1}^{3}\alpha_i^2 + 2\sum_{i<j}^{3}\alpha_i\alpha_j\right] \quad (5.5.20)$$

$$I_Y^Z = k\left[3\sum_{i=1}^{3}\alpha_i^2 - \sum_{i<j}^{3}\alpha_i\alpha_j\right] \quad (5.5.21)$$

Using the tensor invariants defined above (Eqs. 5.5.11 and 5.5.12), we write a spherical polarizability α^2 as the first invariant of the polarizability ensor, and β^2 as the first anisotropy of the polarizability tensor, and obtain

$$I_Y^X = k\{45\alpha^2 + 4\beta^2\} \quad (5.5.22)$$

$$I_Y^Z = k\{3\beta^2\} \quad (5.5.23)$$

If the polarizability tensor is expressed in any system of molecule fixed axes that are not the principal axes of polarizability, the first anistropy is written as

$$\beta^2 = (\alpha_{xx}-\alpha_{yy})^2 + (\alpha_{xx}-\alpha_{zz})^2 + (\alpha_{yy}-\alpha_{zz})^2 + 6(\alpha_{xy}^2 + \alpha_{xz}^2 + \alpha_{yz}^2) \quad (5.5.24)$$

Equations 5.5.22 and 5.5.23 are the take home message from the derivation presented in this section. Starting with a nonrotating molecule defined in a coordinate system of space fixed axes, we obtained the Raman scattered intensity components, I_Y^X and I_Y^Z, expressed in terms of the polarizability components written in space fixed coordinates (Eqs. 5.5.6 and 5.5.7). After a suitable coordinate transformation, we have expressed these same scattered intensity components, I_Y^X and I_Y^Z, in terms of polarizability components written in a coordinate system attached to, and rotating with, the molecule.

These two intensity components, I_Y^X and I_Y^Z, can readily be observed separately by placing a polarization analyzer in the scattered beam. For an isotropic sample, they depend on which elements of the polarizability tensor (i.e., the trace, the first anisotropy, or both) took part in the scattering process.

We define the depolarization ratio as the quotient of the scattered light polarized perpendicularly (I_Y^Z) to the scattered light polarized parallel (I_Y^X), to the incident laser polarization:

$$\rho = \frac{I_Y^Z}{I_Y^X} = \frac{3\beta^2}{45\alpha^2 + 4\beta^2} \quad (5.5.25)$$

This depolarization ratio allows one to assess the symmetry of the vibration giving rise to the scattering. In a totally symmetric vibration, all the Raman scattered intensity is contributed by the spherical part (trace, α^2) of the polarizability, and β^2 is zero. Consequently, the vibration maintains the polarization of the laser beam. Thus I_Y^X of Fig. 5.3 is $45\alpha^2$, I_Y^Z is zero, and therefore ρ is zero.

For a vibration that is not spherically symmetric, α^2 is zero and β^2 contributes all intensity to both I_Y^Z ($3\beta^2$) and I_Y^X ($4\beta^2$), and ρ is 0.75. Thus the polarization properties of the incident laser beam are not maintained, and the vibration acts as a partial polarization scrambler. The limiting values for the depolarization ratios, 0 and 0.75, are reached only for molecules belonging to spherical point groups (T_d, O_h). In CCl_4, for example, the depolarization ratio for the symmetric stretching mode at 459 cm^{-1} (cf. Chapter 7) is smaller than 0.01, where the deviation from exactly zero is mostly due to the isotopic composition of CCl_4. The other three Raman active modes show depolarization ratios very close to 0.75. For molecules with lower symmetry, the observed depolarization ratios may vary typically between 0.01 and 0.6. However, depolarization ratios are very valuable even in molecules of lower symmetry, because the most polarized vibrations originate from the vibration with the most symmetric representation in the point group of the molecule.

The accurate measurement of the depolarization ratios is by no means a trivial problem, because of birefringence in the optics, the large apertures of the collection of Raman scattered light, different polarization responses of monochromator and detector, and so on. Finally, the accurate measurement of any intensity originating from a small volume is difficult, since misalignment of the collection optical system will reduce the light throughput. Some of these aspects will be discussed next, and a few methods of how to observe the depolarization ratios will be introduced.

5.5.1 Raman Scattering Geometries and Depolarization Ratios for Right Angle Scattering

Raman scattering and depolarization ratios are most commonly observed in 90° scattering, as shown in Fig. 5.3. There are a number of ways the depolarization ratios can be measured. In one of them, shown in Fig. 5.3, a polarizer is placed in the scattered beam, and two measurements are carried out with the polarizer transmitting along the Z- and X-directions, respectively. The observed depolarization ratios are given by Eq. 5.5.25 and will vary from 0 to 0.75, depending on the values of α^2 and β^2. For nonspherical point groups, such as C_{2v}, the observed depolarization ratios for a totally symmetric vibration are typically around 0.1 and for an asymmetric vibration above 0.6.

If the laser beam is polarized along the Y-direction, as shown in Fig. 5.4, no differences are observed between I_Z and I_X, since both components are polarized perpendicularly to the incident light vector. Thus no depolarization ratios can be observed. If the laser beam is naturally or circularly polarized (cf. Fig. 5.5), the polarization components shown in Figs. 5.3 and 5.4 need to be added, and the intensities shown in Fig. 5.5 are obtained. Thus, for circularly or naturally (unpolarized) light, the depolarization ratios will vary from 0 to $\frac{6}{7}$.

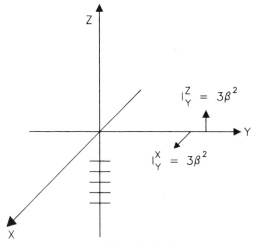

FIGURE 5.4. Scattered Raman intensities for incident light polarized along the Y-direction.

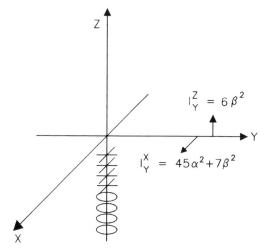

FIGURE 5.5. Scattered Raman intensities for naturally or circularly polarized incident light.

The rotation of a polarizer in the scattered beam bears some disadvantages. First, it is easy to slightly misalign the beam when the polarizer is rotated. Thus not all light might pass the (usually narrow) entrance slit of the monochromator, and this might produce an erroneous depolarization ratio. Second, and more damaging, is the fact that grating-based monochromators, which analyze the scattered light, have a different response for the two light components polarized at right angles to each other. The grating in the monochromator is the main culprit: the diffraction efficiency varies drastically between light components polarized parallel or perpendicularly to the ruling direction of the grating. Thus a polarization scrambler may be placed in front of the monochromator entrance slit (after the polarization analyzer!), which removes the polarization properties of the light.

Alternatively, another scattering geometry is often employed. This geometry involves the rotation of the plane of polarization of the incident laser beam, and observing the depolarization ratios without a polarizer: when the laser beam is polarized as shown in Fig. 5.3, the total intensity observed, $I_x + I_z$, is given by $45\alpha^2 + 7\beta^2$. When the laser polarization is rotated as shown in Fig. 5.4, the total scattered intensity is $6\beta^2$. Thus the depolarization ratio observed by rotating the laser beam is the same as the one observed with naturally polarized light. This latter method lends itself for an automatic measurement of ρ, since the laser plane of polarization may be rotated by 90° with an electro-optic modulator.

5.5.2 Forward and Backscattering

Recently, different scattering geometries for Raman spectroscopy have been explored successfully. Particularly successful is a technique known as Raman backscattering (or 180° scattering), since it allows the collection of Raman data through a microscope, and thus of very small particles and very high spatial resolution. A typical backscattering geometry is shown in Fig. 5.6. The laser beam is reflected by a very small mirror or prism and directed through a lens, which also serves to collect the Raman scattering, often at very large aperture (low f-number). This allows for very efficient collection of the Raman scattered light, and, since the laser beam is very small, spatial resolutions of a few micrometers have been achieved. This geometry is employed in Raman microscopy, where the laser beam is focused through, and the Raman scattered light collected by, the same microscope optics. Backscattering is particularly suitable for solid samples, where the depth of field is of minor importance. The disadvantage of backscattering, particularly in the case of solid samples, is the high level of reflected light at the laser wavelength, which often necessitates the use of triple monochromators.

The depolarization ratios for linearly polarized incident radiation are the

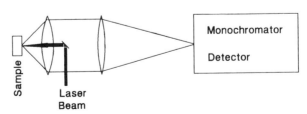

FIGURE 5.6. Geometry employed for the observation of Raman backscattering.

same as in right angle scattering. For unpolarized light, no depolarization ratios can be observed in backscattering, whereas for circularly polarized incident light, the meaning of the depolarization ratio changes somewhat, since a totally symmetric vibration reverses the sense of the circularity in backscattering (i.e., it acts as a mirror in terms of the polarization response). Depolarized vibrations, on the other hand, destroy the circularity partially.

Forward scattering (also known as 0° scattering) is observed by blocking the transmitted beam and analyzing a cone of light scattered forward. Here, totally symmetric vibrations will maintain the sense of circularity in the case of excitation with circularly polarized light.

5.5.3 Computation of Raman Scattered Intensities

The recent progress in the computation of Raman scattered intensities is partially due to the efforts by a few research groups involved in a new spectroscopic technique, known as Raman optical activity (ROA, cf. Chapter 9). This new chiroptical technique measures the difference in Raman scattering between left and right circularly polarized light in chiral (optically active) molecules. Quite similar to the case of modern infrared absorption intensity calculations (Section 3.10), which was partially spurred on by research in infrared circular dichroism, the efforts to interpret ROA spectra required reliable models for the calculation of Raman scattered intensities, which were not tested until recently [Polavarapu, 1990].

The Raman scattered intensity of the jth normal mode of a molecule is determined by the appropriately averaged derivative of the polarizability with respect to the normal coordinates, as given by Eqs. 5.5.22 and 5.5.23:

$$I_Y^T = C[45(\alpha')_j^2 + 7(\beta)_j^2] \tag{5.5.26}$$

where I_Y^T is the total intensity scattered per solid angle along the Y direction. The averaging process was described in Eqs. 5.5.16 to 5.5.18, and each of the tensor elements $\alpha'_{\alpha\beta}$ is the derivative

$$\alpha'_{\alpha\beta} = (\partial \alpha_{\alpha\beta}/\partial Q_j) \tag{5.5.27}$$

C in Eq. 5.5.26 is an expression that contains the exciting frequency to the forth power, a Boltzmann factor, and a number of universal constants.

A quantum mechanical expression for the polarizability tensor elements was given before in Eq. 5.3.5, but this expression would be very hard to implement because it contains a sum over all (electronically and vibrationally) excited states. Thus it is impractical to use Eq. 5.3.5 to determine the polarizability for a given Raman mode, and other ways to do so needed to be worked out.

A better method for the evaluation of the polarizability tensor elements is via an electric field perturbation. For excitation far removed from resonance,

Eq. 5.3.5 can be simplified to

$$\alpha_{\alpha\beta} = 2 \sum_i \frac{\langle \psi_0 | \mu_\alpha | \psi_i \rangle \langle \psi_i | \mu_\beta | \psi_0 \rangle}{E_i - E_0} \tag{5.5.28}$$

where the summation is over all virtual states i. In the presence of a static electric field F along the direction β, the polarizability can be evaluated according to

$$\alpha_{\alpha\beta} = 4 \sum_k \{\phi_k^0 | \mu_\alpha | \phi_k'(F_\beta)\} \tag{5.5.29}$$

where ϕ_k^0 are the unperturbed molecular orbitals, and $\phi_k'(F_\beta)$ the molecular orbitals in the presence of the electric field. Derivatives of these components with respect to normal coordinates are obtained by either calculating the unperturbed and perturbed molecular orbitals for equilibrium geometries and geometries corresponding to displacements along the normal coordinates, or by analytic methods. For a more detailed discussion, including the pertinent references, and comparison of observed and calculated Raman spectra for many small molecules, for example,

$$\text{CO, CO}_2\text{, CH}_4\text{, H}_2\text{C}\overset{\displaystyle O}{\text{———}}\text{CH}_2$$

the reader is referred to an article by Polavarapu [1990].

5.6 RESONANCE RAMAN SPECTROSCOPY

It was pointed out before that resonance enhancement occurs in Raman spectroscopy if the energy of the photon of the exciting laser beam is close to the energy of an electronic transition of the sample. We could argue qualitatively, using Eq. 5.3.5, that the proximity of a real electronic state will make one term of the denominator in the polarizability expression very small and, consequently, the terms with $\omega' \approx \omega$ in the summation very large.

The theory of Raman and resonance Raman intensities is usually not discussed in terms of the polarizability, but the scattering tensor, which has the form

$$(\alpha_{\alpha\beta})_{nm} = \frac{1}{\hbar} \sum_r \frac{\langle n | \mu_\alpha | r \rangle \langle r | \mu_\beta | m \rangle}{(\omega_{rm} - \omega + i\Gamma)} + \frac{\langle n | \mu_\alpha | r \rangle \langle r | \mu_\beta | m \rangle}{(\omega_{rm} + \omega + i\Gamma)} \tag{5.6.1}$$

The scattering tensor contains a damping term $i\Gamma$ in the denominator, which was neglected in the previous discussion of nonresonant Raman spectroscopy. This damping term physically is a lifetime of the intermediate state and prevents the denominator from becoming exactly zero at the resonance condition. In Eq. 5.6.1, r is the intermediate (real or virtual) state, and each element of the scattering tensor is, again, determined by a sum over *all* (vibronic) states of the molecule. The subscripts α and β refer, as previously, to the permutations of the Cartesian coordinates, and the subscripts n and m to the final and original states of the system. At conditions far from resonance, and if the molecule is initially in its ground state, Eq. 5.6.1 is equivalent to the polarizability tensor defined in Eq. 5.3.5 and discussed in the previous section (Eq. 5.5.28). Details on the discussion of the scattering tensor can be found in the literature [e.g., see Koningstein, 1972, Sections 1-4 and 1-5].

For the discussion of resonance enhancement, all states involved are written as the products of vibrational and electronic wavefunctions, and the dipole transition moments are evaluated separately for the purely electronic and vibrational wavefunctions. This allows the scattering tensor to be written as the sum of two terms, referred to as the A and B terms:

$$(\alpha_{\alpha\beta})_{nm} = A + B \tag{5.6.2}$$

with

$$A = \frac{M_e^\alpha M_e^\beta}{\hbar} \sum_v \frac{\langle j|v\rangle\langle v|i\rangle}{(\omega_{iv} - \omega + i\Gamma)} \tag{5.6.3}$$

and a somewhat more complicated expression holding for B. In Eq. 5.6.3, the M_e are pure electronic transition moments, and the terms $\langle j|v\rangle$ are the vibrational overlap integrals (i and j are the vibrational states of the ground electronic state, and v is a vibrational state of an electronically excited state).

The A term is responsible for the resonance enhancement in totally symmetric modes, whereas the B term dominates when the vibrational modes mix the two excited electronic states. The resonance enhancement due to the A and B terms have different frequency responses, which determine the onset of resonance enhancement as the laser wavelength is scanned into an absorption peak [Spiro and Stein, 1977].

The enormous intensity enhancement in resonance Raman spectroscopy has revolutionized the vibrational spectroscopic field of biomolecules, such as enzymes and nucleic acids, since molecules can be studied at concentrations of about 10^{-5} M, whereas in standard Raman spectroscopy, the lower concentration limit is about 10 mM. The solubility (or availability) of many biological molecules is low; consequently, the ability to work with dilute solutions is a prerequisite for these studies. Many proteins and enzymes contain colored reaction centers with chromophores in the visible or accessible ultraviolet region. Hemoglobin and many cytochromes are typical examples of such enzymes, which contain a metal ion (Fe^{2+}) bound in an inorganic ring (porphyrins). Excitation with a laser into such a chromophore will enhance the Raman scattering of vibrations of the chromophore, and a few other vibrations in close contact, such as Fe—S stretching vibrations due to out-of-plane ligands of the central iron atom. Thus the advantage of resonance Raman spectroscopy is not only the increased sensitivity that allows for more dilute solutions to be investigated, but also the selectivity of the effect, which allows for selected regions or groups of the molecule to be resonance enhanced and thus made visible.

With the availability of lasers that can be tuned into the far ultraviolet region, many other chromophores have become available for excitation, such as aromatic amino acid residues, nucleic acid bases, or even $\pi^* \leftarrow \pi$ and $\pi^* \leftarrow n$ transitions of carbonyl groups and the peptide linkage. These measurements have made possible structural elucidation of hundreds of molecular system that cannot be probed otherwise by vibrational spectroscopy. Some application of resonance Raman spectroscopy to biological systems will be discussed in more detail in Chapter 8. Resonance Raman spectroscopy is an extremely busy field of research, and many aspects of the form and symmetry of the scattering tensor, its precise computation, and many other areas are actively being investigated. The reader is referred to one of the many reviews available [Spiro and Stein, 1977; Johnson and Peticolas, 1976].

5.7 TIME-RESOLVED RESONANCE RAMAN SPECTROSCOPY

The ability to create extremely short laser pulses has given rise to a field known as time-resolved resonance Raman spectroscopy (TRRR). In principle, the theory involved in TRRR is not different from resonance Raman spectroscopy discussed in the previous section and can be described by the same equations used for the resonance Raman effect. In practice, however, TRRR has evolved into a very different and sophisticated field, particularly from the viewpoint of the experimental aspects required for observing TRRR. In TRRR, a laser pulse (pump pulse) is used to excite a molecule into an excited electronic state, initiate a photochemical reaction, or create whatever else desired perturbation within a species. A short time interval later, a second pulse (probe pulse) excites the Raman spectrum of the transient species that was created by the first pulse.

Time resolution of nanoseconds was achieved in the 1970s, picosecond resolution in the 1980s, and subpicosecond (50–100 fs) laser pulses have been produced recently, such that excitation on this time scale is possible. Such short time scales enable one to follow dissociation dynamics, deactivation and intersystem crossing mechanisms, and many other fast reactions. A few of those will be introduced later, along with a discussion of the equipment required to perform these experiments.

Lasers that produce very short pulses are commercially available. In gas and solid state lasers, for example, cavity dumping is a well-established procedure for the production of short pulses. Delays between pump and probe pulse are created in two ways. In one of these methods, both pump and probe pulses are derived from one laser pulse: picosecond (ps) pulses from a mode-locked, Nd:YAG laser ($\lambda = 1064$ nm) are split into two beams after frequency doubling to $\lambda = 532$ nm. These beams are used to synchronously pump two dye lasers with cavity dumpers, which are driven by the same synchronization signal. The dye lasers can be tuned to different colors but produce pulses of equal duration. A time delay between the pulses at the sample can be achieved by varying the path the probe and pump pulses travel before impinging on the sample. Optical delay lines are simple to implement at the time scales desired in TRRR, since a laser pulse, traveling at the speed of light, requires 1 ns for 30 cm. Thus a path difference of 3 mm will introduce a time delay between pulses of 10 ps. The apparatus discussed is shown schematically in Fig. 5.7.

Another approach for the creation of short laser pulses for TRRR has been reported by Hester [1987]. In this instrument, two excimer lasers were driven at about 25 Hz repetition rate. In this system, the delay between pump and probe pulses is not produced optically as in the previously discussed instrument, but electronically from one HV pulse, which is split into two paths to trigger the excimer lasers. One of the electronic pulses is delayed with respect to the other to produce a variable, exact delay. The output pulses of the excimer lasers may be directed through dye lasers to produce wavelengths appropriate to the experiments.

Raman spectra are usually collected with double- or triple-stage spectrographs fitted with gated intensified vidicon or reticon detectors (cf. Chapter 6). Numerous laser pump/probe pulse cycles are integrated on the detector elements before readout to improve the signal-to-noise ratio of the Raman spectra. Although integration and readout of the spectra may take many seconds, the time resolution achieved is determined by the time delay between

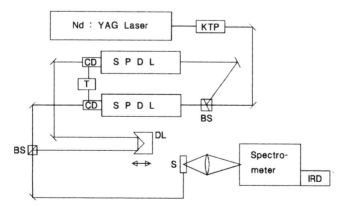

FIGURE 5.7. Apparatus for the observation of time-resolved resonance Raman spectra. KTP, frequency doubler; CD, cavity dumper; SPDL, synchronously pumped dye laser; DL, delay line; T, timing synchronization; BS, beam splitter; S, sample; IRD, intensified reticon detector. (After Atkinson et al. [1987].)

132 INTRODUCTION TO RAMAN SPECTROSCOPY

pump and probe pulses, and the actual Raman acquisition time is determined by the total lengths of the probe pulses.

An enormous amount of molecular dynamics in solution and gaseous states of molecules has been derived from these time-resolved studies. Many of the applications have dealt with the photochemical cycles of rhodopsin, the oxygen, carbon monoxide, and carbon dioxide binding dynamics of hemoglobin and myoglobin, and many of the redox pathways in cytochromes, for which the reaction pathways were elucidated by TRRR spectroscopy [Atkinson et al., 1987]. Some of these applications will be discussed in Chapter 8, in the biophysical applications of vibrational spectroscopy.

Here, we wish to present some other representative results of TRRR to demonstrate the relevance of this field to research in molecular dynamics and structure. A relatively slow time scale study of the photoinduced creation of a free radical anion in a substituted anthraquinone was reported by Moore et al. [1987]. Here, anthraquinone-2,6-disulfonate was illuminated by pump laser pulses at 337 nm. These pulses create a free radical anion, the vibrational spectrum of which is probed with delayed laser pulses at 480 nm. Delay times between pump and probe pulses were varied between 20 ns and 10 μs. The resulting Raman spectra are shown in Fig. 5.8. All Raman bands exhibit the same time dependence and therefore are due to the same species, which was identified to be the radical anion by its vibrational spectrum. After 10 μs, the system has returned to its original state.

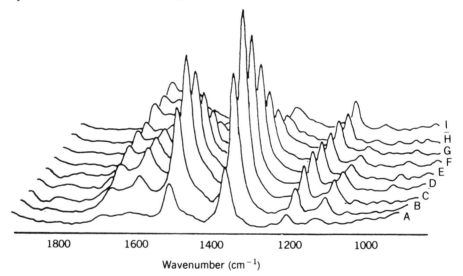

FIGURE 5.8. Time-resolved resonance Raman spectra of the radical anion created by photoexcitation of anthraquinone-2,6-disulfonate. Pump wavelength, 337 nm; probe wavelength, 480 nm. Time delays: A, 20 ns; B, 50 ns; C, 100 ns; D, 500 ns; E, 1 μs; F, 2 μs; G, 5 μs; H, 10 μs; I, probe laser only. (From J. N. Moore, D. Phillips, R. E. Hester, G. H. Atkinson, and P. M. Killough, in *Time Resolved Laser Raman Spectroscopy*, D. Phillips and G. H. Atkinson, Eds., Academic Press, New York, 1987).

In another study, the electronic triplet and singlet states of all-trans diphenylbutadiene (DPB) were detected and identified via TRRR. Although fast fluorescence and (UV) absorption studies had been reported previously, the TRRR studies offer the advantage of characterizing the excited states better and in more detail than many of the other techniques, and, in addition, allow their deactivation dynamics to be followed.

The excitation into the triplet state is symmetry forbidden and thus utilizes an intermediate as follows: biphenyl was excited with pulses at 248 nm, after which nonradiative energy transfer occurs to the triplet state of DPB, which is consecutively sampled with probe pulses at 396 nm.

The DPB singlet state can also be probed by excitation with pump pulses at 308 nm and probing at 640 nm. Lifetime measurements with time resolution as low as 60 ps was achieved in appropriate solvents, which may stabilize or destabilize the excited states. Thus, in this system, resonance Raman techniques can access the ground state, as well as the excited singlet and triplet states, and yield information on their molecular and electronic structure and deactivation dynamics.

5.8 NONLINEAR RAMAN EFFECTS

With the advent of high-power pulsed lasers, several totally novel spectroscopic effects were discovered. These effects all have some principles in common with the classical Raman effects, but the intensities of these effects depend nonlinearly on the field strength of the exciting electromagnetic radiation and are therefore referred to as "nonlinear Raman spectroscopies." The electric field in a pulsed laser can easily exceed 10^{10} V/cm, which is about 100,000 times stronger than the field strength of a CW laser used commonly to excite Raman spectra. At these high laser fields, the induced dipole moment can no longer be represented by Eq. 5.2.1

$$\mu = \alpha E \tag{5.2.1}$$

but has to be replaced by an expansion

$$\mu = \alpha E + \tfrac{1}{2}\beta E^2 + \tfrac{1}{6}\gamma E^3 + \cdots \tag{5.8.1}$$

where β and γ are known as the first and second hyperpolarizability tensors of rank 3 and 4, respectively. Equation 5.8.1 is also often written in terms of macroscopic dielectric susceptibilities:

$$\mathbf{P} = \chi^{(1)}\mathbf{E} + \chi^{(2)}\mathbf{E}\mathbf{E} + \chi^{(3)}\mathbf{E}\mathbf{E}\mathbf{E} + \cdots \tag{5.8.2}$$

where $\chi^{(n)}$ is a tensor of rank $n + 1$, known as the dielectric susceptibility, which relates the inducing electric field $\mathbf{E}$ to the induced macroscopic polarization $\mathbf{P}$. In Eqs. 5.8.1 and 5.8.2, the first term on the right-hand side is responsible for the static dielectric constant, the refractive index, and Rayleigh and Raman scattering. The second term is responsible for hyper-Rayleigh and hyper-Raman scattering, in addition to frequency doubling and the Pockels effect. Finally, the third-order dielectric susceptibility is responsible for a number of effects known as coherent, nonlinear Raman effects.

There is an important distinction between the hyper-Raman effect, which is certainly a form of nonlinear Raman spectroscopy, and the *coherent* forms of nonlinear Raman spectroscopy, such as the stimulated Raman, inverse Raman, Raman gain, and coherent anti-Stokes–Raman effects to be discussed later. In the former technique, the hyper-Raman scattered photons are scattered into all 4π steradians, just like in the case of spontaneous Raman scattering. Also, in both spontaneous and hyper-Raman effects, all normal modes are excited simultaneously when a molecule is exposed to the exciting laser radiation. In the coherent forms, on the other hand, the scattered light exits the sample as a coherent beam with the properties of laser light. Since most of these techniques require two laser input beams, only one normal mode, determined by the frequency difference between the two laser beams, is excited, leading to a much higher efficiency of the scattering process. Thus these effect are very strong (albeit still not easily observed), whereas the noncoherent hyper-Raman effect is so weak that it is still not a particularly practical technique.

In the remainder of this section, some of the theory underlying the coherent nonlinear techniques will be introduced. The equation most commonly used as a starting point of theoretical aspects of coherent Raman effects (CRE) is Eq. 5.8.2:

$$\mathbf{P} = \chi^{(1)}\mathbf{E} + \chi^{(2)}\mathbf{EE} + \chi^{(3)}\mathbf{EEE} + \cdots \quad (5.8.2)$$

which was introduced before. The third-order polarization $P^{(3)}$ induced along the direction of the ith axis, due to the third term in Eq. 5.8.2, may be written as

$$P_i^{(3)} = \sum_{j,k,l} \chi^{(3)}_{i,j,k,l} \cdot E_j E_k E_l \quad (5.8.3)$$

where all three electric fields travel along the Z-axis. This equation is, in principle, similar to Eq. 5.5.1, except that the dimensionality of the process has increased. In Eq. 5.8.3, there are three different electromagnetic fields present, and the third-order susceptibility tensor, which is responsible for their mixing, is therefore of rank 4. Only 21 of the 81 elements of this tensor are nonzero for isotropic media. In fact, only tensor elements for which all four indices are the same (e.g. χ_{xxxx}) and those for which there are two pairs of identical indices (e.g. χ_{xxyy}, χ_{xyyx}, or χ_{xyxy}) are nonzero. For the remaining 21 elements, there are only four different numeric values, which are commonly referred to as χ_{1111}, χ_{1122}, χ_{1221}, and χ_{1212}, where

$$\chi_{1111} = \chi_{xxxx} = \chi_{yyyy} = \chi_{zzzz}, \quad \text{and so on.} \quad (5.8.4)$$

For the further discussion of the third-order nonlinear effects, we shall use the following conventions. The medium is exposed to electromagnetic fields of different frequencies. Let $E(z, t)$ represent an electromagnetic wave with an angular frequency ω traveling along the z-direction according to

$$E(z, t) = E(\omega)(\mathbf{i}\cos\theta + \mathbf{j}\sin\phi\, e^{i\phi})e^{i(\omega t + kz)} \tag{5.8.5}$$

where the term $(i\cos\theta + j\sin\phi)$ determines the orientation of the polarization ellipsoid relative to the unit vector $\mathbf{i}$. ϕ is the phase between the two polarization components along $\mathbf{i}$ and $\mathbf{j}$. $\mathbf{k}$ is the wave vector introduced in Section 2.8. Let the frequencies of the radiation fields be designated by ω_1, ω_2, ω_3, and ω_4, such that the generated photon has a frequency $\omega_4 = \omega_1 + \omega_3 - \omega_2$ (the use of positive and negative signs for the frequency components will become clear in the next section). Then, the third-order polarization along the x-direction for linearly and parallel polarized beams of light is given by

$$P_x^{(3)}(\omega_4) = D[\chi_{1111} + \chi_{1122} + \chi_{1221} + \chi_{1212}]E(\omega_1)E(\omega_2)E(\omega_3)$$
$$\cdot \exp\{i(\mathbf{k}_1 - \mathbf{k}_2 + \mathbf{k}_3 - \mathbf{k}_4)\} \tag{5.8.6}$$

where D is an integer factor between 1 and 6 indicating how often each term must be counted. If the beams are not linearly polarized, and/or if the polarization ellipsoid does not lie along the x-axis, each term in the square bracket in Eq. 5.8.6 needs to be multiplied by the sine and cosine elements of the angle of polarization and by a phase factor. This third-order nonlinear polarization component is subsequently substituted into Maxwell's equations to yield the so-called gain equation, which reveals that the induced electric field at ω_4, $E(\omega_4)$, depends linearly on $P^{(3)}(\omega_4)$ [Tolles and Harvey, 1981; Barrett, 1981].

5.8.1 Coherent Anti-Stokes–Raman Scattering (CARS)

In CARS, the sample is illuminated by two lasers, one of them with a fixed wavelength usually referred to as the pump laser ω_p or ω_1, and a second tunable laser beam, referred to as the Stokes frequency ω_s or ω_2. The energy level diagram depicted in Fig. 5.9 describes the processes giving rise to CARS. In the following paragraphs, the four events necessary for the creation of an anti-Stokes photon are described consecutively, although the four processes do not occur as separate events but are a highly correlated, four-wave mixing phenomenon mediated by the third-order nonlinear susceptibility element χ_{1111}.

An initial photon, $\hbar\omega_p$ is annihilated, promoting the system into a virtual state. A photon from the laser emitting at the (Stokes) Raman frequency $\hbar\omega_s$ causes a transition to the vibrationally excited state. Since both of these

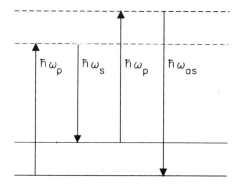

FIGURE 5.9. Energy level diagram and photons participating in the CARS process. The subscripts on the photon energies indicate: as, anti-Stokes; s, Stokes; p, pump.

processes occur with large laser powers, a large population of the vibrationally excited state results. The vibrationally excited state interacts, in turn, with a second laser photon, $\hbar\omega_p$, to populate another virtual state, which emits back to the vibrational ground state. The energy released for this final transition is carried off by a photon for which the frequency is given by

$$\hbar\omega_a = 2\hbar\omega_p - \hbar\omega_s = \hbar\omega_p + \hbar\omega_m \qquad (5.8.7)$$

where $\hbar\omega_m$ is energy of the molecule's vibrationally excited state. Thus the wavelength of this photon is that of an anti-Stokes–Raman process, and the emission of the anti-Stokes photon occurs only if the wavelength of the tunable Stokes laser fulfills the condition

$$\hbar\omega_s = \hbar\omega_p - \hbar\omega_m \qquad (5.8.8)$$

Using Eq. 5.8.7 for the frequency of the generated photon, one can write Eq. 5.8.6 for the CARS process as follows:

$$P^{(3)}(\omega_a) = \chi_{1111}(-\omega_a, \omega_p, \omega_p, -\omega_s)E(\omega_p)E(\omega_p)E(\omega_s)$$
$$\cdot \exp\{i(\mathbf{k}_p - \mathbf{k}_s + \mathbf{k}_p - \mathbf{k}_a)\} \qquad (5.8.9)$$

where the sign associated with the **k** vectors or the angular frequencies indicate a photon is annihilated or created.

The CARS intensity scattered at ω_a is given by

$$I_a = (4\pi^2\omega_a/c^2)^2 I_p^2 I_s |\chi|^2 z^2 \qquad (5.8.10)$$

where z is the distance over which phase matching (*vide infra*) is valid.

The exponential expression in Eq. 5.8.9 needs to be discussed in more detail. It contains the wave vectors of the four interacting electromagnetic fields. For the creation of a CARS photon, the following condition must be fulfilled:

$$(\mathbf{k}_p - \mathbf{k}_s + \mathbf{k}_p - \mathbf{k}_a) = 0 \qquad (5.8.11)$$

This condition is called the phase matching or momentum matching criterion. It implies that the pump and Stokes beams must intersect at an angle given by the vector addition in Fig. 5.10 for CARS photons to be generated. However, as the Stokes laser is scanned in a typical CARS experiment, the lengths of vector $\mathbf{k}_s$ and the refractive index of the medium at ω_s both change. Therefore, phase matching is not a trivial problem, and the detector must be moved as the Stokes laser is scanned to remain at the proper angle with respect to the pump laser beam.

The anti-Stokes photons leave the sample as a collimated, coherent laser beam, for which the spectral resolution is given by the linewidth of the exciting lasers. As pointed out before, spectral information on only one normal mode at a time is obtained as the Stokes laser is scanned to cover the spectral range. However, the lineshapes observed in CARS are rather unusual and often contain dispersive lineshape elements. This is because the third-order susceptibility tensor elements are composed of (complex) resonant and nonresonant terms. Depending on which of these terms contribute, Lorenzian or dispersive lineshapes may be obtained.

CARS is a difficult experiment, since it requires careful matching of the wave vectors of two laser beams. Carreira et al. [1981] have described a fully computerized CARS spectrometer, consisting of two nitrogen laser pumped tunable dye lasers, with a wavelength range from 260 to 740 nm. This instrument is shown schematically in Fig. 5.11. During a CARS experiment, the wavelength of one of the dye lasers is maintained at a fixed value, whereas the other laser is scanned to fulfill Eq. 5.8.7. The CARS beam is detected by a photomultiplier, which is mounted, along with associated detection optics, on a stage that can be rotated under computer control about the point where the pump and Stokes laser beams intersect, and from which the CARS beam emerges. In addition, the intersection angle of the pump and Stokes beams is adjusted under computer control to optimize the phase matching. The instrument described allowed CARS spectra of liquid samples to be collected, in one scan, over a range of many hundreds of wavenumbers, which is difficult in CARS scattering due to the dispersion of the refractive index of the sample. In

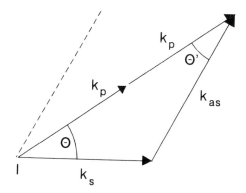

FIGURE 5.10. Phase-matching condition for CARS. Θ denotes the angle between the pump and Stokes beam, which intersect at point I. Θ' is the angle between the pump and the CARS beam, which is emitted from point I along the dotted line. The subscripts used are the same as in Fig. 5.9.

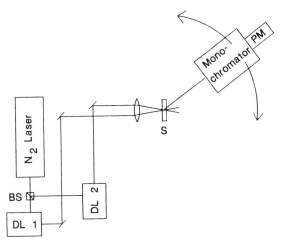

FIGURE 5.11. Apparatus for the observation of CARS. BS, beam splitter; DL1 and DL2, dye lasers. The monochromator/detector may be rotated as indicated to maintain the phase-matching condition. (After Carreira et al. [1981].)

fact, Carreira et al. report methods to adjust the angle of intersection differently for different solvent refractive indices. Other instruments, based on frequency doubled Nd:YAG lasers or ruby lasers, have been reported. Thus it appears that most of the instrumental difficulties associated with CARS have been overcome.

CARS scattering from gases is somewhat easier to observe, since the refractive index does vary more slowly with wavelength in gases than in liquids. The enormous sensitivity of CARS makes it an ideal tool to investigate very low concentrations of gaseous samples. Rotational and rotational–vibrational spectra, with subwavenumber resolution, have been collected at sample pressures of a few torr. Combustion processes, within the exhaust system of combustion and jet engines, or in flames, have successfully been studied via CARS spectroscopy, which is a sensitive, noninvasive techniques to monitor products, reaction mechanisms, and the temperature of gaseous reactions.

5.8.2 Stimulated Raman Scattering

In stimulated Raman scattering, the sample is illuminated with only one laser at the frequency ω_p. If the intensity of this pump laser increases to the levels typically achievable with giant pulse lasers, the intensity of the Stokes–Raman scattering becomes sufficiently large that nonlinear mixing of the Raman radiation field with that of the pump laser causes coherent laser output at ω_s to occur, where $\hbar\omega_s = \hbar(\omega_p - \omega_m)$. This effect can convert up to 50% of the incident photons into Raman scattering, compared to an efficiency of spontaneous Raman spectroscopy on the order of 10^{-12}. Since the effect that

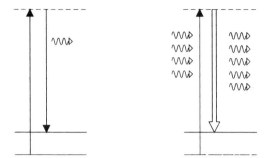

FIGURE 5.12. Principles of stimulated Raman spectroscopy. The diagram on the left is that of spontaneous Raman scattering. On the right, the gain aspect is indicated by the increased number of photons scattered coherently.

generates the Stokes laser beam is a Raman process, the stimulated Raman process is also known as a Raman laser.

Conceptually, we may visualize the stimulated Raman effect as follows: the number of photons temporarily in the virtual state is so large that stimulated emission from this state occurs. This is shown schematically in Fig. 5.12, where three incoming photons of energy $\hbar(\omega_p - \omega_m)$ stimulate a virtual state to emit; consequently, four photons of the same energy leave the sample. The theoretical treatment of the stimulated Raman effect therefore needs to take into account the radiation field of the laser at ω_p as well as the field at $\omega_p - \omega_m$. Whereas CARS is a three-color technique (it involves four photons with three different frequencies), the stimulated Raman effect is a two-color effect, yet it is mediated through the third-order nonlinear susceptibility as well.

5.8.3 Inverse Raman Effect

In the inverse Raman effect, two laser beams are incident at the sample, one pump laser with angular frequency ω_p and another with the frequency of a Stokes photon, ω_s. If the difference of these two laser photon energies match that of a molecular vibrational state, $\hbar\omega_m$,

$$\hbar\omega_p = \hbar\omega_s + \hbar\omega_m$$

and if the intensity of the Stokes laser is above the threshold for stimulated emission, a photon at ω_p must be annihilated, and the virtual state must be populated, before a stimulated emission can occur. The depletion of the laser intensity, monitored at $\hbar\omega_p$, as the Stokes laser is pulsed, is referred to as the inverse Raman effect. Like the stimulated Raman effect, inverse Raman scattering is a two-color phenomenon.

5.8.4 Raman-Induced Kerr Effect

This phenomenon is named after the Kerr effect, which describes the change in optical properties of a medium when the medium is subject to an (static) electric field. In the optical Kerr effect, birefringence is induced in a medium by a radiation field and probed by a light beam at another frequency.

In Raman-induced Kerr effect spectroscopy (RIKES), the electric field is that of a Raman emission. In RIKES, a tunable or broadband probe pulse with the frequency $\hbar\omega_2$ passes through a sample. This beam is linearly polarized and impinges, after passing the sample, onto a second polarizer, which is set at 90° with respect to the first one, such that the probe laser beam is blocked by the second polarizer.

A pulse of a pump laser at frequency $\hbar\omega_1$, which is circularly polarized, intersects the probe pulse in the sample. The pump pulse induces birefringence in the sample, which changes the state of polarization of the probe beam, which is, in turn, transmitted by the second polarizer and detected. This induced birefringence occurs only when the difference between the laser frequencies matches a Raman active mode in the medium: $\hbar\omega_m = \hbar\omega_1 - \hbar\omega_2$.

RIKES is due to different terms of the nonlinear susceptibility tensor than is CARS. In RIKES, the off-diagonal elements $\chi_{1122}(-\omega_2, \omega_1, -\omega_1, -\omega_2)$ and $\chi_{1212}(-\omega_2, \omega_1, -\omega_1, -\omega_2)$ are responsible for the effect.

5.8.5 The Hyper-Raman Effect

Hyper-Raman scattering results from the second term in Eq. 5.8.1, $\frac{1}{2}\beta E^2$, and is a nonlinear, noncorrelated form of Raman spectroscopy. Hyper-Raman and hyper-Rayleigh spectroscopies are three-photon processes depicted schematically in Fig. 5.13. Two photons of frequency ω_L (heavy up arrows) are annihilated by the virtual states, shown by the dashed lines. Simultaneously, a hyper-Rayleigh photon at frequency $2\omega_L$, or a hyper-Raman photon at frequency $2\omega_L - \omega_m$, is created, where ω_L is the frequency of the laser photon and ω_m is the frequency of a molecular vibrational quantum. This is shown by the thin downward arrow in Fig. 5.13. Depending on the symmetry of the scattering medium, the hyper-Raman scattering may or may not be accompanied by hyper-Rayleigh scattering at a frequency of $2\omega_L$, which is, in fact, a frequency doubled photon scattered incoherently.

Hyper-Raman spectroscopy is in many aspects analogous to the standard Raman effect, except that the electric field appears quadratically in the expression for the induced dipole moment. A classical treatment, analogous to Eqs. 5.2.1 through 5.2.8, can be carried out, which yields the hyper-Raman frequencies $2\omega_L \pm \omega_m$. In addition, the hyperpolarizability tensor components can be written in analogy to Eq. 5.3.5, but involving three dipole transition moments and two intermediate, virtual states. Thus it is clear that different selection rules hold for the hyper-Raman effect, since it is a three-photon

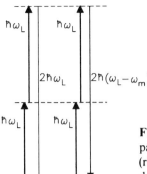

FIGURE 5.13. Energy level diagram and photons participating in hyper-Raman (left) and hyper-Rayleigh scattering (right). Heavy upward arrows, laser photons annihilated; downward arrows, photons scattered; dashed lines, virtual states.

process. In fact, low symmetry vibrations are more allowed, and more intense, in hyper-Raman spectroscopy, and totally symmetric vibrations may be very weak or entirely forbidden. Also, the limiting values of the hyper-Raman depolarization ratios differ from those encountered in Raman spectroscopy and vary from 1/9 to 2/3 for linearly polarized light.

A typical hyper-Raman spectrum appears similar to a Raman spectrum but is shifted from the *harmonic* of the laser frequency: if a ruby laser with $\lambda = 694.3$ nm or 14,403 cm^{-1} is used for excitation of the sample, the (Stokes) Raman spectrum appears red-shifted between 14,403 and about 11,000 cm^{-1}. The hyper-Raman spectrum will appear red-shifted from 28,806 cm^{-1}, but with the same vibrational shift ω_m. CCl$_4$, for example, shows a hyper-Raman spectrum in which the most prominent peak occurs at a shift of 770 cm^{-1}. This peak corresponds to the triply degenerate C—Cl antisymmetric stretching mode, which is weak in the Raman spectrum (cf. Chapter 7). On the other hand, the strong Raman modes at 314 and 459 cm^{-1} are very weak in the hyper-Raman spectrum, and the doubly degenerate deformation at 217 cm^{-1} is inactive in the hyper-Raman effect. It is also noteworthy that the intensity of the hyper-Rayleigh line is somewhat weaker than that of the strongest hyper-Raman peak. This behavior is very different from conventional Raman and Rayleigh scattering, where the intensity of the elastically scattered photon is always many orders of magnitude larger than that of the inelastically (Raman) scattered photons [French, 1981].

As mentioned above, the selection rules for Raman, hyper-Raman, and infrared spectroscopies are all different. Thus the hyper-Raman effect has complemented the two other forms of vibrational techniques in the sense that it has permitted the observation of vibrational states that cannot be excited in Raman or infrared spectroscopies. The torsional vibration in C$_2$Cl$_4$, for example, which transforms as the irreducible representation A_u, is not active in either infrared absorption or Raman scattering, but was observed in hyper-Raman spectroscopy at 110 cm^{-1}.

Finally, most interest in this field has arisen because of second harmonic generation, also known as frequency doubling, in certain crystals. This effect can be viewed as a form of hyper-Rayleigh scattering under phase-matched condition, such that a coherent beam of light at twice the frequency emerges from a frequency doubling crystal. These are used extensively with Nd:YAG lasers, which has an emission wavelength of 1.064 μm. After frequency doubling, green photons at 532 nm are obtained. Because of the symmetry properties of the hyperpolarizability tensor, frequency doubling is possible only in that crystals are devoid of a center of symmetry.

5.9 SURFACE ENHANCED RAMAN SCATTERING (SERS)

When certain molecules, particularly those with free electron pairs on nitrogen atoms, are adsorbed on surfaces of certain metals, the scattered Raman intensities of the adsorbed molecules can be enhanced by several orders of magnitude. In fact, the enhancement is so strong that monolayers of molecules can be observed even in the presence of solvents. Typical examples for which SERS has been reported are aqueous solutions of pyridine, or other amines, on either silver electrodes or silver sol particles. Other metals, notably noble metals, have shown similar surface enhancements, as have a number of metal oxides.

Since a number of different effects contribute to the surface enhancement, the actual mechanism of SERS was not understood initially. Now, it is widely accepted that the underlying effect is an enhancement of the electromagnetic field at the surface. This enhancement can have several reasons. When using a silver sol or colloid, for example, with silver particles with diameters of about 20 nm as the metal surface, visible photons are absorbed into the surface plasmons of the metal spheres. Surface plasmons are oscillations of charges at the surface, and in the case of silver spheres of appropriate size, they can be visualized as spherical harmonic electromagnetic oscillations. The surface plasmon contributes enormously to the local electric field and thus enhances the Raman scattering process.

A silver sol actually appears colored due to the absorption of the light into the plasmons, and the wavelength of maximum absorbance depends on the solvent, the size of the metal spheres, and the nature of the metal. Silver sols can be prepared conveniently by reducing $AgNO_3$ solutions with sodium borohydride.

A similar mechanism for Raman enhancement is proposed for solid silver surfaces as well, although the surface preparation is important. For electrode surfaces, good enhancements are obtained by electrochemically modifying the surface, via repeated reduction–oxidation cycles, which appears to create surface roughness with particles of similar sizes as in silver sols. Island film structures, produced by slowly evaporating Ag atoms onto a substrate, give similar particle size distribution as the sols and produce similar enhancements.

In all cases discussed so far, the SERS effect is due to the enhanced electromagnetic field due to the plasmons.

The adsorbate also can form charge transfer complexes with the metal atoms of the surface, thereby creating an enhancement mechanism that is similar to the resonance Raman effect (cf. Section 5.6). Both mechanisms can also contribute simultaneously, to produce enhancement factors on the order of 10^7 or larger.

Upon binding to the surface, the symmetry of the adsorbate is lowered, resulting in a different vibrational spectrum. Raman bands that are symmetry forbidden in the adsorbate may be allowed in the complex with the surface. Different possible binding structures can often be distinguished from the SERS spectra, and it has been established that many nitrogen-containing molecules such as pyridine bind in such a manner that the ring is perpendicular to the surface.

SERS has been used to follow electrochemical processes on polished electrode surfaces, since it permits a thin layer (monolayer) of adsorbates to be studied. The surface enhancement also can be utilized to increase the Raman scattering efficiency enormously over the (generally unwanted) fluorescence process, which often accompanies Raman scattering. Highly fluorescent dyes have been studied by SERS on silver sol surfaces.

REFERENCES

G. H. Atkinson, T. L. Brack, I. Grieger, G. Rumbles, D. Blanchard, and L. M. Siemanowski, in *Time Resolved Vibrational Spectroscopy*, G. H. Atkinson, Ed., Gordon & Breach Science Publishers, New York, 1987.

R. Aroca and G. J. Kovacs, in *Vibrational Spectra and Structure*, J. R. Durig, Ed., Elsevier, Amsterdam, The Netherlands, Vol. 19, pp. 55–112, 1991.

J.J. Barrett, in *Chemical Applications of Nonlinear Spectroscopy*, A. B. Harvey, Ed., Academic Press, New York, 1981.

L. A. Carreira, L. P. Goss, and T. B. Malloy, Jr., in *Chemical Application of Nonlinear Spectroscopy*, A. B. Harvey, Ed., Academic Press, New York, 1981.

M. J. French, in *Chemical Applications of Nonlinear Spectroscopy*, A. B. Harvey, Ed., Academic Press, New York, 1981.

R. E. Hester, in *Time Resolved Laser Raman Spectroscopy*, D. Phillips and G. H. Atkinson, Eds., Harwood Academic Press, New York, 1987.

B. B. Johnson and W. L. Peticolas, *Annu. Rev. Phys. Chem.*, **27**, 465 (1976).

W. Kauzmann, *Quantum Chemistry*, Academic Press, New York, 1957.

J. A. Koningstein, *Introduction to the Theory of the Raman Effect*, D. Reidel Publishing, Dordrecht, The Netherlands, 1972.

J. N. Moore, D. Phillips, R. E. Hester, G. H. Atkinson, and P. M. Killough, in *Time Resolved Laser Raman Spectroscopy*, D. Phillips and G. H. Atkinson, Eds., Harwood Academic Press, New York, 1987.

P. L. Polavarapu, *J. Phys. Chem.*, **94**, 8106 (1990).

T. G. Spiro and P. Stein, *Annu. Rev. Phys. Chem.*, *28*, 501 (1977).

W. M. Tolles and A. B. Harvey, in *Chemical Applications of Nonlinear Spectroscopy*, A. B. Harvey, Ed., Academic Press, New York, 1981.

E. B. Wilson, J. C. Decius, and P. C. Cross, *Molecular Vibrations*, McGraw-Hill, New York, 1955.

6

INSTRUMENTATION FOR THE OBSERVATION OF VIBRATIONAL SPECTRA

In this chapter, the instrumentation necessary for the observation of vibrational spectra will be introduced. Since there are two significantly different techniques in vibrational spectroscopy, Raman scattering and infrared absorption, very different instruments are needed for the observation of these effects. However, both techniques have in common that polychromatic radiation from a source needs to be analyzed, in terms of the intensity of its components, and detected. The light source to be analyzed in Raman spectroscopy is the focused laser beam in the sample, whereas the source for infrared absorption spectroscopy is the actual image of the infrared source. Light emitted from either source is collected and focused into the wavelength sorting device, which can be a monochromator in a dispersive instrument or an interferometer in Fourier transform methods, for both Raman and infrared measurements. Thus this chapter will begin with a discussion of dispersive monochromators, along with their entrance optics, and interferometers. Subsequently, the specific aspects of the instrumental requirements for the two techniques, as well as sampling methods, will be introduced.

6.1 DISPERSIVE SYSTEMS

6.1.1 Monochromators

Monochromators are devices that separate polychromatic radiation from the source spatially into monochromatic components. In nearly all modern dispersive instruments, this step is achieved via diffraction gratings. A grating is a reflective surface that is ruled, or scratched, with parallel lines that are spaced

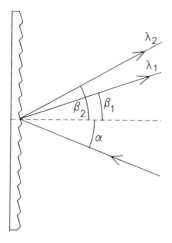

FIGURE 6.1. Schematic of a diffraction grating. Polychromatic rays of light are incident at angle α. The rays diffracted at angles β_1 and β_2 have wavelengths λ_1 and λ_2.

by a distance about the same as the wavelength of the light to be analyzed. This is shown in Fig. 6.1.

Polychromatic radiation incident on a grating is diffracted from it according to the grating equation, Eq. 6.1.1:

$$d(\sin \alpha + \sin \beta) = n\lambda \qquad (6.1.1)$$

where λ is the wavelength of the incident light, α is the angle of incidence, and β is the angle of diffraction. n is an integer known as the diffraction order, and d is the spacing between the rulings. Often, the grating groove density g is specified, usually given in mm^{-1}, where $d = 1/g$.

Thus if polychromatic and parallel light impinges on a grating as shown in Fig. 6.1, the different wavelengths are diffracted off at different angles. A grating is incorporated into a monochromator as shown in Fig. 6.2. This particular

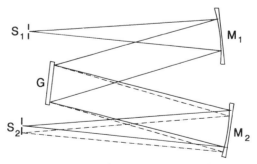

FIGURE 6.2. Czerny–Turner monochromator. S_1 and S_2, entrance and exit slits, respectively; G, grating; M_1 and M_2, spherical mirrors. Different wavelengths are diffracted off the grating at different angles and reach the exit focal plane at different locations. Only one wavelength is transmitted by the exit slit S_2.

design, with two spherical mirrors and a plane grating, is known as a Czerny–Turner monochromator. The grating may be rotated, and depending on the angle of the grating, different wavelengths are focused onto, and transmitted by, the exit slit.

In a Czerny–Turner monochromator, the incident light is diffracted into its component "colors" according to

$$\frac{d\lambda}{dx} = \frac{\cos\beta}{Fng} \quad (6.1.2)$$

where $d\lambda/dx$ is known as the linear dispersion of a monochromator and F is the focal length of the monochromator. The linear dispersion determines the spatial separation of the different wavelengths of light in the exit plane of the monochromator and thus determines the optical resolution achieved at a given slit width. Equation 6.1.2 can be derived from the grating equation, 6.1.1, by differentiating it with respect to β, which yields the angular dispersion ($d\lambda/d\beta$). Substituting $dx = F\,d\beta$ to relate angular deviation with the focal length, we obtain Eq. 6.1.2.

This equation indicates that the linear dispersion of a monochromator is not constant but varies with the cosine of the diffracted angle. Thus the dispersion of a monochromator changes with the wavelength of light being measured, since the angle of diffraction is given by Eq. 6.1.1. Thus in order to maintain constant spectral bandpass, the exit slit needs to be adjusted as the wavelengths are scanned.

In dispersive infrared (IR) instruments, 20–50 cm focal length monochromators are often used with gratings ruled at 50–300 lines/mm. This produces optical resolutions on the order to 0.5–5 cm^{-1} at a mechanical slit width of 0.1–1 mm. Typical IR spectral measurements are made over a range of 400 to nearly 4000 cm^{-1} (25–2.5 μm). Inspection of Eq. 6.1.2 shows that the linear dispersion of the monochromator depends on the cosine of the angle of diffraction. Consequently, one finds that it is impossible to achieve a reasonably constant resolution over the wavelength range with one and the same grating. Instrument designers avoided this problem by using the grating in first and second order, depending on the wavelength investigated, which changes the dispersion by a factor of $\frac{1}{2}$. Also, many dispersive instruments use two gratings mounted on a turret to allow for automatic grating change at a certain wavelength.

In dispersive Raman instrumentation, the design requirements are more stringent. First, due to the presence of Rayleigh scattering, which may be orders of magnitude more intense than the Raman scattered intensities, the rejection of stray light must be very high. Second, since the scattered light is still visible, with relatively small frequency shift, the resolution of a monochromator system for Raman spectroscopy must be higher than that for infrared monochromators.

A simple example is given here to illustrate this point. To resolve two infrared peaks at 2995 and 3000 cm^{-1} (3.3389 and 3.3333 μm, respectively), the resolving power needs to be about one part in a thousand. Two Raman lines at 2995 and 3000 cm^{-1} shift from an exciting Ar ion laser line (514.5 nm, or 19,436 cm^{-1}) occur at 16,441 and 16,436 cm^{-1}, respectively (608.23 and 608.42 nm). To separate these lines, a resolving power of three parts in 10^4 is needed.

Thus monochromators for Raman spectroscopy typically use longer focal lengths (50–100 cm), much higher grating groove densities (1000–2400 lines/mm), and narrower slits (50–200 μm). The linear dispersion of such a monochromator is typically on the order 5 Å/mm.

6.1.2 Light Collection Optics for Monochromators

In a dispersive system, the actual throughput of the monochromator depends on the square of the mechanical slit width, the optical aperture, and the efficiency of the grating. Thus one wishes to work at very low f-numbers and at very wide slit width. These two conditions, however, produce a low resolution; thus a trade-off has to be made between these parameters.

One of the most important aspects determining the performance of a spectrometer is the transfer of the light from the source into the monochromator. This is shown schematically in Fig. 6.3. In principle, one wishes to collect the light from the source with low f-number optics, since this corresponds to sampling the largest solid angle. However, monochromators typically have apertures (f-numbers) between $f/4$ and $f/8$. This aperture is determined by the size of the grating that can be manufactured at a reasonable cost and limits the throughput of the entire spectrometer. If light from the source is collected, for example, at $f/1$, it can be focused into a $f/8$ monochromator. However, this process will cause an increase of the size of the source image by a factor given by the ratio of the f-numbers, in this case 8. Thus an infrared source of 1 mm diameter cannot be focused to an image size less than 8 mm wide if the light is collected at $f/1$ and focused at $f/8$.

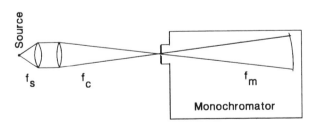

FIGURE 6.3. Matching light collection and monochromator apertures. f_s, f_c, and f_m are the f-numbers of sampling, collection, and the monochromator, respectively. Ideally, f_c and f_m should be equal. The ratio f_c/f_s determines the magnification of the source image at the entrance slit.

For most spectral applications, 8 mm slits are unacceptable due to the loss of resolution. If the slits are closed below 8 mm, however, the size of the image is cut, and the advantage of collecting at $f/1$ is lost.

Thus there are a number of factors that must be considered when designing a light collection system: the mechanical slit width that is acceptable for the instrument, the size of the souce, and the f-number of the system desired. The mechanical slit width depends, of course, on the monochromator local length and grating, as given by Eq. 6.1.2. It is advantageous in the design of a spectrometer to start with the question: What is the widest slit that produces an acceptable spectral resolution? The answer to this question narrows the choice of a monochromator and grating. The monochromator f-number, in turn, determines the focal length of the focusing optics (cf. Fig. 6.3). The mechanical slit width and the size of the source determine what magnification factors are acceptable in the collection optics and, in turn, the light collection f-number. A number of these arguments are presented in a review article by the author [Diem, 1991].

6.2 INTERFEROMETRIC METHODS

6.2.1 General Aspects of Fourier Transform Infrared Spectroscopy

In a dispersive systems incorporating a grating and an exit slit, only one spectral element is sampled by the detector at a time. In contrast, in Fourier transform spectroscopy, one examines all wavelengths arriving at the detector simultaneously. In order to obtain the desired spectral distribution, $I(v)$ versus v, one varies the interference pattern of the light reaching the detector, $I(v)$, interferometrically and takes the Fourier transform of the signal at the detector. The principle of the central apparatus in Fourier transform spectroscopy, the interferometer, will be discussed next, followed by a short review of the interference process taking place in such an interferometer, and the mathematical foundation to understand the process.

6.2.2 The Michelson Interferometer

A schematic of a Michelson interferometer is shown in Fig. 6.4. Light from a source is collimated by a lens and enters the interferometer as a parallel beam. This beam of light impringes on a beam splitter (BS) such that half the light intensity is reflected and half is transmitted. The transmitted light is reflected by a mirror (fixed mirror) and impinges on the beam splitter again.

The portion of the light that was originally reflected by the beam splitter is reflected by a mirror marked "movable mirror" in Fig. 6.4, which may be scanned in the direction indicated by the arrow. The light reflected by this movable mirror reaches the beam splitter with a phase shift, or path difference,

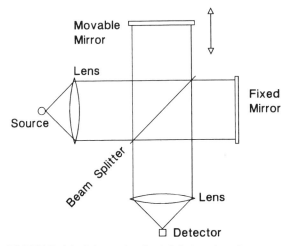

FIGURE 6.4. Schematic of a Michelson interferometer.

depending on the position of the movable mirror. The beams are recombined at the beam splitter and undergo constructive or destructive interference, depending on their path difference, and are focused through a second lens onto the detector.

In order to understand the function of the interferometer, let us assume the movable mirror is in a fixed position, and the light entering the interferometer is purely monochromatic. The two waves that interfere constructively or destructively are described by

$$E_1 = E_0 \sin(\omega t - \alpha_1)$$
$$E_2 = E_0 \sin(\omega t - \alpha_2) \quad (6.2.1)$$

where the α's are phase angles and ω is the frequency of the waves (which is exactly equal for both waves since they are derived from one beam of light via a beam splitter). E_1 and E_2 are the instantaneous values of the electric vectors of each wave, E_0 is the amplitude, which is equal for both waves, since we assumed that exactly one-half of the total intensity is reflected and one-half transmitted by the beam splitter.

The total amplitude E_T of the wave resulting by adding the two components of Eq. 6.2.1 can be written as

$$E_T = E_0[\sin(\omega t - \alpha_1) + \sin(\omega t - \alpha_2)] \quad (6.2.2)$$

which can be rewritten, using the identity

$$\sin(\alpha - \beta) = \sin\alpha \cos\beta - \cos\alpha \sin\beta \quad (6.2.3)$$

$$E_T = (\cos\alpha_1 + \cos\alpha_2)E_0 \sin(\omega t) - (\sin\alpha_1 + \sin\alpha_2)E_0 \cos(\omega t) \quad (6.2.4)$$

The sum of two harmonics of equal frequency and along the same line is another harmonic with amplitude A:

$$A \cos \Theta = E_0(\cos \alpha_1 + \cos \alpha_2) \qquad (6.2.5)$$

$$A \sin \Theta = E_0(\sin \alpha_1 + \sin \alpha_2) \qquad (6.2.6)$$

Squaring these time-independent amplitudes and using Eq. 6.2.3 again yield

$$A^2 = 2E_0^2 + 2E_0^2 \cos(\alpha_1 - \alpha_2) \qquad (6.2.7)$$

$$= 2E_0^2(1 + \cos \delta) \qquad (6.2.8)$$

where δ is the phase shift, $\delta = \alpha_1 - \alpha_2$. The derivation of Eqs. 6.2.5 through 6.2.8 can be found in texts on classical optics [e.g., see Jenkins and White, 1957].

Since the intensity of light is proportional to the square of the amplitude, we substitute $I(\delta)$ for A^2, and I_0 for E_0^2:

$$I(\delta) = 2I_0(1 + \cos \delta) \qquad (6.2.9)$$

If the movable mirror is scanned, we can write the phase difference δ in Eq. 6.2.8 in terms of the wavelength λ of the light and the mirror position x as follows:

$$\delta = 2\pi x/\lambda = 2\pi v x \qquad (6.2.10)$$

where v is the wavenumber of the radiation, and Eq. 6.2.8 becomes

$$I(x) = 2I_0^2\{1 + \cos(2\pi v x)\} \qquad (6.2.11)$$

If the light source is not monochromatic but has a spectral distribution $S(v)$, we obtain instead of Eq. 6.2.11

$$I(x) = \int 2S(v)\{1 + \cos(2\pi v x)\}\,dv \qquad (6.2.12)$$

$$= \int 2S(v)\,dv + \int S(v) \cos(2\pi v x)\,dv \qquad (6.2.13)$$

where the integration is from zero to infinite wavenumber. The first term on the right-hand side of Eq. 6.2.13 can be evaluated for zero path difference, $x = 0$:

$$I(0) = 4 \int_0^\infty S(v)\,dv \qquad (6.2.14)$$

Combining Eqs. 6.2.13 and 6.2.14 yields

$$I(x) - \tfrac{1}{2}I(0) = 2\int_0^\infty S(v)\cos(2\pi vx)\,dv = J(x) \qquad (6.2.15)$$

Extending the integral symmetrically in frequency space yields

$$J(x) = \int_{-\infty}^\infty S(v)\cos(2\pi vx)\,dv \qquad (6.2.16)$$

and

$$S(v) = \int_{-\infty}^\infty J(x)\cos(2\pi vx)\,dx \qquad (6.2.17)$$

Equation 6.2.16 and 6.2.17 are said to be a Fourier pair. In order to get from the interferogram $J(x)$, which is an intensity distribution as a function of the position x of the movable mirror, to the spectrum $S(v)$, one has to take the Fourier transform (FT) of $J(x)$. Fourier series and Fourier transforms will be introduced in the next section, and methods to perform the FT computationally will be discussed thereafter.

6.2.3 Fourier Series and Fourier Transform

Any function $f(x)$, which is periodic, can be approximated in an interval from $-L$ to L by a Fourier series. Such an expansion in terms of the Fourier components is also known as *harmonic analysis*, since it extracts the appropriately weighted harmonic components from a general waveform. An example of this procedure will be discussed below. The Fourier series expansion of $f(x)$ is defined by

$$f(x) = \sum_{n=-\infty}^{\infty} c_n e^{in\pi x/L} \qquad (6.2.18)$$

where the expansion coefficients c_n are given by

$$c(n) = \frac{1}{2L}\int_{-L}^{L} f(x) e^{in\pi x/L}\,dx \qquad (6.2.19)$$

In the case of real functions, the expansion given in Eq. 6.2.18 takes the form

$$f(x) = a_0 + \sum_{n=-\infty}^{\infty} a_n \cos(n\pi x/L) + b_n \sin(n\pi x/L) \qquad (6.2.20)$$

with the real expansion coefficients given by

$$a_n = \frac{1}{2L} \int_{-L}^{L} f(x) \cos(n\pi x/L) \, dx \qquad (6.2.21)$$

and

$$b_n = \frac{1}{2L} \int_{-L}^{L} f(x) \sin(n\pi x/L) \, dx \qquad (6.2.22)$$

The harmonic analysis of a square wave, given by Eq. 6.2.23,

$$f(x) = \begin{cases} +1 & \text{for } 0 \leqslant x \leqslant \pi \\ -1 & \text{for } \pi \leqslant x \leqslant 2\pi \end{cases} \qquad (6.2.23)$$

may serve as an example. Substituting Eq. 6.2.23 into the equations for the coefficients a_n and b_n (Eqs. 6.2.21 and 6.2.22) and integrating from 0 to 2π, one finds that all the terms a_n will be zero, and the terms b_n assume the values of $1/n$. Thus the square wave can be expanded into an infinite series of all odd harmonics, scaled by $1/n$:

$$f(x) = \frac{1}{n} \sum_{n=1,3,5\ldots}^{\infty} \sin nx \qquad (6.2.24)$$

This is shown in Fig. 6.5 for the first few terms of the expansion.

A logical connection between the principles of Fourier expansion and Fourier transform can be made by substituting

$$k = n\pi/L$$

into Eqs. 6.2.18 and 6.2.19 and letting the interval, L, over which the function is expanded, go to infinity. Thus k gets very small, and we are justified to

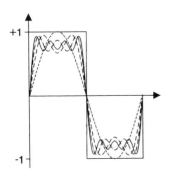

FIGURE 6.5. Fourier series expansion of square wave. Short dashes, first term; long dashes, first and second term; dot-dashes, first to or through third terms; solid line, first to or through fourth terms.

substitute the sum in Eq. 6.2.18 by an integral:

$$f(x) = \frac{1}{C} \int_{-\infty}^{\infty} c(k) e^{ikx} \, dx \qquad (6.2.25)$$

and

$$c(k) = \frac{1}{C} \int_{-\infty}^{\infty} f(x) e^{ikx} \, dx \qquad (6.2.26)$$

where C is a normalization constant, $C = \sqrt{2\pi}$.

We can view the process of taking a Fourier transform as a harmonic analysis with infinitely small increments of the frequency intervals. Consider the simplest example of a Michelson interferometer, with the movable mirror at its zero position and at rest. All frequencies of the spectral distribution $S(v)$ arrive at the detector at the same time. However, the detector experiences a superposition of all amplitudes, and a unique decomposition of this signal into the harmonic components is indeterminate and meaningless. Moving the mirror by an amount x gives a different superposition of all amplitudes, since now a different set of wavelengths or frequencies undergo destructive or constructive interference. If we collect an interferogram at sufficiently closely spaced mirror positions, the spectral distribution $S(v)$ and the interferogram $J(x)$ are just the FT of each other (cf. Eqs. 6.2.16 and 6.2.17).

Next, some properties of Fourier transforms will be discussed. The Fourier transform, denoted by the symbol $\mathcal{f}\{f(x)\} = F(k)$, of a Gaussian function is another Gaussian function:

$$\mathcal{f}\{e^{-\alpha x^2}\} = F(k) = (2\alpha)^{-1/2} e^{-k^2/4\alpha} \qquad (6.2.27)$$

such that a strongly peaked (narrow) Gaussian produces a broad Gaussian in FT space, and vice versa. The Fourier transform of a shifted δ-function, for example, for an infinitely narrow-band (monochromatic) light source, is

$$\mathcal{f}\{\delta(v - v_0)\} = e^{iv_0 x} \qquad (6.2.28)$$

which is in agreement with our original analysis (cf. Eq. 6.2.11) of monochromatic light entering a Michelson interferometer. We reasoned then that monochromatic input into the interferometer should yield a simple cosine function interferogram. This is confirmed by the result above, since the real part of the expression on the right-hand side of Eq. 6.2.28 is a simple cosine function.

6.2.4 Discrete and Fast Fourier Transform Algorithms

The equations given so far for Fourier transform pairs describe the process and concepts but not a practical method to obtain Fourier transforms of a set of data. The transformation from an interferogram to frequency space is presently performed by an algorithm known as FFT (fast Fourier transform).

This is based on the principles of the discrete Fourier transform: since the interferogram is not sampled continuously, but at short intervals (i.e., at discrete times), the interferogram is obtained as a one-dimensional matrix of digital values. For such a situation, the process of taking the Fourier transform can be written as

$$g(kT) = \frac{1}{N} \sum_{n=0}^{N-1} G(n/NT) e^{2\pi i nk/N} \qquad (6.2.29)$$

and

$$G(n/NT) = \sum_{k=0}^{N-1} g(kT) e^{-2\pi i nk/N} \qquad (6.2.30)$$

where N is the total number of data points, T is the sampling interval, and n and k are running indexes in g and G space, respectively. Equation 6.2.29 and 6.2.30 are the discrete (point-by-point) versions of Eqs. 6.2.25 and 6.2.26. Setting $n/NT = m$ and $kT = p$, Eq. 6.2.30 can be written as

$$G(m) = \sum_{k=0}^{N-1} g(p) W^{mp} \qquad (6.2.31)$$

where

$$W^{mp} = e^{-2\pi i mp} \qquad (6.2.32)$$

Equation 6.2.31 can be cast into matrix notation:

$$\mathbf{G} = \mathbf{W}^{mp} \mathbf{g} \qquad (6.2.33)$$

Thus the computation of a discrete Fourier transform from G to g space reduces to computing a (complex) transformation matrix and multiplying the vector of discrete points with this matrix. For modern spectroscopic applications, a typical sample set may be 8K data points (i.e., 8192 points in the $g(p)$ vector). The Fourier transform operation, according to Eq. 6.2.32 or 6.2.33, requires for each of the 8192 points in G space an 8K × 8K matrix to be multiplied by a 8K vector. Such a matrix manipulation would be prohibitively slow, since for each data point 8K multiplications and 8K additions are required.

This problem was alleviated by the FFT algorithm, developed originally by Cooley and Tukey [1965], which avoids the problem of the large number of multiplications and additions by factoring the W matrix into product matrices that are sparse; that is, they have many zero elements. It can be shown that such a factoring is possible, but the factoring will require reordering the entries in the G and g vectors. A detailed discussion of the proofs involved is beyond the scope of this chapter, and the reader is referred to the literature [e.g., see

Brigham, 1974]. However, a simple example, namely, that of a four-point FFT, will be presented here. Incidentally, the original FFT algorithm works only for data sets for which the size is given by an integer power of 2. In the case of a four-point FT, $N = 4$, $mp = nk$, and

$$W^{mp} = e^{-2\pi i mp} = e^{-\pi i nk} \tag{6.2.34}$$

where the superscript mp refers to the product of these two indices. Thus the W matrix in Eq. 6.2.33 is given by

$$\begin{bmatrix} W^0 & W^0 & W^0 & W^0 \\ W^0 & W^1 & W^2 & W^3 \\ W^0 & W^2 & W^4 & W^6 \\ W^0 & W^3 & W^6 & W^9 \end{bmatrix} \tag{6.2.35}$$

which can be rewritten as

$$\begin{bmatrix} 1 & 1 & 1 & 1 \\ 1 & W^1 & W^2 & W^3 \\ 1 & W^2 & W^4 & W^6 \\ 1 & W^3 & W^6 & W^9 \end{bmatrix} \tag{6.2.36}$$

For this step, the relation $W^0 = 1$ was used. This matrix can be further simplified, using $W^4 = W^0$, $W^6 = W^2$, and $W^9 = W^1$. These relationships will be discussed in more detail below. Thus the matrix manipulation that needs to be carried out can be written as

$$\begin{bmatrix} G(0) \\ G(1) \\ G(2) \\ G(3) \end{bmatrix} = \begin{bmatrix} 1 & 1 & 1 & 1 \\ 1 & W^1 & W^2 & W^3 \\ 1 & W^2 & W^0 & W^2 \\ 1 & W^3 & W^2 & W^1 \end{bmatrix} \begin{bmatrix} g(0) \\ g(1) \\ g(2) \\ g(3) \end{bmatrix} \tag{6.2.37}$$

This process requires 16 multiplications and 16 additions. The FFT algorithm now breaks the W matrix into sparse partial matrices, the product of which is identical to the W matrix. The actual scheme according to which this matrix decomposition is carried out is complicated, and only the results will be given. In the case of a four-point transform, the two partial matrices are

$$\begin{bmatrix} G(0) \\ G(2) \\ G(1) \\ G(3) \end{bmatrix} = \begin{bmatrix} 1 & W^0 & 0 & 0 \\ 1 & W^2 & 0 & 0 \\ 0 & 0 & 1 & W^1 \\ 0 & 0 & 1 & W^3 \end{bmatrix} \begin{bmatrix} 1 & 0 & W^0 & 0 \\ 0 & 1 & 0 & W^0 \\ 1 & 0 & W^2 & 0 \\ 0 & 1 & 0 & W^2 \end{bmatrix} \begin{bmatrix} g(0) \\ g(1) \\ g(2) \\ g(3) \end{bmatrix} \tag{6.2.38}$$

The product of these two matrices can be evaluated easily, and yields (before unscrambling) the following matrix:

$$\begin{bmatrix} 1 & W^0 & W^0 & W^0 W^0 \\ 1 & W^1 & W^0 & W^0 W^2 \\ 1 & W^1 & W^2 & W^1 W^2 \\ 1 & W^3 & W^2 & W^2 W^3 \end{bmatrix} \quad (6.2.39)$$

The product of two terms, $W^n W^k$, is given by the remainder of dividing $n + k$ by the dimension of the transform, N. Since $N = 4$ for the example chosen, we find that $W^1 W^2 = W^3$, $W^2 W^3 = W^1$, and so on.

Although Eq. 6.2.38 requires that two matrix multiplications be carried out, and although the resulting G vector is scrambled, this computation needs fewer multiplications and additions than the one described by Eq. 6.2.33. The time savings is particularly impressive for large transforms, which can be carried out in a few seconds on a modern personal computer, using the FFT algorithm.

For more details, the reader is referred to specific reference books, in particular, *The Fast Fourier Transform* by E. O. Brigham [1974], which treats the subject very thoroughly yet understandably. The approach presented here is taken from this book.

6.3 INSTRUMENTATION FOR INFRARED SPECTROSCOPY

6.3.1 General Experimental Considerations

Infrared absorption, or IR spectroscopy for short, has been utilized in structural chemistry, spectroscopy, and as a qualitative tool by a number of other branches of sciences. Its theory, namely, that of dipole allowed vibrational transitions, was treated before. However, even in a field as widely applied and relatively well understood as IR spectroscopy, there are still a number of areas where a lot of research may be carried out, in particular, in the IR spectroscopy of macromolecules and oriented samples and in the measurement and computation of absolute IR intensities. Furthermore, enormous progress has been made, particularly over the past decade, on the instrumental front. New IR spectrometers allow spectra to be collected from samples that have long been considered "impossible," such as dilute aqueous solutions or nontransparent samples such as rubber and fat.

The reasons that IR absorption spectroscopy is such a widespread technique are numerous. First, the technique has been around much longer than other techniques such as NMR. Second, sample preparation is easy, and materials can be examined in the gaseous, liquid, or solid states. Sampling techniques will be discussed later in this chapter (cf. Section 6.3.4). Third, a qualitative interpretation is often possible very quickly by inspection of the spectral features. Finally, data acquisition takes but a few seconds on a modern Fourier

158 INSTRUMENTATION FOR THE OBSERVATION OF VIBRATIONAL SPECTRA

transform spectrometer, and the price of these instruments is very reasonable, particularly compared to other available techniques such as NMR.

One of the disadvantages of IR spectroscopy is that glass does not transmit in the infrared region. Thus all infrared sample cells must be equipped with windows that are transparent in the spectral region of interest, which typically are made of materials such as NaCl, KBr, AgCl, AgBr, CsI, ZnSe, CaF_2, or KRS-5 (a mixed thallium halide), depending on the wavelength range desired and the chemical reactivity between sample and window material. In general, the suitable optical materials for the infrared region are ionic compounds that do not contain polyatomic ions, since any polyatomic species—molecular or ionic—exhibits vibrational bands, which would interfere with the vibrations of the sample. Thus salts such as carbonates, sulfates, or phosphates are unsuitable as optical materials in the infrared. The transmission range, along with the refractive index and some chemical properties, of several common materials is presented in Table 6.1. The index of refraction is significant, since it determines the light losses incurred at the air/window and the window/sample interface.

In infrared absorption spectroscopy, one observes the absorbance A of the

TABLE 6.1. Infrared Transmitting Materials

Material	Wavenumber Range[a] [cm^{-1}]	Refractive Index[b]	Compatibility
CaF_2	1100–5000	1.40	Chemically inert, water insoluble
BaF_2	870–5000	1.45	Slightly water soluble; soluble in ammonium salts
NaCl	625–5000	1.52	Cannot be used with water or methanol
KBr	400–5000	1.54	Cannot be used with water or methanol; hygroscopic
CsI	200–5000	1.74	Cannot be used with water or methanol; soft and hygroscopic
AgCl	435–5000	2.0	Water insoluble; reacts with amines and metals; UV sensitive
KRS-5	275–5000	2.38	Nearly water insoluble; chemically inert; poisonous!!!
ZnSe	550–5000	2.4	Chemically inert; water insoluble; high reflection losses

[a] The upper wavenumber range is the limit of normal infrared instrumentation. Most of the materials listed (with the exception of KRS-5 and ZnSe) are clear in the visible range. Thus for many of them the transmission range extends past 25,000 cm^{-1}.
[b] At 2000 cm^{-1} (5 μm).

INSTRUMENTATION FOR INFRARED SPECTROSCOPY

sample (cf. Chapter 5)

$$A = \varepsilon Cl = \log(I_0/I) \tag{5.1.1}$$

where C is the concentration of the sample [mol/L], l is the path length [cm], and ε is the molar extinction coefficient [L/mol·cm]. I and I_0 are the light intensity emerging from, and incident on, the sample, respectively.

The integral of a infrared peak, plotted in ε units versus wavenumber,

$$\int (\varepsilon/v) dv = D = \mu^2 \tag{3.10.1}$$

can be related to the quantum mechanical observable and is thus the relevant quantity. In infrared spectroscopy, ε values for strong absorptions are typically about 1000 L/mol·cm and are much smaller than ε values encountered in electronic absorption spectroscopy for symmetry allowed transitions, for which ε may be over 100,000 L/mol·cm.

6.3.2 Dispersive Infrared Instrumentation

Like most other absorption spectrophotometers, a dispersive infrared instrument consists of an infrared source, a device to sort the light by wavelength, the sample, a suitable detector, and the necessary electronics to amplify the signal. These element will be discussed in turn.

Since IR light is heat radiation, any hot source is an emitter of IR, according to the law of black body radiation:

$$dW = \frac{c_1}{\lambda^5(e^{c_2/\lambda T} - 1)} d\lambda \tag{6.3.1}$$

where dW is the energy radiated per unit area and time into the wavelength band $d\lambda$, T is the absolute temperature, and c_1 and c_2 are constants. The total (integrated) intensity is given by the Stefan–Boltzmann law:

$$W = \sigma T^4 \tag{6.3.2}$$

Thus the higher the temperature of the source, the more intense is the total radiation emitted, and the lower the wavelength of the peak emission. The blue-shift of the peak emission can easily be observed: at room temperature, the peak of the emission of a black body is well in the infrared and cannot be seen with the naked eye. At about 1000 K, a black body glows red. Between 1000 and about 1800 K, the color changes from red to orange and yellow, and above 2200 K the emission appears white, since all visible colors are represented in the emitted spectrum. Higher temperature arc sources appear distinctly blue, since the emission maximum is shifted past the blue wavelengths into the ultraviolet range.

Thus it may appear at first that a lower temperature (~ 1000 K) source would be ideal for IR spectroscopy since its peak emission occurs in the infrared spectral region. However, since the total energy radiated increases with the fourth power of temperature, a source as hot as possible is advantageous as an infrared source, and the light emitted in the visible or even ultraviolet range is discarded.

Furthermore, the source material should be as close as possible to a black body; that is, its emissivity should be close to unity. Finally, the dimensions of a source are important, and from a viewpoint of light collection and focusing the image of the source on a slit, a small source with the highest possible surface brightness is desirable. The aspect of source size and the design of the light collection optics was discussed before in Section 6.1.

Sources used most often are globars (SiC), Nichrome wires, or ceramic Nernst glowers, which are heated electrically to between 1600 and 2400 K. These sources may be operated in a standard atmosphere, since they are not susceptible to oxidation. Nichrome sources are found in many Fourier transform (*vide infra*) instruments where the requirements for maximum source brightness are not quite as stringent as in dispersive instrumentation.

A few very sophisticated infrared absorption experiments use tunable infrared lasers, or frequency subtraction techniques with tunable visible lasers, for extremely high-resolution studies, but these experiments are outside the intended coverage of this chapter.

The light from the source is collected most of the time via mirrors (because one need not worry about optical materials in reflective optics) and focused into the monochromator. These were discussed before in Section 6.1. Filters are employed after the monochromator to remove unwanted radiation, such as higher order diffractions of the grating. The monochromatic radiation is then focused through the sample and onto a detector. The sample cell materials are the salts introduced above, in addition to polyethylene for the far-infrared region. The position of monochromator and sample may be exchanged: thus the sample is illuminated with polychromatic light if it is mounted before the monochromator, and the light is analyzed subsequently. Alternatively, the light may first be rendered monochromatic and then be allowed to interact with the sample.

Detectors in the infrared region can be classified as thermal and semiconductor detectors. Thermal detectors measure a heating effect due to the IR radiation. These include thermocouples, bolometers (thermal resistor), and pyroelectric detectors. Deuteriated triglycine sulfate (DTGS) is an example of a pyroelectric material: in these materials, the dielectric constant depends on the temperature. Thus DTGS is used as a capacitor material, and the change in capacitance, induced by the temperature change, is measured as a voltage across the capacitor. It has a flat wavelength response into the far-infrared region and is reasonably sensitive and can handle frequencies well into the kilohertz range. It is probably the most frequently used, low cost, high performance detector and can be found in many commercial instruments.

Among the semiconductors, photoconductive HgCdTe (also known as MCT for mercury–cadmium–telluride) and photoconductive and photovoltaic InSb have been used for the highest sensitivity applications. In photoconductive semiconductors, the infrared photon promotes an electron across the band gap between valence and conductivity band and creates in this manner mobile electrons in the conductivity band. This is measured as a change in current across the semiconductor detector, which is biased at constant voltage. In photovoltaic detectors, electrons are also promoted across the band gap, the structure of which is such that they directly induce a potential change proportional to the light intensity. InSb is used at wavelengths above 5 μm, whereas HgCdTe can be produced, by varying the HgTe/CdTe ratio, to detect wavelengths as long as 12.5 μm. Other semiconductor detectors used include PbS, PbSe, InGaAs, and Ge.

Since the sensitivity of a given detector, the grating, and the source efficiencies vary enormously over the wavelength range covered by a typical infrared instrument, it is most advantageous to carry out the IR absorption measurement in a dual-beam mode to compensate for the instrument response function. In this mode of operation, the beam is divided into two identical paths via a rotating mirror chopper. One-half the light is passed through the sample path, and the other half through the reference path. The signal at the detector is amplified by a phase sensitive detector (lock-in amplifier) tuned to the chopping frequency. In this way, the actual difference between the sample and reference beam is measured, and effects such as source, detector, and grating efficiency may be compensated. Dual-beam mode also helps to remove solvent or atmospheric water absorption features. In the early designs of commercial instrumentation, a variable beam attenuator in the reference beam was used to null the signal at the lock-in amplifier, and the position of the variable beam attenuator was coupled to the recorder pen.

With modern electronics, the recording of sample and background spectra is often done sequentially, and the difference is formed mathematically on stored data. This approach was pioneered in the early years of Fourier transform infrared spectroscopy (*vide infra*) since data stations with sufficient memory for storage were required for the Fourier transform anyway, and since a true dual-beam experiment is difficult to perform with an interferometer.

The instrumentation described so far was the standard instrumental setup for infrared absorption spectroscopy for a number of decades. However, the design principles of these instruments leave a lot to be desired. The main shortcoming is that as the grating is scanned, only a small fraction (<1%) of the total infrared radiation actually passes through the exit slit of the monochromator and reaches the detector, the remainder of the light being rejected by the exit slit jaws and wasted. Thus only a single spectral element with a bandpass of 0.5–5 cm^{-1} is examined at a given time, and the time to acquire a spectrum of 3000 cm^{-1} was often measured in hours. Obviously, this measuring time can be reduced enormously if all wavelengths could be examined simultaneously. This *multiplexing advantage* available with Fourier transform spectrometers made the acquisition of spectral data via dispersive instrumentation obsolete. In the next section, the background of Fourier transform absorption spectroscopy will be introduced.

6.3.3 Principles of FT Spectrometers

After having investigated in Section 6.2 how interferometric data can be efficiently transformed into the frequency domain, a few words on the operation of standard FT infrared spectrometers are in order. Sources and detectors are more or less the same as the ones discussed before for dispersive instruments. Care has to be taken that the frequency response of the detector material is sufficiently high to follow the Fourier frequencies, which result from the rapid scanning of the movable mirror. Typical scan speeds are on the order of several centimeters/second, and typical Fourier frequencies are in the low kilohertz range.

Data acquisition of the interferometer is synchronized to the fringes produced by passing a He–Ne laser beam through a portion of the interferometer (typically the corners or periphery of the mirrors). A diode detector measures the laser intensity variations produced when the interferometer is scanned. These laser fringes are used as a trigger for the data acquisition, and by counting these fringes, the path difference that the light experiences in the two arms of the interferometer can be determined.

Since the wavelength of a He–Ne laser is very stable, and very accurately known, data can be acquired at very precise path differences, even if there is some jitter in the velocity of the moving mirror. Thus the accuracy of the interferometric data is very high, allowing for repetitive scans to be very well aligned with respect to each other. This is known as the Conne advantage of Fourier transform spectroscopy. Thus accurate coaddition of scans and subtraction of background spectra are possible.

In addition to the multiplexing advantage, discussed earlier, and also known as the Fellgett advantage, there is one other major (optical) advantage of Fourier transform over dispersive instrumentation. This advantage, known as the Jacquinot advantage, is the absence of beam restricting elements such as slits. Instead, large diameter beams are used, and the intensity losses due to the slits, which are necessary in dispersive systems to define the optical resolution, are avoided. In FT instrumentation, the resolution is not determined by the size of the beam, but strictly by the stroke (travel) of the movable mirror and the number of data collected during a stroke. Thus Fourier transform spectroscopy offers major optical advantages, which account for an enormously reduced data acquisition time for a spectrum obtained with similar resolution on a scanning instrument. These advantages more than compensate for the disadvantage that data have to be transformed mathematically to obtain the desired spectra. With the recent increase in computational power of personal computers, it is possible to carry out an 8 K transform in less than 2 seconds via software FFT, and in fractions of a second if a dedicated, hard wired FFT accelerator board, is used.

6.3.4 Sampling Techniques in Infrared Spectroscopy

6.3.4.1 Transmission Spectroscopy In the discussion of both FT and dispersive infrared spectroscopy, we assumed that the sample to be examined is relatively transparent in the spectral range of interest such that the vibrational spectrum can be collected via standard absorption spectroscopy. Here, the attenuation of the infrared beam within the sample follows the Lambert–Beer law (Eq. 5.1.1). The most accurate spectral information is obtained if the sample path or concentration is adjusted such that the peaks of interest have absorbances between 0.01 and 1.

Gases with 10–50 mm Hg vapor pressure at room temperature can be studied at these pressures at about 50–100 mm path lengths. For samples with lower vapor pressures, correspondingly longer path lengths need to be employed. Commercial gas cells with up to 10 m path lengths are available. For these long paths, sample cells with internal mirrors to fold the optical path are utilized.

Gas phase spectra give the most accurate spectral data, because the molecules are normally monomeric units with no solvent interactions. The rotational fine structure (cf. Sections 2.10 and 7.2) can sometimes be used to obtain very accurate structural data and can aid in the vibrational assignment of small molecules. If the rotational structure obscures vibrational peaks, addition of a few atmospheres of inert gas often sufficiently broaden the rotational–vibrational lines into broad envelopes to permit observation of the vibrational peaks.

Neat liquids are sampled most often with sample path lengths between 15 and 100 μm, between windows of the materials described in Table 6.1. For liquids and solutions, one has to ascertain that no chemical interaction occurs between the sample and window materials: amines and silver halides are incompatibles, as are NaCl or KBr windows with aqueous solutions. Cells with variable path lengths (0.01–5 mm) are commercially available to accommodate a large range in sample concentration.

Even a very short path length such as 15 μm often will produce very intense absorption spectra. Therefore many routine spectral data acquisitions are carried out with the sample dissolved in a suitable solvent. In dilute solution spectra, monomeric solute spectra can be obtained, and interactions between solute molecules (e.g., hydrogen-bonded dimers) or between solvent and solute molecules can be observed. Vibrational frequencies of gaseous and solution phase species are usually shifted due to solvent interactions. Low solubilities of the solute in a given solvent may require long path lengths to be used, which may create problems, since even for a 0.5 mm path, most organic liquids have enormously strong overtone and combination bands in the mid-infrared spectrum. Therefore many organic solvents, such as acetone, DMSO, or acetonitrile, can be used only in very narrow spectral regions. Acceptable solvents above about $1200\,\text{cm}^{-1}$ are CCl_4, CS_2, chloroform-d_1 ($CDCl_3$), and bromoform-d_1 ($CDBr_3$).

Biological molecules and ionic compounds often are insoluble in any solvents except water and deuteroxide. Both these solvents have enormous infrared absorption cross sections, and it is virtually impossible to utilize path lengths of more than 50 μm with H_2O and 100 μm with D_2O. Therefore data acquisition from these solvents in transmission spectroscopy is challenging.

Due to problems in dissolving samples, many routine observations of infrared spectra are carried out in the solid phase. Most applications use polycrystalline samples (to minimize interface effects) pressed into pellets of an inert salt (such as KBr or KCl) or solid samples suspended in a transparent oil (mull).

Solid phase spectra are the most difficult to interpret, since a number of effects perturb the spectral features of the isolated molecules. Among them are strong interactions, such as hydrogen bonding, and site interaction. This latter effect is due to the fact that more than one molecule, occupying different sites, may be found in a unit cell. Since the unit cell, and not the molecule, is the smallest repeating unit, the observed vibrational spectrum is that of the unit cell. Thus some vibrations may be split into multiplet structures depending on the number of molecules per unit cell. In principle, the site splitting can be analyzed by group theoretical methods, since the sites are related by symmetry. However, most qualitative infrared work totally ignores these effects.

6.3.4.2 Attenuated Total Reflection and Diffuse Reflectance Although absorption methods are by far the most common way to collect infrared vibrational data, there are other methods to collect the same spectral information from light reflected from surfaces. This method is particularly useful if the sample is opaque in the spectral region of interest.

The reflection and refraction of a beam of light are summarized in Fig. 6.6. Figure 6.6A shows a typical reflection process when a beam of light from a less dense medium strikes an interface of an optically more dense medium. Here, the angle of incidence and angle of reflection are equal. The reflected beam contains spectral information on the sample interface, and this phenomenon is used in diffuse reflectance spectroscopy. A small portion of the light is also refracted into the medium.

When a beam of light traveling in a more dense medium strikes the interface, as shown in Fig. 6.6B, a number of phenomena may occur, depending on the angle of incidence Θ_i, and the ratio of refractive indices, n_1/n_2. If

$$\sin \Theta_i > n_1/n_2 \tag{6.3.3}$$

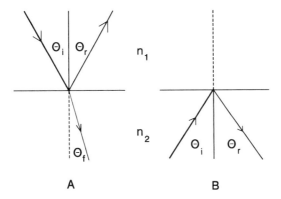

FIGURE 6.6. (A) Reflection and refraction. (B) Total internal reflection. Θ_i, angle of incidence; Θ_r, angle of reflection; Θ_f, angle of refraction. Medium 2 is optically denser than medium 1: $n_2 > n_1$.

Θ_t is undefined and total reflection occurs within the dense medium. If

$$\sin \Theta_i = n_1/n_2$$

then Θ_i is known as the critical angle, and the refracted beam disappears into the surface. If medium 2 is a crystal material with high refractive index, such as ZnSe or KRS-5, and medium 1 is a typical organic liquid, it is easily seen that internal reflection takes place at moderate angles of incidence.

In order to understand the phenomenon of attenuated total reflection, the refractive index of medium 1 must be written in its most general form, which is a complex number:

$$\eta = n + ik \tag{6.3.4}$$

The real part, n, of Eq. 6.3.4 is the regular refractive index, which holds outside a region of absorption. If there are electronic or vibrational transitions at the wavelength of the light incident on the interface, the absorption due to this transition affects the refractive index via the imaginary part in Eq. 6.3.4. k is the extinction coefficient, which is related to ε used in the Lambert–Beer law (Eq. 5.1.1) as follows:

$$\varepsilon = (4\pi\tilde{v}/C)k \tag{6.3.5}$$

where C is the concentration and $\tilde{v}$ the wavenumber of the light.

Equation 6.3.4 implies that the real index of refraction, which governs reflection and refraction processes of everyday life, needs to be replaced by the expression in Eq. 6.3.5 in the neighborhood of a transition. This can be visualized as follows. It was pointed out in the introduction that interactions of light and matter can occur on and off resonance. In the discussion of the polarizability tensor (Eq. 5.3.3) responsible for Rayleigh and Raman scattering,

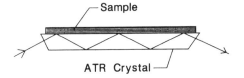

FIGURE 6.7. Experimental arrangements for ATR spectroscopy: multiple internal reflections in trapezoidal crystal.

we introduced the concept that matter responds to electromagnetic radiation with frequencies that are *not* absorbed. The polarizability of a medium, in turn, is related to the index of refraction according to Eq. 5.3.4. Thus the refractive index is a measure of the off-resonance interaction of matter and light.

It is also well established that at frequencies of dipole allowed transitions of the medium, the refractive index unergoes an effect known as anomalous dispersion, which is described by Eq. 6.3.4. Thus a beam of polychromatic light, reflected from a surface, will experience regular reflection, if there is no transition at a given wavelength; but at the wavelength of an absorption, the reflected beam will be attenuated. Conceptually, one can visualize that reflection not only occurs at the outermost layer of the surface, but that interaction occurs in a few layers of molecules inside the medium. This layer will absorb the wavelength where transitions occur, just as in standard absorption spectroscopy.

Attenuated total reflection (ATR) spectroscopy uses this principle to allow infrared spectral data to be collected from samples where transmission spectroscopy does not work. In ATR, the IR beam is totally reflected within a transparent crystal (typically ZnSe or KRS-5), as shown in Fig. 6.7. The sample, which may be solid or a solution, is in contact with this crystal surface. Since the infrared data are collected from the reflected, and not the transmitted, light, samples that are opaque can be investigated. Thus infrared data can be observed for samples such as a piece of black rubber tire or suspensions such as milk, grease, and fat. Intensities can be enhanced by using multiple internal reflections in a trapezoidal crystal as shown in Fig. 6.7.

Diffuse reflectance spectroscopy works in a similar fashion. A beam of IR radiation is reflected directly from a polycrystalline surface of the sample, and the reflected light is collected and analyzed as usual. Diffuse reflectance works better than ATR when it is difficult to obtain good physical contact between the ATR crystal and a sample, such as powdery substances. Diffuse reflectance spectroscopy has been used to differentiate, for example, between rye and wheat flour.

6.4 INSTRUMENTATION FOR RAMAN SPECTROSCOPY

6.4.1 Dispersive Single Detector Raman Instrumentation

In the early 1970s, a number of spectroscopic instrument manufacturers began introducing Raman instrumentations, which could compete with the infrared instrumentation available at that time in terms of convenience and ease of data acquisition. These instruments were made possible by the availability of commercial gas lasers and photomultiplier detectors that had been optimized for low light levels, mostly for astronomic research.

Originally, this instrumentation utilized either He–Ne or Ar ion lasers for excitation, 90° scattering geometry (vide supra), double or triple scanning monochromators, photomultiplier tubes (PMT) for detectors, and either DC amplification of the PMT output or photon counting. Many Raman spectrometers of this generation are still in use, and the design is sufficiently general to justify a discussion of them. Their components will be introduced briefly, followed by a more detailed discussion of more modern instrumentation.

Sufficient literature is available about lasers that they will not be discussed here. It is sufficient to state that they emit between a few milliwatts and a few watts of continuous wave (CW), or average pulse power at a bandwidth of about a wavenumber or less. Visible laser lines used commonly are the 632.8 nm line of He–Ne lasers, 488.0 and 514.5 nm of Ar ion lasers, and 647 nm of Kr ion lasers. Often, the powerful lines of Ar or Kr ion lasers are used to pump tunable dye lasers, the output of which can be tuned from the ultraviolet into the near-infrared range.

The laser beams, whether continuous wave or pulsed, are collimated, low diameter beams, which often are highly polarized. Thus they may be directed over distances of many meters without significant increase in beam diameter. Specially coated, high reflection optics need to be employed with lasers of more than 500 mW of power; otherwise, destruction of a mirror is possible by the few percent light losses on the surface.

The samples, contained in melting point capillaries, cuvettes, spinning cells (to avoid sample photolysis), or even as free sample jets, are illuminated by the laser beam focused into the sample. The minimum possible beam waist w of the laser at the focal point in the sample is determined by the focal length f of the focusing lens as follows:

$$w = \lambda f / \pi d \qquad (6.4.1)$$

where λ is the wavelength of the laser emission, and d is the $1/e^2$ radius of the laser beam (i.e., the diameter of a pinhole that transmits a fraction of $1/e^2$ of the total laser power). For a typical focusing geometry for the beam of an Ar ion laser in the visible region ($\lambda = 0.5\,\mu m$, $d = 0.7\,mm$, and $f = 50\,mm$), the beam waist is about $10\,\mu m$. Note that a wider beam waist of the focused beam is obtained with a longer focal length lens.

For right angle scattering, the light is collected by a large aperture lens and collimated, and focused into the monochromator. Aspects of the light collection optics were discussed in Section 6.1.2. In the early 1970s instrumentation, the monochromators used were nearly exclusively scanning double monochromators with a focal length of about 1 m in each of the monochromators, and grating groove densities of about 1200/mm. Double monochromators were necessary to discriminate against Rayleigh scattering, which is much more intense than Raman scattering, or even light reflected at the laser wavelength. With modern holographic gratings, single monochromators can be used since these gratings reject stray light better than ruled gratings.

The dispersion of the two monochromators can be additive, such that in the focal plane of the second monochromator (where the detector is mounted), the dispersion is double that of each monochromator. This is tantamount to having one monochromator with twice the focal length, but one gains the added stray light rejection offered by the intermediate slit. The Spex Industries Model 1403-series monochromators operated on this principle.

One also can arrange the second monochromator such that the dispersion is zero at the exit slit; that is, the dispersions of the two monochromators cancel, and the wavelengths at the exit plane of the second monochromator are scrambled. The Jarrell–Ash Model 25-100 used this subtractive dispersion.

The gratings, usually with between 600 and 1800 grooves/mm, are commonly driven by a stepping motor, and different colors of the dispersed light are scanned by the exit slit and the detector.

A number of commercial photomultipliers (PMT) are available, which have single-photon sensitivity, and a dark count of a few counts/second when they are cooled thermoelectrically. The quantum efficiency, however, is relatively low (about 12–15%) for most photomultipliers.

There are basically two avenues to collect and process the signal from the photomultiplier. One of these is a simple DC amplification and the other is digital photon counting. In DC amplification, the anode of the PMT, which is at ground potential, is connected directly to the input of a current-to-voltage converter with a variable gain of about 10^7–10^{10}.

A simple calculation will help clarify the magnitude of signals expected. A single photon detected by the PMT will be converted along the dynode chain of the PMT to a pulse of electrons. Typically, the gain of a high-quality PMT is in excess of 10^6 electrons/photon. A pulse of 10^6 electrons has a charge of about 1.6×10^{-13} coulomb. Thus a dark count of 1 photon/s translates into an anode current of about 10^{-13} A. A strong Raman signal may produce 10^6 photons at the detector, giving an anode current of $0.1\,\mu A$ which can be displayed on a strip chart recorder after suitable amplification. DC detection is not used much any more, since it does not discriminate against secondary dynode emission in the PMT. Since the dynodes are at high potential with respect to ground and with respect to each other, spontaneous emission of electrons from these dynodes occurs frequently. If this emission occurs in one of the early dynodes, a significant electron pulse will arrive at the anode, which in a DC amplification system would be amplified just like a pulse due to a photon and contribute to the observed dark current.

In photon counting, on the other hand, one makes use of pulse shaping and discriminator circuitry, which only allows pulses above a certain threshold value to pass, thus discriminating against dynode emission. These pulses are then counted digitally and processed by a computer. Since a PMT is a device that detects digital events (photons), it is advantageous to process the data digitally in the electronic system.

The same arguments cited before in the comparison between dispersive and FT infrared spectroscopy apply here as well: most of the light diffracted from the grating will be rejected by the slit jaws, and only a small fraction will pass the slit and be registered by the detector. Since Raman scattering is such a weak effect, discarding most of the spectral information is not advisable. Thus one attempts to improve the S/N ratio, or reduce the data acquisition time, by utilizing the multiplexing advantage. There are two avenues to achieve that: multichannel detectors or Fourier transform methods. These will be discussed in the following sections.

6.4.2 Multichannel Dispersive Raman Spectrometers

In the mid- and late 1970s, multichannel visible detectors became available commercially. A multichannel, or array, detector consists of a large number (typically between 500 and a few thousand) of solid state detector elements, which are arranged in close proximity and are placed into the focal plane of the monochromator without an exit slit. Each detector element simultaneously detects a diffeent color of the diffracted light, and the resolution of the optical system is determined (among other factors) by the width of the detector element.

The first commercially available array detectors were called Vidicon detectors which were photodiodes being read by a scanning electron beam. These Vidicons were soon substituted by solid state photodiode array detectors, and later by charge coupled devices (CCDs). We shall restrict our discussion of array detectors to diode array detectors and CCDs, since Vidicon cameras are no longer very common. The discussion of the photodiode array detector will use the Reticon chip, produced by Princeton Applied Research, for an example.

A diode array detector consists of 512 or 1024 silicon photodiodes, which are charged to a certain level electronically and subsequently discharged by electron holes created by the absorption of a photon. The remaining charge is measured by special circuitry, which connects each diode consecutively to a charge-to-voltage converter/amplifier and a D/A converter, which measures the height of the pulse that corresponds to the remaining charge of the diode. During the read cycle, the diode is also recharged.

Each diode in the Reticon measures about 0.025 mm wide and 2.5 mm high. Thus the total light sensitive area of the array measures about 25×2.5 mm^2. The rectangular shape of each diode makes them ideally suited for slit and grating based spectrometers, since the entrance slit of a monochromator has a similar aspect ratio of 1:100 (typically 0.1 mm wide and 10 mm high). The image of this entrance slit is projected the exit plane of the spectrometer with suitable image demagnification. Thus the image of the slit ideally fills each diode completely.

Although the quantum yield is high in silicon, the S/N ratio is relatively poor in Si photodiode array detectors; that is, the relative change in charge in a diode brought on by the absorption of a photon is small compared to the readout noise. Thus single-photon detection is not practical in photodiode arrays. To alleviate this problem, image intensifiers were used commonly with diode arrays. An image intensifier is a true photon multiplier: an incoming photon produces a photoelectron, which is amplified via a microchannel plate electron amplifier. The resulting pulses of electrons are converted back to light by making them impinge on phosphorous screens. In this way, photon gains of between 10^3 and 10^4 could be achieved. With these amplified light levels, diode array detectors work satisfactorily.

Image intensifiers also offer the possibility of "gating," or rapidly turning the detector on and off. This is accomplished by pulsing the high voltage on the electron amplifier in response to an external signal, such as a trigger from a laser pulse. Gating offers the ability to synchronize the detector to external events, which is particularly important when it is necessary to discriminate against fluorescent background scattering: since fluorescence is a much slower process than Raman scattering, it is possible to collect Raman photons in the first few hundred nanoseconds after an exciting laser pulse and gate out the fluorescent intensity, which occurs thereafter. The time-resolved Raman experiments discussed in the previous chapter often employ gated detectors to discriminate Raman scattering created from the pump laser from that created by the probe pulse.

Since the mid-1980s, a new multichannel technology has emerged for use in Raman spectroscopy. This is the so-called charge coupled device (CCD) technology, which was originally developed for video cameras. The photosensitive area in the CCD is typically divided into $30\,\mu\text{m} \times 30\,\mu\text{m}$ large picture elements (pixels), composed of Si photodetectors. As mentioned before the quantum yield of Si is high, and charge due to electron holes can be accumulated very efficiently. The difference between CCDs and diode arrays is in the readout logic. The readout in a CCD proceeds as follows: the first element in the bottom row is read by converting its charge, via a charge-to-voltage converter, and stored digitally. Now, the charges of all consecutive elements in the bottom row are shifted by one position by draining the electrons into the next pixel, and the first element is read again. Once the entire bottom row has been read, the charges in each pixel element in the next row up are drained into the pixel elements in the bottom row, and the readout process continues, until the entire array has been read. The efficiency of the charge transport (the "charge coupling") must be very high, on the order of 0.99999, since arrays often contain about 600×300 or even 1000×300 pixels, and the charge in a pixel may have to be shifted over 1000 times until it is read.

CCDs offer the advantage of a rectangular detector, with resolution in two perpendicular directions, whereas diode arrays were strictly linear arrays. This can be used advantageously in either of two ways. First, one may actually perform Raman imaging experiments in real time. This is done in Raman microscopy in material science, where the variation of a surface composition can be studied by monitoring the Raman scattering with a spatial resolution of a few micrometers.

Second, if CCDs are used in a standard spectrograph configuration, it is advantageous to "bin," that is, co-add, the charge in pixels of the same column of the array. If the light is dispersed in the horizontal direction of the array, which typically contains between 600 and 1000 pixels over a distance of 18–30 mm, then the pixels in a column experience the same wavelength (and presumably intensity) of the light. The vertical dimension of the array may be between 5 and 9 mm. Thus, in the binning process, the pixels in a vertical column are co-added and read out together, which significantly speeds the reading process and increases the sensitivity of the measurement by the square root of the number of pixels binned together.

The sensitivity of a CCD chip can further be enhanced by a process known as back-thinning, in which the layer of silicon, onto which the diode has been deposited, is mechanically thinned. Back-thinned CCDs can have quantum efficiencies of over 50% and single-photon detectivity. They have further increased the sensitivity of Raman spectroscopy and have allowed a number of experiments to be carried out which were previously impossible.

An interesting problem arises about the calibration of the wavelength in an array detector. Any detector, be it a photomultiplier or an array detector, does not distinguish the actual color of the light but only determines the light intensity. In the case of the dispersive, single-channel detection system, the grating is scanned at a constant rate and varying intensities are swept through the exit slit to the detector. The sine of the grating angle (in linear wavelength spectrometers) or the cosecant of the grating angle (in linear wavenumber spectrometers) determines the abscissa of the spectrum, and the ordinate is just the signal sensed by the detector.

In multichannel Raman instruments, the grating is left stationary during the data acquisition, and dispersed light of different wavelengths impinges onto different detector elements, depending on the angle of diffraction. The detector elements in an array detector are linearly spaced; typical element widths are between 20 and 40 μm. Unfortunately, an inspection of the equation governing the linear dispersion of a monochromator,

$$\frac{d\lambda}{dx} = \frac{\cos\beta}{Fng} \qquad (6.1.2)$$

reveals that in the exit plane of the monochromator, the dispersion changes with the cosine of the diffracted angle. Thus the diode positions are linear in neither wavelength nor wavenumber. Therefore mathematical procedures need to be used to present array detector data in linear wavelength or wavenumber scales [Diem, et al., 1986].

6.4.3 Fourier Transform Raman Instrumentation

The enormous advantages realized in infrared spectroscopy when utilizing FT versus dispersive instrumentation had long tempted spectroscopists to perform FT–Raman experiments. However, interferometers work best in the infrared region of the spectrum for a number of reasons. First, the motion of the movable mirror must be parallel to the fixed mirror to better than the wavelength of light, or the interference patterns will be distorted. Second, the wavelength of the reference beam (typically a He–Ne laser) should be much shorter than the wavelength of the light to be analyzed to ensure that the interferogram is sufficiently well defined. Thus interferometry was commercialized in the infrared region.

However, the observation of Raman spectra in the infrared is hampered by the fact that Raman scattering depends on the v^4 factor; consequently, the use of long wavelength lasers to excite the Raman process seriously degrades the scattered intensities. Recently, methods have been developed to scan the mirror of a Michelson or modified Michelson interferometer more precisely, evenly, and free of jitter such that observation of interferometric data in the near-IR (0.7–1.2 μm) range has become practical. Thus it is now possible to collect interferometric data with the required precision for Raman spectroscopy in the red end of the visible and the near-IR portion of the visible spectrum.

In addition, high-power lasers are now available in the near-infrared region. YAG lasers, for example, are available which produce many watts of CW power output at 1.064 μm. In addition, diode lasers with a wavelength of about 780 nm are available at low cost. The combination of near-infrared lasers, improved interferometers, and solid state detectors for the near-IR range (such as InGaAs) have made the observation of FT–Raman spectra possible.

Excitation in the infrared bears another major advantage. Raman spectra are often obscured by fluorescence of the sample or fluorescence by minor impurities in the sample. "Burning" off this fluorescence by exposing the sample to hours of high-intensity visible light works sometimes for small amounts of impurities; yet real fluorescence can only be averted by time-resolved techniques, which rely on the fact that Raman scattering emerges faster than fluorescence. Excitation with an infrared laser can circumvent the problem completely, since the infrared photons normally do not have sufficient energies to reach the vibronic states that cause fluorescence. In fact, FT–Raman spectra of highly fluorescent dyes, such as Rhodamine G, have been recorded free of fluorescence interference using IR laser excitation. Thus FT–Raman offers the major optical advantages discussed earlier in FT IR spectroscopy, and the advantage of reduced fluorescence—while suffering from the disadvantage of a lower scattering cross section due to infrared excitation.

FT–Raman, however, is by no means a trivial experiment. The main problems arise from laser noise during data acquisition. This noise is sampled via the Rayleigh scattering and is Fourier transformed into high-intensity noise throughout the entire spectral range. It can be so strong that it masks the

Raman spectrum completely. This problem is sometimes referred to as the "multiplex disadvantage" [Schrader, 1990]. Optical filters, or a premonochromator, are used to remove the Rayleigh scatter and, with it, the source of this noise.

Commerical FT–Raman instruments use mostly YAG lasers, notch filters at the laser wavelength to reduce the Rayleigh scatter, high-quality interferometers, and detectors such as InGaAs, which peak in the near-infrared. By switching a few optical elements, it is possible to use the instrument as a FT infrared absorption spectrometer; thus the possibility of utilizing one instrument to collect Raman and IR data has become a reality. This offers the opportunity to digitize spectra in the same data format, which is, of course, necessary for the automatic search for fitting spectral data.

6.5 HADAMARD TRANSFORM SPECTROSCOPY

An interesting alternative to Fourier transform spectroscopy is Hadamard transform spectroscopy, pioneered mostly by Fateley and co-workers [Tilotta et al., 1987]. In Fourier transform spectroscopy, a pattern of interferences is recorded as a function of a path retardation, which is caused in most cases by a moving mirror but can also be due to the different paths as the light travels through refractory wedges. Both these methods rely on an accurate movement of the device that produces the retardation. Such a device may not be suitable under very harsh conditions involving lateral accelerations or very rough and bumpy motions. Hadamard transform spectroscopy offers an advantage here since it can be implemented without any moving parts. In Hadamard transform spectroscopy, the polychromatic beam of light emerging from the sample is passed through a dispersive monochromator as discussed before with the dispersive instrumentation. Instead of rotating the grating and using a slit to select a bandpass, a mask of many transparent and nontransparent areas is inserted between the grating and the detector, and all the spectral elements transmitted by the mask are simultaneously focused on the detector.

This mask serves as an encoder, which allows certain wavelengths to impinge on the detector, while others are rejected. By moving the mask laterally, a different pattern of wavelengths is allowed to impinge on the detector. The spectrum can be recovered from the encoded intensities (also referred to as an "encodegram") via the Hadamard transformation, which can be written in a matrix notation similar to Eq. 6.2.33. However, the main difference in Fourier transform and Hadamard transform lies in the fact that the former is a harmonic analysis of the signal, whereas the latter expands the signal in terms of rectangular waves. The transformation matrix, also known as the simplex matrix, is dependent on the shape and size of the mask and contains 0 and 1 elements representing transparent or opaque spots.

The advantage of this method is, as pointed out before, that masks can be constructed that are devoid of any moving parts. This can be accomplished using electro-optic materials or liquid crystalline elements, for which the transmission can be varied by applying suitable voltages to the optical elements of the mask. In this case, an extremely rugged spectrometer may be constructed that still has the multiplexing advantage of Fourier transform spectroscopy, while avoiding the rigorous motion of the movable mirror and the precise location of the mirror during data acquisition. Hadamard transform instruments have been applied to both infrared and Raman spectroscopies.

REFERENCES

E. O. Brigham, *The Fast Fourier Transform*, Prentice-Hall, Englewood, NJ, 1974.

J. W. Cooley and J. W. Tukey, *Math. Comput.*, *19*, 297 (1965).

J. W. Cooper, *Introduction to PASCAL for Scientists*, Wiley, New York, 1981.

M. Diem, in *Vibrational Spectra and Structure*, Vol. 19, J. R. Durig, Ed., Elsevier Science Publishers, Amsterdam, The Netherlands, 1991, p. 1.

M. Diem, F. Adar, and R. Grayzel, *Computer Enhanced Spectroscopy*, *3*, 29 (1986).

F. A. Jenkins and H. E. White, *Fundamentals of Optics*, McGraw-Hill, New York, 1957.

B. Schrader, in *Practical Fourier Transform Infrared Spectroscopy*, J. R. Ferraro and K. Krishnan, Eds., Academic Press, New York, 1990.

D. C. Tilotta, R. D. Freeman, and W. G. Fateley, *Appl. Spectrosc.*, *41*, 1280 (1987).

7
VIBRATIONAL SPECTRA OF SELECTED SMALL MOLECULES

So far, we have discussed the quantum mechanical foundations of vibrational spectroscopy, the symmetry arguments, computational aspects of vibrational frequencies, and methods to observe vibrational data. In this chapter, the vibrational spectra observed for a number of polyatomic molecules will be presented, and the detailed arguments necessary for a consistent vibrational assignment will be given. The importance of interpreting and assigning vibrational spectra *before* attempting any normal mode calculation cannot be overemphasized: there are many papers in the literature where calculations have been performed on incomplete data sets, data sets that ignored isotopic information, or just poorly assigned vibrational spectra. The sentence frequently found in the literature—"normal coordinate calculations were carried out to confirm the vibrational assignment"—epitomizes this fallacy: for most molecules, the calculations are so indeterminate that the calculations at best confirm the input assumptions. Thus it is absolutely essential that the input assumptions are correct, and this implies, in the case of vibrational spectroscopy, that isotopic data, band shape analyses, intensity, and polarization data are taken into account before calculations are carried out.

In this chapter, some examples will be introduced that demonstrate the arguments necessary for a proper assignment of the vibrational modes of a number of small molecules. For some of these, normal mode calculations were carried out in the author's research group following the detailed vibrational assignment. Some other molecules chosen were taken from the literature, mostly for their pedagogical value or to demonstrate how vibrational spectroscopy can be used to solve structural problems. No effort was made to secure the most recent and most accurate value of the vibrational frequencies. Rather, the data reported here are meant to convey some feeling for typical vibrational

176 VIBRATIONAL SPECTRA OF SELECTED SMALL MOLECULES

frequencies, and the emphasis in the following discussion is on the logic behind an assignment and not on the presentation of the most recent and exhaustive data. Finally, the reader is herewith reminded that the following three chapters are not intended to be reviews of the material, but are intended to illustrate the concepts using a few selected examples.

7.1 TRIATOMIC MOLECULES

Most of the triatomic molecules have been studied in great detail, and many of them are discussed in detail in Herzberg [1945, Chapter III]. Here, only those that affect vibrational spectroscopists frequently will be summarized. Two of them, water and carbon dioxide, can actually be quite a nuisance because their absorptions can be found in just about every infrared spectrum collected. CS_2 is included since it is an excellent solvent for vibrational studies, if one ever gets accustomed to its rather unpleasant smell.

In the previous chapters, the vibrational spectra of water, carbon dioxide, and carbon disulfide were introduced as examples of various topics (cf. Sections 3.9, 2.10, and 5.4, respectively), and their vibrational frequencies are summarized in Table 7.1. These molecules are certainly interesting from a theoretician's viewpoint, since they are sufficiently simple to allow very good calculations to be made. Furthermore, their low number of vibrational modes makes them desirable solvents for vibrational spectroscopy. For water and heavy water, this is certainly true in Raman spectroscopy, whereas the enormous absorption cross section of these molecules makes them poor solvents for infrared absorption spectroscopy.

Water, being a very light and polar molecule, exhibits a broad and very intense rotational–vibrational spectrum that is ubiquitous. Although most infrared absorption spectra are collected in a double-beam mode, these water lines often do not subtract out completely and are observed as derivative shape peaks between 1400 and $1750\,\text{cm}^{-1}$ and between about 3400 and $3900\,\text{cm}^{-1}$.

TABLE 7.1. Vibrational Frequencies [cm^{-1}] of Selected Triatomic Molecules

Mode Designation	H_2O (vapor)	D_2O (vapor)	CO_2 (gas)	CS_2 (liquid)
Point group	C_{2v}	C_{2v}	$D_{\infty h}$	$D_{\infty h}$
δ	1595	1179	667	397
ν_s	3652	2666	1286^a	657
			1388^a	
ν_{as}	3756	2789	2349	1523

^{a}Fermi resonance doublet between ν_s and 2δ.

Since water is such a light molecule, its rotational–vibrational bands cover two very broad regions.

Similarly, the rotational–vibrational envelope of CO_2 is observed at approximately $2350\,cm^{-1}$ in most infrared spectra. However, since the oxygen atoms are so much heavier than the hydrogen atoms in H_2O, the rotational–vibrational envelope is much narrower in CO_2. As mentioned before, the water and carbon dioxide contamination of the vibrational spectra can be quite a nuisance. Typical interference patterns observed are pointed out in the caption of Fig. 7.2.

The vibrational spectrum of CO_2 also contains one of the classic examples of Fermi resonance. According to group theoretical considerations, the Raman spectrum of a linear triatomic molecule with $D_{\infty h}$ symmetry should exhibit only one Raman active mode (the symmetric stretching vibration) and two infrared active modes (the antisymmetric stretching and the bending vibrations). This is the case in an analogous molecule such as CS_2. However, in CO_2, two strong Raman modes are observed at 1286 and $1388\,cm^{-1}$. They are due to the interaction of the symmetric stretching mode, estimated to occur at about $1337\,cm^{-1}$ and the overtone of the deformation, expected somewhat below $1334\,cm^{-1}$. The Fermi resonance nearly equalizes their intensities and shifts them apart by over $100\,cm^{-1}$.

7.2 TETRATOMIC MOLECULES

We begin this section with the discussion of the vibrations of formaldehyde, $O{=}CH_2$. This molecule has C_{2v} symmetry, and its six degrees of vibrational freedom fall into the following symmetry species: $3A_1$, $2B_1$, and $1B_2$. All of these are active in infrared and Raman spectroscopies. The observed frequencies are listed in Table 7.2.

The two high-frequency modes are the antisymmetric (B_1) and symmetric

TABLE 7.2. Vibrational Frequencies of Formaldehyde $O{=}CH_2$ and $O{=}CD_2$

Observed Frequencies [cm^{-1}]			
$O{=}CH_2$	$O{=}CD_2$	Symmetry	Assignment
2874	2156	B_1	v_{as}(CH) or (CD)
2780	2056	A_1	v_s(CH) or (CD)
1743	1700	A_1	v_s(CO)
1503	1106	A_1	δ_s(CH) or (CD)
1280	990	B_1	δ_{as}(CH) or (CD)
1167	938	B_2	δ_{oop}

(A_1) C—H stretching modes. As far as CH (and CD) stretching motions are concerned, these frequencies are extremely low and nearly a bit atypical. This is due to the presence of the electron withdrawing character of the carbonyl oxygen, which weakens the C—H bonds. We observe a typical behavior of these two identical groups attached to a central atom: the vibrations of these two groups couple and form symmetric and antisymmetric combination modes, which can be discussed in terms of symmetry coordinates introduced in Chapter 3.

The mode at 1743 cm^{-1} is strong in infrared absorption, which is typical for C=O groups. Its shift upon substitution of the two hydrogens by deuterium atoms is rather large and indicates that the C=O stretching motion is accompanied by a relatively large amplitude H motion to maintain the center of mass. This observation demonstrates that the concept of group frequencies, although useful, needs to be invoked carefully when precise descriptions of the molecular motions are required.

The two modes at 1503 and 1280 cm^{-1} contain two O=C—H and the H—C—H deformation coordinates. These three internal coordinates are not linearly independent, which means they cannot all increase simultaneously. Thus only two normal modes are observed, which all contain contributions of the three deformations. The lower frequency deformation at 1280 cm^{-1} (B_1) consists mainly of the two H—C=O internal coordinates, whereas the symmetric deformation at 1503 cm^{-1} contains a large amount of the H—C—H deformation, in addition to the H—C=O deformation coordinate. These modes can be identified and assigned by their shift upon deuteriation.

The mode at 1167 cm^{-1} is due to the out-of-plane deformation (wagging mode) of formaldehyde. This frequency depends strongly on the masses attached to the central atom, as can be seen by the large deuterium isotopic shift. In molecules such as acetone, it occurs at even lower frequency. The assignment presented so far was originally performed using group frequencies and group theoretical arguments, as well as Raman depolarization ratios. The latter data are very important in deciding which of the vibrations belong to the totally symmetric representation and, therewith, helping to distinguish between symmetric and antisymmetric vibrations of a given group.

Another tetratomic molecule to be discussed here is N_2F_2, since the vibrational data demonstrate very nicely how the *cis-* and *trans-*isomer of this molecule can be distinguished. The data also show vibrational group frequencies of compounds not containing any carbon atoms. One can see from Table 7.3 that the N=N double bond stretch occurs at slightly lower frequency than a C=C stretching mode, due to the higher masses. Similarly, the N—F stretching frequency is about 100 cm^{-1} lower than that of a C—F bond.

The structure of *trans-*N_2F_2 was shown in Chapter 4. It belongs to the C_{2h} point group, which is centrosymmetric. Therefore the mutual exclusion principle holds, but all six vibrational degrees of freedom are accessible in either infrared (A_u and $2B_u$) or Raman spectroscopy ($3A_g$). These data are taken from King and Overend [1966, 1967] and are summarized in Table 7.3. *Cis-*N_2F_2,

TABLE 7.3. Vibrational Frequencies [cm^{-1}] on N$_2$F$_2$

cis-N$_2$F$_2$ (C_{2v})		trans-N$_2$F$_2$ (C_{2h})	
$\nu_{NN}(A_1)$	1525 Ra, IR	$\nu_{NN}(A_g)$	1522 Ra
$\nu_{NF}(A_1)$	892 Ra, IR	$\nu_{NF}(A_g)$	1010 Ra
$\delta_{NNF}(A_1)$	341 Ra, IR	$\delta_{NNF}(A_g)$	600 Ra
$\tau_{N=N}(A_2)$	550? Ra	$\tau_{N=N}(A_u)$	364 IR
$\nu_{NF}(B_1)$	952 Ra, IR	$\nu_{NF}(B_u)$	990 IR
$\delta_{NNF}(B_1)$	737 Ra, IR	$\delta_{NNF}(B_u)$	423 IR

on the other hand, belongs to C_{2v}. Its $3A_1$ and $2B_1$ vibrations are both Raman and infrared active, whereas the torsional vibration about the N=N bond belongs to A_2 and is active in the Raman spectrum only.

7.3 PENTATOMIC METHANE DERIVATIVES

The point group of pentatomic molecules, where the central C atom is surrounded tetrahedrally by four other atoms, can vary from T_d, C_{3v}, C_{2v}, C_s, to C_1, depending on the nature of the atoms attached. Although all these molecules have nine degrees of vibrational freedom, the actually observed number of vibrational modes is less than nine in the most highly symmetric (T_d and C_{3v}) molecules due to the occurrence of degenerate modes. The number of vibrations allowed in Raman and infrared spectra of pentatomic molecules of varying symmetry is given in Table 7.4. In terms of the normal coordinate calculations discussed in Chapter 3, it appears that the lack of observable frequencies will complicate the calculations. However, due to symmetry of the

TABLE 7.4. Number of Infrared and Raman Active Fundamentals, and Their Symmetry, for Pentatomic Molecules

Symmetry	Example	Raman	Infrared	Total
T_d	CCl$_4$	4	2	4
		A_1, E, T_1, T_2	T_1, T_2	
C_{3v}	CHCl$_3$	6	6	6
		$3A_1$, $3E$	$3A_1$, $3E$	
C_{2v}	CH$_2$Cl$_2$	9	8	9
		$4A_1$, A_2, $2B_1$, $2B_2$	$4A_1$, $2B_1$, $2B_2$	
C_s	CH$_2$FCl	9	9	9
		$6A'$, $3A''$	$6A'$, $3A''$	
C_1	CHFClBr	9	9	9
		$9A$	$9A$	

180 VIBRATIONAL SPECTRA OF SELECTED SMALL MOLECULES

molecule, the number of independent force constants needed to describe the molecular potential energy is reduced to an even greater degree. Thus the more highly symmetric molecules are more easily described in terms of normal mode calculations, and consequently, the literature on symmetric species far outnumbers that of low symmetry molecules. We shall start the discussion with two examples of tetrahedral molecules, CCl_4 and CH_4.

7.3.1 Pentatomic Molecules with T_d Symmetry

CCl_4 does not contain fundamental vibrations above $800\,cm^{-1}$ and thus is a suitable solvent for vibrational spectroscopy if the low frequency spectral range is unimportant. As shown in Table 7.4, it exhibits four Raman and two infrared active vibrations. Figure 7.1 shows the well-established Raman spectrum of CCl_4. However, a cursory inspection of the Raman spectrum shows (at least) five bands, with overtones and combination bands contributing to the spectrum. Since the peak at $459\,cm^{-1}$ is highly polarized, it is assigned to the C—Cl symmetric stretching mode of A_1 symmetry. The low frequency modes at 218 and $314\,cm^{-1}$ fall in the area of C—Cl deformations and are assigned to the E and T_2 modes, respectively. That leaves one of the peaks at $762/790\,cm^{-1}$ to be the antisymmetric stretching modes of T_1^- symmetry. The other member of

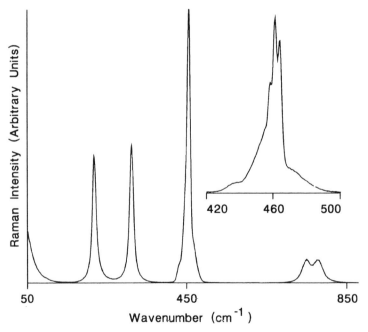

FIGURE 7.1. Raman spectrum of liquid CCl_4. Inset: The v_1 mode at $459\,cm^{-1}$ at $1\,cm^{-1}$ resolution, showing the isotopic components.

this double peak is an enhanced overtone or combination. Inspection of the possible combinations yields that $459 + 314 \text{ cm}^{-1}$ is close to 770 cm^{-1}. Consequently, the peak at 790 cm^{-1} is assigned to the antisymmetric C—Cl stretching mode of T_1 symmetry, and the peak of 762 cm^{-1} to a combination band, $\delta^{\text{C—Cl}}(T_2) + v^{\text{C—Cl}}(A_1)$. The intensity enhancement experienced by the (normally weak) combination band is due to Fermi resonance, which was discussed in Section 3.3. Fermi resonance is due to a breakdown of Eqs. 3.3.3 and 3.3.4, caused by the proximity of two energy levels $\delta^{\text{C—Cl}}(T_2) + v^{\text{C—Cl}}(A_1)$ and $v^{\text{C—Cl}}(T_1)$, which allows for mixing of the wavefunctions of the modes involved; consequently, the corresponding energy levels interact. This interaction has two major consequences: the intensities of the two peaks tend to equalize (i.e., the weaker combination band borrows intensity from the stronger fundamental), and the two peaks split further apart than they would in an unperturbed situation. These manifestations of Fermi resonance were discussed in detail in Section 3.4 and will be mentioned in a number of cases below, since it is a common phenomenon and particularly prevalent in low symmetry molecules.

There is one further remarkable spectral feature about the Raman spectrum of CCl_4. When the bandpass of the spectrometer is decreased (i.e., the resolution is improved), the peak at 459 cm^{-1} appears with a distinct multiplet structure, shown in the inset of Fig. 7.1. The origin of the multiplet is due to isotopic effects. Naturally occurring Cl consists of roughly 25% ^{37}Cl and 75% ^{35}Cl. Thus the most probable CCl_4 molecule found will be C ^{37}Cl ^{35}Cl$_3$. The stretching frequency of this molecule is the most dominant peak in the spectrum, at about 459 cm^{-1}. The next most likely species is C ^{35}Cl$_4$, which will have a slightly higher vibrational frequency (461 cm^{-1}) due to the lower mass of the molecules. The highest frequency component of the multiplet corresponds to this species. The shoulder at 457 cm^{-1} is due to the C ^{37}Cl$_2$^{35}Cl$_2$ species, and the other two isotopomers are not easily observed at room temperature, although they appear clearly at 77 K. The broad features at lower frequency in the inset of Fig. 7.1 are due to a Fermi resonance enhanced overtone of the E mode at 218 cm^{-1}.

We now turn to the discussion of the infrared spectra of methane. Gaseous CH_4 can be studied very conveniently at moderate pressures (10–50 mm Hg) in an infrared gas cell with a path length of a few centimeters. It exhibits two infrared allowed transitions of T_1 and T_2 symmetry. The observed spectrum is shown in Figs. 7.2 and 7.3 at low and high resolution.

The band shapes in the low resolution spectrum (Fig. 7.2) show the typical behavior of unresolved rotational–vibrational transition introduced in Section 2.10: the sharp peak in the center is the actual $v = 0$ to $v = 1$ vibrational transition, whereas the envelopes to higher and lower frequencies of the center peak are due to the vibrational–rotational transitions.

Since methane is a spherical top rotor (cf. Chapter 2), its rotational energy is given by

$$E = BJ(J + 1) \tag{2.2.9}$$

182   VIBRATIONAL SPECTRA OF SELECTED SMALL MOLECULES

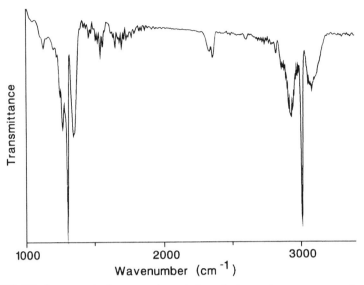

FIGURE 7.2. Low resolution gas phase infrared absorption spectrum of methane. There are two allowed fundamentals in the absorption spectrum, both with pronounced P, Q, and R branches. Note also the water rotational–vibrational bands centered at $1620\,\text{cm}^{-1}$ and at $3000\,\text{cm}^{-1}$, and the CO_2 band at $2350\,\text{cm}^{-1}$.

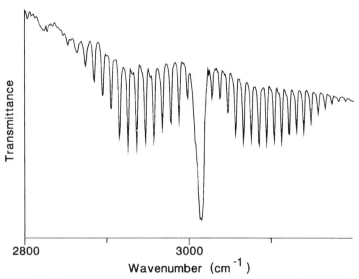

FIGURE 7.3. High resolution rotational vibrational spectrum of the v_3 vibration of methane.

where B is the (only) rotational constant. Since CH_4 has no permanent dipole moment, its pure rotational spectrum is not allowed in absorption. Thus its rotational constant cannot be measured directly in microwave spectroscopy. However, the rotational–vibrational interaction spectrum allows this quantity to be determined readily. The broad rotational–vibrational envelopes, observed under low resolution (Fig. 7.2), can be resolved into the distinct rotational–vibrational transitions under higher resolution, shown in Fig. 7.3. For the v_3 band (T_1), the vibrationally excited state is distorted from T_d symmetry and thus gives rise to simultaneous $\Delta v = \pm 1$ and $\Delta J = \pm 1$ transitions. Thus the vibrational transitions show rotational fine structure, which allows the determination of the rotational constant and, consequently, the moment of inertia and structure of methane. The spacing of the rotational lines is not constant, as can be seen in Fig. 7.3, due to the interaction of rotational and vibrational energies, as discussed previously in Section 2.11.

The band envelope observed at low resolution, or in the presence of an inert gas at high pressure to broaden the individual rotational lines, is typical for a spherical top rotor, with pronounced P, Q, and R branches. Due to the low masses attached to the central atom, the moment of inertia of methane is small, and consequently, the rotational constant is large. A splitting of $10\,\mathrm{cm}^{-1}$ is observed only for such small and light molecules, and rotational energy levels of larger molecules containing heavy atoms are often in the subwavenumber range.

7.3.2 Pentatomic Molecules with C_{3v} Symmetry

These molecules will be discussed in the most detail, since an analysis of their vibrational spectra reveals an enormous amount of principles of vibrational spectroscopy. Figures 7.4 and 7.5 show infrared and Raman data of neat $CHCl_3$ and $CDCl_3$. These molecules have C_{3v} symmetry and exhibit nine degrees of vibrational freedom. Following the procedures given in Chapter 4, the reader may verify that these nine vibrations fall into three A_1 and three E symmetry classes; that is, only six vibrational modes are observed. The observed frequencies for $CHCl_3$ are tabulated in Table 7.6. We introduce in this table the standard nomenclature of denoting stretching vibrations by v, with a superscript denoting the vibrating group, and the symmetry in parentheses. δ denotes deformation vibrations, ρ rocking vibrations, and τ torsional vibrations. Thus $v^{CH_3}(E)$ implies the antisymmetric stretching mode of E symmetry of a methyl group.

The assignment of the undeuteriated parent molecule, $CHCl_3$, proceeds through the use of group frequencies and by analogy with the two molecules discussed in the previous section. An inspection of a table of group frequencies shows that most molecules that contain a single C—Cl bond will show a strong vibration in both Raman and IR spectra at a little above $700\,\mathrm{cm}^{-1}$ due to the carbon–chlorine stretching motion.

184 VIBRATIONAL SPECTRA OF SELECTED SMALL MOLECULES

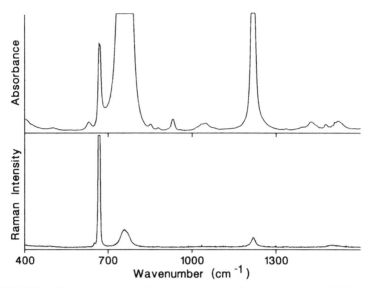

FIGURE 7.4. Infrared absorption (top) and Raman (bottom) spectra of chloroform in the C—Cl stretching and methine deformation region.

A —CCl_3 group with local C_{3v} symmetry will exhibit three C—Cl stretching motions. The linear combination of three C—Cl stretching coordinates R_1, R_2, and R_3 will produce three symmetry coordinates S_i (cf. Section 4.5):

$$S_1(A_1) = 3^{-1/2}(R_1 + R_2 + R_3)$$
$$S_2(E) = 6^{-1/2}(2R_1 - R_2 - R_3) \quad (7.3.1)$$
$$S_3(E) = 2^{-1/2}(R_2 - R_3)$$

and, consequently, three normal modes of vibration, two of which are degenerate. Thus two vibrations are observed, a symmetric stretching motion and an antisymmetric stretching motion. The symmetric stretching mode appears at 667 cm^{-1} and is extremely sharp and strong in the Raman spectrum, whereas the antisymmetric stretching mode at 760 cm^{-1} is broad and weak. In the infrared spectrum, on the other hand, the antisymmetric stretching mode is more prominent (cf. Fig. 7.4). Here, the general rule that symmetric modes are strong in the Raman spectra and antisymmetric modes are strong in infrared absorption immediately points toward the proper assignment of the two bands in terms of the symmetric and antisymmetric modes.

In addition to the two stretching group frequencies of the —CCl_3 group, there will be other group frequencies due to the three Cl—C—Cl deformation coordinates, which will combine in a similar fashion as the stretching coordinates to form a symmetric (A_1) deformation (the "umbrella" mode) that is observed at 364 cm^{-1} and a degenerate antisymmetric deformation (E) at

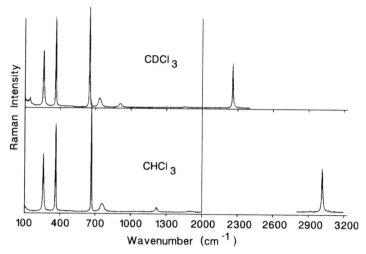

FIGURE 7.5. Raman spectra of $CDCl_3$ (top) and of $CHCl_3$ (bottom). All six fundamentals are observable for both species.

$260\,cm^{-1}$. Both of these are strong in the Raman spectra. Group frequencies for some common groups with C_{3v} symmetry are summarized in Table 7.5.

Returning to the spectra of chloroform and deuteriochloroform in Figs. 7.4 and 7.5, we see that with the arguments above the observed frequencies at 260, 364, 667, and $760\,cm^{-1}$ can be uniquely assigned. Raman depolarization values, or a comparison between infrared and Raman intensities, are used to distinguish between symmetric and antisymmetric vibrations. The highest frequency mode, observed at about $3030\,cm^{-1}$, obviously is due to the C—H stretching motion, since group frequencies of carbon–hydrogen stretching motions all are in the range from 2850 to $3050\,cm^{-1}$. The frequencies of C—H stretching modes are influenced by the mass and electronegativity of the groups attached to the C and the hybridization of the C atom.

A comparison of the vibrational spectra of chloroform and deuteriochloroform very impressively demonstrates the isotopic shift upon substitution of the

TABLE 7.5. Group Frequencies of Some Common—CX_3 Groups

Frequency	—CH_3	—CD_3	—CCl_3	—CBr_3
$\nu_{as}(E)$	3020	2240	760	655
$\nu_s(A_1)$	2955	2125	665	541
$2\delta_{as}$	2890[a]	—	—	—
$\delta_{as}(E)$	1460	1040	260	155
$\delta_s(A)$	1360	1060	365	222

[a] Fermi resonance enhanced overtone of $2\delta^{as}$ and $\nu^s(A_1)$; see Section 3.4.

C—H by a C—D bond. Such a substitution, of course, confirms the assignment of the C—H(C—D) stretching motion. Since the four low frequency vibrations listed above are affected very little by deuterium substitution, they do not appear to contain much H motion.

The H versus D isotopic shift can be estimated by using Eq. 2.6.16

$$v = (1/2\pi)\sqrt{f/M_R} \qquad (2.6.16)$$

and assuming that the force constants for the C—H and the C—D bonds are the same. This is a fair assumption since the additional neutron in D should not change the electronic distribution of the bond to a first approximation. Furthermore, let us assume that the mass of the remainder of the molecule is much larger than that of the H or D atom. Then we can approximate the ratio of the stretching frequencies of the C—H and the C—D bonds by

$$\frac{v_{(C-H)}}{v_{(C-D)}} = \sqrt{m_H/m_D} = 1.4 \qquad (7.3.2)$$

This factor is indeed very close to the observed H/D isotopic shift: the C—D stretching frequency in $CDCl_3$ occurs at 2255 cm^{-1}, and the observed ratio is $3030/2256 = 1.3$. The discussion so far has left out the remaining fundamental, observed for $CHCl_3$ at 1216 cm^{-1} and for $CDCl_3$ at 908 cm^{-1}. Its large isotopic shift suggests a motion with sizable H amplitudes; this last mode is, indeed, assigned as the H—C—Cl deformation. This mode is doubly degenerate (E) and can be described as the linear combination of the three H—C—Cl deformation internal coordinates, α_1, α_2, and α_3, according to $2\alpha_1 - \alpha_2 - \alpha_3$ and $\alpha_2 - \alpha_3$ (cf. Fig. 3.5) in analogy to the stretching symmetry coordinates described in Eq. 7.3.1. This methine deformation mode is important in the discussion of low symmetry species containing a single C—H group.

The molecule methyl chloride, CH_3Cl, has the same symmetry as chloroform, but its vibrational spectrum is very different (cf. Table 7.6). Here, the lowest vibrational frequency is the C—Cl stretching motion at 705 cm^{-1}, with A_1 symmetry. Now, there are three C—H stretching vibrations interacting with each other, and one observes a number of molecular vibrations with group frequencies typical for the methyl group.

The two high-frequency modes, at 2966 and 3042 cm^{-1}, are the symmetric (A_1) and antisymmetric (E) methyl stretching modes, for which we can write the same linear combinations (Eq. 7.3.1) of internal coordinates as we did before for the —CCl_3 group. The methyl deformations, observed at 1355 cm^{-1} for the symmetric deformation (umbrella) mode and at 1455 cm^{-1} for the antisymmetric deformation (E) mode, are also very characteristic for a methyl group (cf. Table 7.5). In addition, methyl groups exhibit an overtone of the antisymmetric deformation mode (expected around 2900 cm^{-1}), which is in Fermi resonance with the symmetric stretching mode and typically observed at 2880 cm^{-1} in both Raman and infrared spectra (cf. Section 3.4).

So far, we have accounted for seven (fundamental) degrees of vibrational freedom of CH_3Cl, and the remaining mode is an E mode. The corresponding vibration is referred to as the methyl rocking vibration, which we define in terms of linear combinations of the Cl—C—H deformation coordinates as follows:

$$2\beta_1 - \beta_2 - \beta_3$$
$$\beta_2 - \beta_3 \quad (7.3.3)$$

In methyl chloride, this mode occurs at $1015\,cm^{-1}$. In a low symmetry environment, the formerly degenerate states split strongly, and often two separate CH_3 rocking vibrations are observed at about 1000 and $925\,cm^{-1}$. Their identification is often difficult because of extensive mixing of these modes with skeletal (e.g., C—C stretching) vibrations. Substitution of a —CH_3 by a —CD_3 group is often necessary to assign these vibrations unambiguously.

The —CD_3 group has, in general, vibrations that are analogous to those of the —CH_3 group (cf. Table 7.5). However, there exists one difference in the order of the deformation modes. Whereas in the methyl group the (degenerate) antisymmetric deformation vibrations (ca. $1460\,cm^{-1}$) occur always at higher wavenumber than the symmetric one (ca. $1380\,cm^{-1}$), this order may or may not invert in the case of the deuteriomethyl group, depending on the mass of the atom(s) to which the group is attached. Thus it is obviously necessary to collect complete vibrational data, including Raman depolarization ratios, for a proper and complete assignment of the vibrational spectra, even for very small molecules.

7.3.3 Pentatomic Molecules with C_{2v} Symmetry

So far, we have compared two molecules of C_{3v} symmetry. It is instructive to investigate the molecule that logically occupies the place between $CHCl_3$ and CH_3Cl, namely, methylene chloride, CH_2Cl_2. The vibrational frequencies and assignments of this molecule are contained in Table 7.6 as well. Here, two C—Cl bonds emerge from the same C atom, and these stretching coordinates couple to form a symmetric and an antisymmetric stretching mode under local C_{2v} symmetry. The motion corresponding to the symmetric (A_1 under C_{2v}) stretching combination occurs at $705\,cm^{-1}$, and the antisymmetric (B_2) stretching combination occurs at $735\,cm^{-1}$. It is again typical that antisymmetric combinations occur at higher wavenumber than the symmetric ones (*vide infra*). They may be distinguished by their depolarization ratios, or sometimes even by a comparison of the Raman and infrared intensities. Of course, the C—H coordinates couple similarly to form group frequencies of the methylene group. Under C_{2v} symmetry, all vibrations of pentatomic molecules are Raman active, but the A_2 mode, which is a torsion of the CCl_2 group with respect to the CH_2 group, is not active in the infrared.

TABLE 7.6. Comparison of Vibrational Frequencies [cm^{-1}] of CHCl$_3$, CH$_2$Cl$_2$, and CH$_3$Cl

CHCl$_3$		CH$_2$Cl$_2$		CH$_3$Cl	
3030	$\nu^{C-H}(A_1)$	3040	$\nu^{C-H_2}(B_1)$	3040	$\nu^{C-H_3}(E)$
		3000	$\nu^{C-H_2}(A_1)$	2965	$\nu^{C-H_3}(A_1)$
1205	$\delta^{C-H}(E)$	1465	$\delta^{C-H_2}(A_1)$	1455	$\delta^{C-H_3}(E)$
		1265	$\delta^{C-H_2}(B_2)$	1355	$\delta^{C-H_3}(A_1)$
		1155	$\tau^{C-H_2}(A_2)$		
		900	$\rho^{C-H_2}(B_2)$	1015	$\rho^{C-H_3}(E)$
760	$\nu^{C-Cl_3}(E)$	755	$\nu^{C-Cl_2}(B_2)$	730	$\nu^{C-Cl}(A_1)$
665	$\nu^{C-Cl_3}(A_1)$	715	$\nu^{C-Cl_2}(A_1)$		
365	$\delta^{C-Cl_3}(A_1)$				
260	$\delta^{C-Cl_3}(E)$	285	$\delta^{C-Cl_2}(A_1)$		

Table 7.6 demonstrates the principle of group frequencies impressively. It also shows the number of observable fundamentals when the symmetry changes. Most of the assignments can be made in a straightforward fashion, using the vibrational frequencies of CH$_3$Cl and CHCl$_3$ and Raman depolarization ratios to distinguish symmetric and antisymmetric vibrations.

7.3.4 Pentatomic Low Symmetry Species

The assignment of lower symmetry species is carried out according to the same principles. In CH$_2$FCl (a member of the freon family of refrigerants) with C_s symmetry, all vibrations are active in both infrared and Raman spectra, yet there can still be symmetric (A') and antisymmetric (A'') vibrations, and thus the methylene group still exhibits two vibrations that may be classified as a symmetric and an antisymmetric stretching vibration.

In the author's laboratory, one of the major efforts has been the understanding of the vibrations of small, asymmetric molecules. The interest in these molecules resulted, of course, from the emergence of a new field of spectroscopy, namely, vibrational optical activity (infrared circular dichroism and Raman optical activity, cf. Chapter 9), and the measurement of natural optical activity requires asymmetric or dissymmetric molecules. Vibrational literature on these molecules has been extremely sparse, due to the inherent difficulties of performing normal mode analyses on molecules without symmetry. In the following section, some details of the vibrational spectra of small, asymmetric molecules will be presented. The assignments were done in great detail, using isotopic data, and normal mode calculations were carried out using different force fields.

One of these molecules studied in detail was HCFClBr and its isotopomer, DCFClBr [Diem and Burow, 1976]. An asymmetric species such as this one teaches an enormous amount about the vibrations of small molecules and gives good insight into the complex coupling pattern of internal coordinates. As pointed out in Table 7.4, all nine fundamentals are active in both Raman and infrared spectroscopies. In order to obtain an unambiguous vibrational assignment for the normal mode calculations, isotopic species had to be studied as well. Thus DCFClBr was used for the vibrational assignment, and it was found that nearly all vibrations of this molecule shift upon substitution of the H for D. Further isotopic data could be obtained by a detailed interpretation of the gas phase Raman spectra: some modes showed a distinct $2\,\text{cm}^{-1}$ splitting, with intensity ratios of about 3:1. This splitting was attributed to the 75%:25% isotopic ratio of ^{35}Cl to ^{37}Cl. Such isotopic splitting leads to a detailed understanding of which normal modes involve sizable Cl atom motion.

The observed nine fundamentals for each of the four isotopic species are listed in Table 7.7. In addition to these fundamentals, about a dozen overtones and combinations can be observed. Some of these are enhanced by Fermi resonance, and the amount of enhancement is very sensitive to isotopic substitution, since the isotopic shift can move bands apart and prevent Fermi resonance from happening.

The C—H and C—D stretching vibrations occur at frequencies typical of these groups in trihalomethanes (3026 and $2264\,\text{cm}^{-1}$), and the pattern of isotopic shifts is very similar to that observed for CHCl_3 and CDCl_3. The halogen–carbon–hydrogen deformations show up as two distinct peaks of medium intensity in both Raman and infrared spectra. These two peaks are interesting, since they originate from a doubly degenerate (E) mode (at $1216\,\text{cm}^{-1}$ in a molecule such as chloroform, *vide supra*). Under C_1 symmetry, the degeneracy is lost, and two vibrations are observed at 1311 and $1205\,\text{cm}^{-1}$. Both of these bands shift, upon deuteration, to 974 and $919\,\text{cm}^{-1}$. One might expect three distinct hydrogen–carbon–halogen deformation modes, since

TABLE 7.7. Vibrational Frequencies of Fluorochlorobromo Methane and Isotopomers in the Gas Phase

CHF^{35}ClBr	CHF^{37}ClBr	CDF^{35}ClBr	CDF^{37}ClBr	Frequency
3026	3026	2264	2264	ν(C—H)
1311	1311	974	974	δ(X—C—H)
1205	1205	919	919	δ(X—C—H)
1078	1078	1084	1084	ν(C—F)
788	784^a	750	746	ν(C—Cl)
664	664	621	621	ν(C—Br)
427	423	424	421	δ(F—C—Cl)
314	314	313	313	δ(F—C—Br)
226	223	224	221	δ(Cl—C—Br)

^{a}Estimated from solid state splitting.

there are three different halogen atoms in this molecule. However, since this particular mode is doubly degenerate in symmetric trihalomethanes such as chloroform, a splitting into only two components is observed.

The C—F stretching vibration is by far the weakest fundamental in the Raman spectrum and the strongest in the infrared spectrum. This is because the highly polar C—F stretching motion produces an enormous change in the molecular dipole moment. On the other hand, the electrons are held so tightly by the electronegative F atom that the motion of the fluorine produces very little change in polarizability, which was qualitatively defined above as the ease with which electrons can be set oscillating by the exciting light. The C—F stretching mode is the only vibration in this molecule which shifts to higher frequencies upon deuteriation. This can be explained in terms of the quite different mixing pattern of the C—F stretching coordinate with a number of other motions between the hydrogenated and deuteriated compound.

The C—Cl and C—Br stretching vibrations are typical for such modes if only one of each are present. Each of the three low frequency vibrations can be associated with one of the three heavy atom deformation coordinates. Two of them show distinct Cl isotopic splitting and thus may be assigned to the Br—C—Cl and the F—C—Cl deformations. Therefore the remaining deformation is the F—C—Br deformation mode. The normal coordinate calculations reveal, of course, that these internal coordinates mix to some extent to produce the normal mode of vibration. The determination of the amount of mixing is difficult and requires the availability of isotopic data. It is amazing that even these low frequency and relatively well localized vibrations depend on the presence of H versus D: these modes show a small, but reproducible, isotopic shift upon deuterium substitution.

In order to carry out normal mode calculations on such a molecule, the choices of force fields and coordinate systems become very important. The definitions of the coordinates will be discussed first. One could carry out the normal mode calculations using nine nonredundant internal coordinates. The most logical choices for these would be the three halogen–carbon–halogen deformations, the two halogen–carbon–hydrogen deformation coordinates, and the four stretching internal coordinates. This choice is reasonable since these coordinates correspond directly to the assignment of the observed normal modes.

The other choice for the internal coordinates is to utilize the four stretching and all six deformation internal coordinates. Thus all three hydrogen–carbon–halogen deformation coordinates will be used, in addition to the F—C—Cl, F—C—Br, and Cl—C—Br deformation coordinates. One of these six deformation coordinates is redundant (cf. Fig. 3.5). Both definitions have their inherent advantages and disadvantages. Using only five deformation coordinates eliminates the redundancy (with the related argument of linear force constants), but the choice of which force constant to ignore presents an ambiguity, which affects the transferability of the obtained force field. Using six deformation force constants creates a redundant coordinate and a zero calculated frequency but results in computed numerical values for the three H—C—X deformation force constants.

Independent of the choices, the two observed hydrogen deformation modes at 1311 and 1205 cm^{-1} will be described as mixtures of two or three halogen–carbon–hydrogen deformation internal coordinates, but the description in terms of the force constants will be different. In view of the desired transferability of diagonal force constants, the use of the redundant set of internal coordinates is somewhat preferable.

The second major decision in such calculations is the choice of the force field. A GVFF description of the force field requires 55 distinct force constants if all 10 (redundant) internal coordinates are used. Such computations are indeterminate and require that some of the off-diagonal force constants be arbitrarily ignored, even with the large number of isotopic data available The UBFF calculations require only 17 independent force constants and appear less arbitrary. This course was taken in the original normal mode calculation. However, even with all available isotopic data, the description of the two methine deformations at 1311 and 1205 cm^{-1} remained somewhat ambiguous and did not reproduce the direction of the dipole change properly, which was determined via a detailed analysis of the band shapes. Thus one can see the difficulties inherent in a detailed vibrational analysis of an asymmetric molecule such as HCFClBr, although this species is one of the simplest conceivable asymmetric species.

7.4 ETHANE AND ETHANE DERIVATIVES

Ethane, in a staggered conformation (cf. Fig. 4.4), has D_{3d} symmetry, and 18 degrees of vibrational freedom. The vibrations belong to the following symmetry classes: three A_{1g} (Raman), one A_{1u} (inactive), two A_{2u} (infrared), three E_u (infrared), and three E_g (Raman). The inactive vibration is the torsion about the C—C bond.

This molecule has been studied in such detail that its vibrational spectrum need not be elaborated on any further in this text. One aspect, however, is important for the following discussion. Although its torsional vibration is inactive in both infrared and Raman spectra, it has been measured in substituted ethane derivatives and is typically about 250 cm^{-1}, corresponding to a thermal energy of about 100°C. (The approximation 200 cm^{-1} ≈ 600 cal was used here.) The torsional barrier has been established via thermodynamic methods and is about 2.93 kcal/mol for ethane, and nearly twice as high in substituted ethanes such as CH$_3$—CCl$_3$. Considering that room temperature is about 0.6 kcal/mol, one finds that the picture of a freely rotating methyl group, often proposed by organic chemists, is erroneous.

Interestingly, the prevalent picture of a freely rotating methyl group results from the misinterpretation of a number of experimental data. Due to the slow time scale of the experiment (about six orders of magnitude slower than vibrational spectroscopy), NMR results, for example, seem to indicate a free rotation, and so do a number of chemical observations. Indeed, when a —CH_2D group is attached to a molecule, all three possible orientations of the C—D bond within the molecule's framework are found equally populated, suggesting that the methyl group rotates freely. However, the potential function and energy levels discussed in Section 3.3 show clearly that there are three distinct states of internal rotation in which the methyl or —CH_2D group is trapped, and the torsional vibration can be described as a small amplitude, back-and-forth libration, or hindered internal motion, without crossing the energy barrier.

At room temperature, according to the Boltzmann energy distribution, a sizable fraction of the hindered internal rotors have sufficient energy to jump over the barrier and arrive at one of the other energy minima. This accounts for the occurrence of all three possible orientations of a substituted methyl group, as discussed before. However, the difference here lies in principle: in a freely rotating system, the kinetic energy of the motion is larger than the potential energy barrier, which is clearly not the case here. Durig and Sullivan [1985] have reviewed the quantum mechanical foundations of hindered internal rotation and have published numerous papers on the determinations of the barriers of internal rotations in substituted ethanes.

We shall discuss here one asymmetrically substituted derivative, namely, the substituted ethane CH_3—CFClBr. There are two justifications for using this particular system: first, we may investigate the vibrations of a methyl group under the perturbation of a group with no symmetry; and second, we may verify the statements about internal rotations of the previous paragraphs.

When a —CH_3 group is placed into an environment that lowers the overall molecular symmetry to C_1, degenerate modes are no longer possible, and the two antisymmetric stretching motions corresponding to the internal coordinate combinations $2R_1 - R_2 - R_3$ and $R_2 - R_3$ occur with a small frequency difference. Similarly, the corresponding methyl antisymmetric deformation modes may show such splitting as well.

Vibrational data for CH_3—CFClBr are presented in Table 7.8. In this table, we have used symbols appropriate for molecules with higher symmetry, for example, $v_{as}(CH_3)$, although there is no distinction between symmetric and antisymmetric vibrations. Yet a detailed normal coordinate calculation on similarly perturbed methyl groups [Nafie et al., 1980] has shown that the stretching vibrations of the methyl group may still be described by a $R_1 + R_2 + R_3$ combination of internal coordinates, which may be classified as a "symmetric" mode, and two antisymmetric modes with the internal coordinate combinations $2R_1 - R_2 - R_3$ and $R_2 - R_3$. The "symmetric" mode is observed in CH_3—CFClBr at 2940 cm^{-1}, with a depolarization ratio of $\rho = 0.02$. The two (no longer degenerate) "antisymmetric" stretching modes are observed at 3017 and 2999 cm^{-1} and are depolarized ($\rho = 0.6$ and 0.7, respectively). Thus the relatively uncoupled methyl stretching vibrations main-

TABLE 7.8. Vibrational Frequencies of Fluorochlorobromo Ethane

CH_3—CFClBr	Assignment	CH_2D—CFClBr	Assignment
3017	$\nu_{as}(CH_3)$	3006	$\nu_{as}(CH_2)$
2999	$\nu_{as}(CH_3)$	2991	$\nu_s(CH_2)$
2940	$\nu_s(CH_3)$		
		2205	$\nu_s(CD)$
1437	$\delta_{as}(CH_3)$		
		1417	$\delta_{as}(CH_2)$
1382	$\delta_s(CH_3)$		
		1280	$\delta_s(CH_2)$
1150	ν(C—F, C—C)	1139	ν(C—F, C—C)
1105	$\rho(CH_3)$	960	$\rho(CH_2)$
1078	ν(C—C, C—F)	1063	ν(C—C, C—F)
915	$\rho(CH_3)$	808	$\rho(CH_2)$
715	ν(C—Cl)	710	ν(C—Cl)(DtF)[a]
		692	ν(C—Cl)(DtCl)
		676	ν(C—Cl)(DtBr)
555	ν(C—Br)	551	ν(C—Br)(DtF)
		539	ν(C—Br)(DtCl)
		519	ν(C—Br)(DtBr)
412	δ(F—C—Cl)	407	δ(F—C—Cl)
365	δ(C—C—F, C—C—Cl)	357	δ(C—C—F, C—C—Cl)
318	δ(F—C—Br)	307	δ(F—C—Br)
279	δ(C—C—Br)	275	δ(C—C—Br)
216	δ(Cl—C—Br)	211	δ(Cl—C—Br)

[a]DtF, DtCl, and DtBr denote conformations where the deuterium of the —CH_2D group is *trans* to F, Cl, and Br, respectively.

tain their symmetry properties in spite of the fact that the overall molecular symmetry is reduced to C_1.

This molecule serves as an example of how valid the group frequency approach is for a methyl group placed into an asymmetric molecule. A comparison of the vibrational frequencies of the methyl group under C_{3v} and C_1 symmetries are listed in Table 7.9.

This molecule shows interesting spectral behavior upon monodeuteriation at the —CH_3 group. In liquid CH_2D—CFClBr, the C—Cl and C—Br stretching modes at 715 and 555 cm^{-1} are split at room temperature into triplet peaks of near equal intensities due to the presence of three conformers with the deuterium atom *trans* to either F, Cl, or Br. These observed results demonstrate the conceptual limits of group frequencies: although the heavy atom stretching modes certainly deserved to be called "group frequencies," the spectral results demonstrate that even in a C—Br stretching motion, some H

TABLE 7.9. Comparison of Methyl Group Frequencies Under C_{3v} and C_s Molecular Symmetry

C_{3v}			C_1
3040	$\nu^{C-H_3}(E)$	3017	$\nu^{C-H_3}(A)$
		3000	$\nu^{C-H_3}(A)$
2965	$\nu^{C-H_3}(A_1)$	2940	$\nu^{C-H_3}(A)$

motion contributes significantly (most likely a C—C—H deformation) such that the C—Br stretching depends on the isotopic nature of the CH_3 group. Furthermore, the temperature dependence of the observed triplet structure of the C—Cl and C—Br stretching modes demonstrated that from 10 K to above room temperature, the intensity ratio of the three components remained constant, and only at elevated temperatures in the gas phase does the splitting disappear. This was interpreted as nearly equal energies of the three conformers and a barrier of internal rotation much higher than the energy associated with room temperature. These results dispel the idea of a freely rotating methyl group, as discussed above.

7.5 EXAMPLE OF AN AMINO ACID: ALANINE

To facilitate the transition from smalll molecules to larger and biological molecules, we shall present in this section a detailed analysis of the vibrational spectra of alanine, which is an asymmetric molecule with a chiral center to which locally symmetric groups are attached. When this study was published, it represented the most in-depth study of an amino acid and required five isotopomers for a detailed vibrational assignment. This study was the starting point for a similarly thorough study on small peptides, to be discussed in Chapter 8. Most previous studies either dealt with the infrared data of the solid only, omitted the low frequency range, or presented calculations based on incomplete data sets. Consequently, some of the vibrational analyses prior to the mid-1970s are totally unreliable.

Alanine is the simplest naturally occurring chiral amino acid. Its structure (**I**) converts in aqueous solution to the zwitterionic form (**II**):

$$CH_3-\underset{NH_2}{\overset{H}{\underset{|}{\overset{|}{C}}}}-CO_2H \rightarrow CH_3-\underset{NH_3^+}{\overset{H}{\underset{|}{\overset{|}{C}}}}-CO_2^-$$

(I)　　　　(II)

The hydrogen atoms on the nitrogen are labile; that is, they exchange rapidly with hydrogen atoms of the solvent. Thus when alanine is dissolved in D_2O, structure **II**, incorporating a $-ND_3^+$ group instead of a $-NH_3^+$ group, is obtained.

With 13 atoms and 33 vibrational degrees of freedom and no symmetry, alanine presents a very difficult, if not impossible, problem for a vibrational assignment unless isotopic data are available. In order to provide these data, all hydrogen atoms in alanine were exchanged for deuterium in several stages. Thus the following species were examined: alanine, alanine-N-d_3, alanine-C-d_3, alanine-C^*-d_1, and alanine-C^*-d_1-C-d_3, where N, C, and C^* denote deuteration at the amine, methyl, or methine group.

The zwitterionic character of alanine was established unequivocally via vibrational spectroscopy. In the vibrational spectra, a carboxylic acid functional group (**III**) shows distinct peaks due to the C=O and the C—O stretching modes at 1760 and 1440 cm^{-1}, in addition to the C—O—H deformation at 1253 cm^{-1}. When the acid group is deprotonated, the two C—O bonds become equivalent with a bond order of about 1.5, and the locally symmetric R—CO_2^- group of C_{2v} symmetry (**IV**)

$$-R-\overset{O}{\underset{\|}{C}}-OH \qquad R-C\begin{matrix}\diagup O \\ \diagdown O\end{matrix}^-$$

(III)   (IV)

exhibits a strong symmetric stretching motion in the Raman spectrum at about 1410 cm^{-1}, and a strong antisymmetric stretching mode in the infrared spectrum at 1610 cm^{-1}. These modes can actually be used to study the degree of dissociation of an organic acid, and the vibrational spectra of alanine indicate that this amino acid is mostly in the zwitterionic form.

Thus, in solution, one would expect to observe the vibrations of a $-NH_3^+$ or $-ND_3^+$ group as well. However, the vibrations of H_2O or D_2O mask these vibrations. Since the amine protons are labile and exchange rapidly with the solvent protons, the $-NH_3^+$ cannot be observed in D_2O, and vice versa. In the solid state, on the other hand, alanine is not zwitterionic, and the vibrations of the $-NH_2$ and $-CO_2H$ groups are observed. Normal mode calculations (*vide infra*) indicate that the antisymmetric and the symmetric $-NH_3^+$ vibrations occur around 3070 and 3020 cm^{-1}.

The carbon–hydrogen stretching vibrations present an interesting assignment problem, which only could be solved correctly using isotopic data. There are four C—H bonds in alanine, as shown in structure **V**:

$$X-\underset{Y}{\overset{H^*}{\underset{|}{\overset{|}{C}}}}-\underset{H}{\overset{H}{\underset{|}{\overset{|}{C}}}}-H$$

(V)

In alanine, the solution phase Raman spectra show strong peaks at 3003, 2993, 2949, and 2893 cm^{-1}. Upon substitution of the methine proton (marked with an asterisk in structure V) with a deuterium atom, virtually the same spectrum is observed [Diem et al., 1982] as in the totally protonated case. Thus these four vibrations are all associated with the methyl group, the lowest frequency being a Fermi resonance enhanced overtone of the methyl deformation mode, and the other three the two antisymmetric and the symmetric methyl stretching motions. Previous workers had assigned the methine stretching mode to the lowest frequency band.

The C—H stretching peak could only be identified in alanine-C-d_3, that is, the isotopomer where the methyl group in structure V was deuteriated and the methine proton was left unchanged. In this species, the C*—H stretching motion was found to be a broad, weak peak at 2962 cm^{-1}. This shows very clearly that isotopic data are essential for detailed vibrational assignments, and without them, one arrives at an incorrect conclusion that had indeed been reached in all previous studies on alanine.

Most of the remainder of the assignment is relatively straightforward and is summarized in Table 7.10. However, there are two interesting aspects that need to be emphasized. One of them is the methine deformation region, which shows, in analogy to the situation in HCFClBr, two distinct peaks at 1351 and 1301 cm^{-1}, which shift to 959 and 880 cm^{-1} upon deuteration of the methine group. Although these vibrations are typical C—H deformation group frequencies, they are influenced by deuteration of either the —CH$_3$ or the —NH$_3^+$ groups and shift between 10 and 15 cm^{-1} toward lower wavenumber.

The other interesting point about the vibrations of alanine is that the

$$\text{CH}_3-\underset{|}{\overset{\overset{\displaystyle H}{|}}{\text{C}}}-\text{NH}_3^+$$

group behaves, in terms of its vibrations, more like a

$$\text{CH}_3-\underset{|}{\overset{\overset{\displaystyle H}{|}}{\text{C}}}-\text{CH}_3$$

group: when either the methyl or amine functions are deuteriated, the entire group responds by vibrational shifts more characteristic of a locally symmetry and coupled group than noninteracting moieties.

Detailed normal coordinate analyses, based on the solution and solid phase Raman data, and the available infrared data for all isotopomers listed above were carried out. Also included in this data set were limited heavy atom isotopic data. The choice of the force field is important in a molecule of this

TABLE 7.10. Frequencies for Zwitterionic Alanine

Assigned Mode	Ala-d_0		Ala-C^*-d_1		Ala-C-d_3		Ala-C-d_4		Ala-N-d_3	
	Observed	Calculated	Observed	Calculated	Observed	Calculated	Observed	Calculated	Observed	Calculated
ν^a NH$_3^+$	3080	3068	3080	3068	3080	3068	3080	3068	2290	2263
ν^a NH$_3^+$	3060	3068	3060	3068	3060	3068	3060	3068	2230	2262
ν^a NH$_3^+$	3020	3020	3020	3018	3020	3020	3020	3018	2160	2162
$\nu^a_{CH_3}$	3003	2999	3008	2999	2251	2228	2256	2228	3003	2999
$\nu^a_{CH_3}$	2993	2998	2991	2998	2236	2226	2245	2226	2993	2998
ν_{CH}	2962	2966	2210	2170	2962	2964	2194	2170	2962	2968
$\nu^s_{CH_3}$	2949	2946	2947	2949	2126	2116	2129	2114	2949	2946
δ^a NH$_3^+$	1645	1635	1645	1634	1645	1635	1645	1634	1190	1186
δ^a NH$_3^+$	1625	1633	1625	1632	1625	1633		1632	1180	1170
ν^a CO$_2^-$	1607	1610	1607	1605	1607	1610	1607	1604	1607	1609
δ^s NH$_3^+$	1498	1507	1498	1505	1498	1507	1495	1505	1145	1135
δ^a CH$_3$	1459	1462	1456	1462	1038	1050	1055	1056	1461	1463
δ^a CH$_3$	1459	1463	1456	1461	1038	1050	1055	1053	1461	1462
ν^s CO$_2^-$	1410	1405	1407	1404	1402	1405	1401	1404	1409	1405
δ^s CH$_3$	1375	1383	1373	1371	1050	1066	1055	1070	1375	1382
δ_{CH}	1351	1349	959	975	1347	1355	947	968	1337	1345
δ_{CH}	1301	1294	880	892	1291	1301	887	893	1291	1296
$\zeta_{NH_3^+}$	1220	1191	1211	1263	1220	1187	1248	1256	874	896
$\zeta_{NH_3^+}$	1145	1155	1158	1174	1165	1147	1165	1170	840	849
ν^s_{CN}	1110	1126	1079	1144	1109	1116	1110	1132	1148	1163
$\nu_{CC(O_2)}$	1001	1017	1010	1033	941	999	1015	1017	1097	1078
ζ_{CH_3}	995	964	1000	958	820	791	823	779	1055	1045
ζ_{CH_3}	922	934	899	934	758	747	751	743	920	937
ν^a_{CCN}	850	879	823	852	921	883	880	856	812	798
$?_{CO_2^-}$	775	764	747	761		727		723	778	757
$\delta_{CO_2^-}$	640	656	635	645	610	619	610	613	613	634
$\zeta_{CO_2^-}$	527	519	520	514	509	498	520	495	513	493
$\tau_{NH_3^+}$	477	473		471	478	468	476	465	335	341
δ_{skel}	399	397		396	374	378	372	377	377	378
τ_{CH_3}	296	292		292	191	201		200	273	288
δ_{skel}	283	267		266	297	259	293	258	258	263
δ_{skel}	219	226		226	220	223		223	211	214
$\tau_{CO_2^-}$	184	192		191	184	185	187	185	184	187

size and was made in favor of a Urey–Bradley type force field, using 38 nonzero force constants. However, normal mode calculations need to be carried out very cautiously, because they tend to converge to physically nonrealistic results, to be discussed next.

In order to derive a physically meaningful force field for alanine, which reproduced the isotopic data sets, it was found necessary to refine the molecule piecewise. For example, by setting the force constants of the CH_3 and NH_3^+ groups to zero, the force field that reproduces the

$$H-C-CO_2^-$$

moiety could be refined. Physically, such a procedure "decouples" certain group vibrations from the rest of the molecule. Subsequently, the force constants of other groups are refined by decoupling them from the rest of the molecule. In such a fashion, a description of the molecule was obtained in which the normal modes are less mixed than in previous reports, and the isotopic data sets were reproduced exceedingly well. Thus it appears that this approach is preferable, and the extensive mixing of all coordinates reported in many normal mode calculations in the 800 to 1200 cm^{-1} region may be artificial.

7.6 CYCLIC MOLECULES: BENZENE

We shall conclude this chapter with a short discussion of the interpretation and computational procedures of a highly symmetric, cyclic system such as benzene. In this case, two entirely different procedures can be adopted for the normal mode calculations. One of them, published and discussed in detail by Wilson et al. [1955, Chapter 10] uses symmetry coordinates to reduce the size of the matrix to be diagonalized, whereas the other uses present day computational power to achieve the same results. The latter approach, though less elegant, has the advantage that it is more easily adapted to benzene derivatives of low symmetry.

Both approaches start by defining a set of redundant internal coordinates. These are six C—C R_{CC} and six C—H R_{CH} stretching coordinates, 12 inplane CCH deformations R_{CCH}, six in-plane CCC deformations R_{CCC}, six

$$H-C\genfrac{}{}{0pt}{}{\diagup C}{\diagdown C}$$

out-of-plane deformations R_{CCHC}, and six

CC—CC torsional coordinates R_{CCCC}, for a total of 42 internal coordinates. These internal coordinates are represented schematically in Fig. 7.6. Since there are 30 vibrational degrees of freedom for benzene, 12 of the internal coordinates are redundant. For computations using the procedures outlined in Section 3.9, the B matrix has dimensions of 42×36, which still can be handled conveniently on a personal computer. GVFF or UBFF force constants can be used to calculate the four allowed infrared and the seven allowed Raman frequencies (*vide infra*).

If symmetry coordinates are desired, they can be defined as described in Wilson et al. [1955]. First, linear combinations of adjacent CCH in-plane deformations are defined as follows:

$$\alpha_1 = -[R_{CCH}(1) + R_{CCH}(2)]$$
$$\beta_1 = \tfrac{1}{2}[R_{CCH}(1) - R_{CCH}(2)]$$

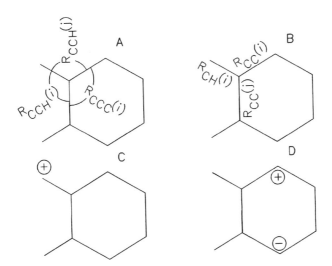

FIGURE 7.6. Definition of internal coordinates of benzene. (A) In-plane deformation coordinates: $\alpha = -(R_{CCH}(i) + R_{CCH}(j)) = R_{CCC}$ and $\beta = \tfrac{1}{2}(R_{CCH}(i) + R_{CCH}(j))$. (B) stretching coordinates. (C) Out-of-plane CCHC deformation (δ). (D) CC—CC torsion (τ).

The α's are equivalent to the R_{CCC} coordinates defined above, which therefore may be dropped, and only 36 redundant internal coordinates are left. Their totally symmetric combination will be a redundant symmetry coordinate, since not all six CCC angles can increase simultaneously.

Next, in-plane and out-of-plane symmetry coordinates are defined such that the total number of nonzero symmetry coordinates, and their symmetry species, agree with the 30 vibrational degrees of freedom of benzene. These can be shown to be, using the procedure outlined in Chapter 4:

$$2A_{1g} + A_{2g} + 4E_{2g} + 2B_{1u} + 2B_{2u} + 3E_{1u}$$

for the in-plane vibrations, and

$$2B_{2g} + E_{1g} + A_{2u} + 2E_{2u}$$

for the out-of-plane vibrations.

The in-plane symmetry coordinates are:

$S_1 = 6^{-1/2}[R_{CH}(1) + R_{CH}(2) + R_{CH}(3) + R_{CH}(4) + R_{CH}(5) + R_{CH}(6)]$ $\{A_{1g}\}$
$S_2 = 6^{-1/2}[R_{CC}(1) + R_{CC}(2) + R_{CC}(3) + R_{CC}(4) + R_{CC}(5) + R_{CC}(6)]$ $\{A_{1g}\}$
$S_3 = 6^{-1/2}[\alpha(1) + \alpha(2) + \alpha(3) + \alpha(4) + \alpha(5) + \alpha(6)]$ $\{A_{1g}\}$
$S_4 = 6^{-1/2}[\beta(1) + \beta(2) + \beta(3) + \beta(4) + \beta(5) + \beta(6)]$ $\{A_{2g}\}$
$S_5 = 12^{-1/2}[2R_{CH}(1) - R_{CH}(2) - R_{CH}(3) + 2R_{CH}(4) - R_{CH}(5) - R_{CH}(6)]$ $\{E_{2g}\}$
$S_6 = 12^{-1/2}[R_{CC}(1) - 2R_{CC}(2) + R_{CC}(3) + R_{CC}(4) - 2R_{CC}(5) + R_{CC}(6)]$ $\{E_{2g}\}$
$S_7 = 12^{-1/2}[2\alpha(1) - \alpha(2) - \alpha(3) + 2\alpha(4) - \alpha(5) - \alpha(6)]$ $\{E_{2g}\}$
$S_8 = \frac{1}{2}[\beta(1) + \beta(2) + \beta(3) + \beta(4)]$ $\{E_{2g}\}$
$S_9 = 6^{-1/2}[R_{CH}(1) - R_{CH}(2) + R_{CH}(3) - R_{CH}(4) + R_{CH}(5) - R_{CH}(6)]$ $\{B_{1u}\}$
$S_{10} = 6^{-1/2}[\alpha(1) - \alpha(2) + \alpha(3) - \alpha(4) + \alpha(5) - \alpha(6)]$ $\{B_{1u}\}$
$S_{11} = 6^{-1/2}[R_{CC}(1) - R_{CC}(2) + R_{CC}(3) - R_{CC}(4) + R_{CC}(5) - R_{CC}(6)]$ $\{B_{2u}\}$
$S_{12} = 6^{-1/2}[\beta(1) - \beta(2) + \beta(3) - \beta(4) + \beta(5) - \beta(6)]$ $\{B_{2u}\}$
$S_{13} = 12^{-1/2}[2R_{CH}(1) + R_{CH}(2) - R_{CH}(3) - 2R_{CH}(4) - R_{CH}(5) + R_{CH}(6)]$ $\{E_{1u}\}$
$S_{14} = \frac{1}{2}[R_{CC}(1) - R_{CC}(3) - R_{CC}(4) + R_{CC}(6)]$ $\{E_{1u}\}$
$S_{15} = 12^{-1/2}[2\alpha(1) + \alpha(2) - \alpha(3) - 2\alpha(4) - \alpha(5) + \alpha(6)]$ $\{E_{1u}\}$
$S_{16} = \frac{1}{2}[\beta(1) + \beta(2) - \beta(3) - \beta(4)]$ $\{E_{1u}\}$

These 16 coordinates contain three redundant degrees of freedom: S_3, as pointed out above, and one E_{1u} coordinate. The out-of-plane symmetry coordinates are defined as combinations of the six R_{CCHC} and the six CC—CC torsional coordinates R_{CCCC}, henceforth referred to as $\delta(1)$ to $\delta(6)$ and $\tau(1)$ to

$\tau(6)$, respectively. In terms of these internal coordinates, the out-of-plane symmetry coordinates are:

$$S_{17} = 6^{-1/2}[\delta(1) - \delta(2) + \delta(3) - \delta(4) + \delta(5) - \delta(6)] \quad \{B_{2g}\}$$
$$S_{18} = 6^{-1/2}[\tau(1) - \tau(2) + \tau(3) - \tau(4) + \tau(5) - \tau(6)] \quad \{B_{2g}\}$$
$$S_{19} = 12^{-1/2}[2\delta(1) + \delta(2) - \delta(3) - 2\delta(4) - \delta(5) + \delta(6)] \quad \{E_{1g}\}$$
$$S_{20} = 12^{-1/2}[\tau(1) + 2\tau(2) + \tau(3) - \tau(4) - 2\tau(5) - \tau(6)] \quad \{E_{1g}\}$$
$$S_{21} = 6^{-1/2}[\tau(1) + \tau(2) + \tau(3) + \tau(4) + \tau(5) + \tau(6)] \quad \{A_{1u}\}$$
$$S_{22} = 6^{-1/2}[\delta(1) + \delta(2) + \delta(3) + \delta(4) + \delta(5) + \delta(6)] \quad \{A_{2u}\}$$
$$S_{23} = 12^{-1/2}[2\delta(1) - \delta(2) - \delta(3) + 2\delta(4) - \delta(5) + \delta(6)] \quad \{E_{2u}\}$$
$$S_{24} = \tfrac{1}{2}[\tau(1) - \tau(3) + \tau(4) - \tau(6)] \quad \{E_{2u}\}$$

There are, again, three redundant degrees of freedom contained in these, which are one E_{1g} and the A_{1u}.

Benzene exhibits seven allowed transitions in the Raman spectrum, two A_{1g}, one E_{1g}, and four E_{2g}. Infrared allowed transitions are one A_{2u} and three E_{1u} modes. These are listed in Table 7.11, along with the symmetry coordinates responsible for the modes. Thus inspection of the symmetry coordinates yields a reasonable picture of the form of the normal modes. The details of the assignment may be found in Wilson et al. [1955, Chapter 10]. Data for C_6D_6 and extensive use of the Teller–Redlich product rule were necessary for the original assignment.

TABLE 7.11. Vibrational Frequencies [cm^{-1}], Symmetry, and Corresponding Symmetry Coordinates of Normal Vibrations of Benzene

Observed Frequency		Symmetry Species	Internal Coordinate	Activity
C_6H_6	C_6D_6			
606	577	E_{2g}	S_7	Ra
671	503	A_{2u}	S_{22}	IR
849	661	E_{1g}	S_{19}	Ra
992	945	A_1	S_2	Ra
1037	813	E_{1u}	S_{24}	IR
1178	876	E_{2g}	S_8	Ra
1485	1333	E_{1u}	$S_{14} + S_{15}$	IR
1595[a]	1559	E_{2g}	S_6	Ra
3047	2264	E_{2g}	S_5	Ra
3061	2292	A_1	S_1	Ra
3080	2292	E_{1u}	S_{13}	IR

[a]In Fermi resonance with vibrational combination band $v(992 + 606)$.

7.7 OUTLOOK

At the writing of this book, it is safe to state that the vibrational spectra of any known, stable or unstable, small molecule (<10 atoms) have been examined and reexamined, and even for molecules of up to 25 atoms, good vibrational data are available. In general, as the symmetry of molecules decreases, both the quality and quantity of vibrational data decrease enormously. Some of the examples discussed above show this trend very clearly: whereas there are hundreds of papers on methane, methylene chloride, and other highly symmetric species, there were less than half a dozen papers on HCFClBr.

As the trend in modern vibrational spectroscopy is toward larger molecules, such as models for biochemical compounds or even these compounds themselves, the strategies of obtaining vibrational information get more difficult. Some groups have opted for the same (albeit slow) approach the author has pursued: utilizing detailed isotopic data for a vibrational assignment, followed by thorough normal mode calculations. Enormous efforts were required to elucidate, for example, the molecular vibrations of retinal, the prosthetic group (chromophore) of the visual pigment rhodopsin. In order to properly assign the *cis* and the *trans* isomers of this conjugated polyene, dozens of isotopomers had to be synthesized and analyzed. The results obtained by the research group of R. Mathies [1980], and to be discussed in Chapter 8, showed very clearly that many previous studies were incorrect, and that results obtained on this complex system could only be explained after a detailed vibrational study of the isotopomers. It follows that a thorough chemical approach, namely, *experimental* efforts that include isotopic substitutions, is more valuable than a number of *computational* studies based on incomplete data, which cannot stand up to detailed scrutiny.

At the other extreme of this discussion are the efforts to describe the molecular vibrations of large organic molecules and biological macromolecules by vibrational techniques, using group frequencies and qualitative correlations for the interpretation. These efforts are certainly worthwhile but have to be viewed as strictly a qualitative tool of structural chemistry. Attempts to corroborate vibrational interpretations on such large systems by "normal mode calculations" are usually so indeterminate that the calculations do not corroborate anything, since the state-of-the-art of these computations is still at a stage where the computational results are unreliable.

REFERENCES

M. Diem and D. F. Burow, *J. Chem. Phys.*, *64*, 5179 (1976); *J. Phys. Chem.*, *81*, 476 (1976).

M. Diem, P. L. Polavarapu, M. Oboodi, and L. A. Nafie, *J. Am. Chem. Soc.*, *104*, 3329 (1982).

J. R. Durig and J. F. Sullivan, in *Chemical, Biological and Industrial Applications of Infrared Spectroscopy*, J. R. Durig, Ed., Wiley, New York, 1985.

G. Herzberg, *Molecular Spectra and Molecular Structure. II. Infrared and Raman Spectra of Polyatomic Molecules*, Van Nostrand Reinhold, New York, 1945.

S. T. King and J. Overend, *Spectrochim. Acta*, 22, 689 (1966); *23A*, 61 (1967).

R. Mathies, G. Eyring, B. Curry, A. Bloek, I. Palings, R. Fransen, and J. Lugtenburg, in *Proceedings of the Seventh International Conference on Raman Spectroscopy*, North-Holland, Amsterdam, 1980, p 546.

L. A. Nafie, P. L. Polavarapu, and M. Diem, *J. Chem. Phys.*, 73, 3530 (1980).

E. B. Wilson, J. C. Decius, and P. C. Cross, *Molecular Vibrations: The Theory of Infrared and Raman Vibrational Spectra*, McGraw-Hill, New York, 1955.

8

BIOPHYSICAL APPLICATIONS OF VIBRATIONAL SPECTROSCOPY

Vibrational spectroscopy has become one of the common qualitative techniques to determine the structure and dynamics of biological molecules and to monitor their conformational changes. In particular, Raman spectroscopy has become an increasingly popular technique to study biological molecules for a variety of reasons. One of these is the fact that water is an excellent solvent for Raman spectroscopy, because of its low light scattering cross section. Water is the preferred solvent for studying biological molecules, which often are insoluble in all other solvents, and whose function often depends on the presence of water as a solvent. Furthermore, the time resolution and selectivity of the resonance Raman technique have allowed hundreds of studies to be carried out which would have been impossible with any other technique.

In this chapter, we shall explore a number of different applications of vibrational spectroscopy to biomolecules. First, the vibrational features of the peptide moiety will be introduced and the conformational information obtainable from the peptide backbone vibrations will be discussed. These studies have been carried out in nonresonance Raman as well as in infrared spectroscopy. Subsequently, resonance Raman results on protein prosthetic groups and reactive centers, including the visual pigments, will be presented, which have been used to follow the dynamics and mechanism at the site of enzymatic activity.

Vibrational data have also been used to investigate DNA conformation and the packing of nucleic acids in viruses. Again, Raman studies have been particularly successful, allowing various regions (of a virus) to be probed with the very finely focused laser beam and with different excitation wavelengths. Raman and infrared spectroscopic results have also been used to probe local order and structures in lipids and membranes. The discussion will take quite different forms for these classes of biomolecules, since the data available are quite different.

This is the first general text on vibrational spectroscopy that devotes a chapter to the biophysical applications of vibrational spectroscopy. This is due to the author's own inclinations and research interests, which are in the application of vibrational optical activity (infrared circular dichroism and Raman optical activity, cf. Chapter 9) to biological systems. Although the outlook of this book has been, so far, relatively unbiased and has dealt with theoretical principles and the interpretation of vibrational data for small molecules, the author feels that the future of vibrational spectroscopy belongs to applications to biological systems. This field presents entirely different challenges but also enormous rewards in terms of the amount of information that is available from vibrational techniques.

Previous texts in vibrational spectroscopy did not devote much space to the subject of biophysical applications, for a good reason: one has to remember that this field has developed only over the last two decades, and many measurements and methods described (such as infrared spectroscopy in aqueous solution or time-resolved spectra in the picosecond regime) were still unheard of three decades ago.

We shall start this chapter with a discussion of the infrared and nonresonant Raman spectroscopy of peptides, which follows closely the discussion on alanine presented in Chapter 7.

8.1 PEPTIDE VIBRATIONAL SPECTROSCOPY

Peptides and proteins are key biological molecules, which are responsible for a wide variety of biological functions, such as enzymatic activity, transport of reactants and products, and repair and replication of DNA. The distinction between peptides and proteins is one based mainly on the size of the molecules: peptides are the smaller versions of proteins, but both consist of amino acids linked together by amide linkages. The amino acids, in general, are molecules with the following structures:

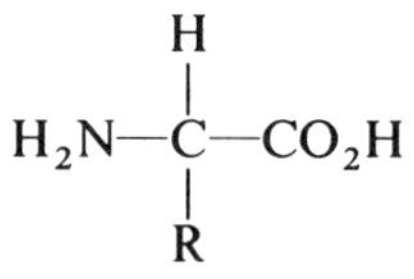

$$H_2N-\underset{\underset{R}{|}}{\overset{\overset{H}{|}}{C}}-CO_2H$$

(I)

where R is a side group, for example, $-H$, $-CH_3$, $-CH_2OH$, $-(CH)_2CO_2H$, or $-(CH)_4NH_2$. There are 20 naturally occurring amino acids, each one with a distinct side group, which can be hydrophobic, hydrophilic, or even ionic. In peptides and proteins, the sequence and spatial arrangement of amino acids can create active sites, or pockets, in the protein structure where a given substrate may bind and undergo a chemical reaction.

The amino acids in a peptide or protein are linked together according to the following scheme:

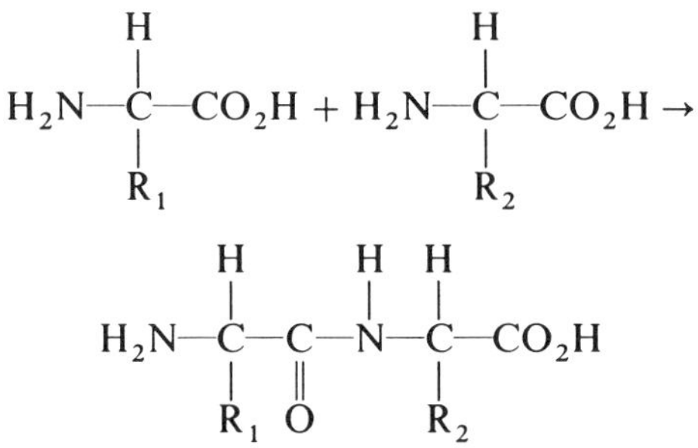

$$H_2N-\underset{R_1}{\overset{H}{C}}-CO_2H + H_2N-\underset{R_2}{\overset{H}{C}}-CO_2H \rightarrow$$

$$H_2N-\underset{R_1}{\overset{H}{C}}-\underset{O}{\overset{\|}{C}}-\overset{H}{N}-\underset{R_2}{\overset{H}{C}}-CO_2H$$

(II)

Compound (II) above consists of two amino acids and would be referred to as a dipeptide. The amino acid residues are connected by a linkage (III)

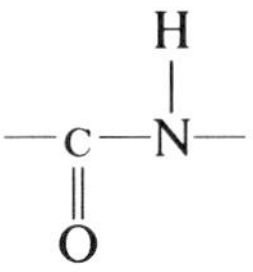

$$-\underset{O}{\overset{\|}{C}}-\overset{H}{N}-$$

(III)

which is known as the amide or peptide moiety. In a peptide or protein, hundreds or thousands of amino acids are linked together according to a sequence that is specific and different for every protein. This sequence of amino acids in a naturally occurring peptide or protein is determined by a DNA molecule (*vide infra*) via the genetic code and is referred to as the protein *primary structure*.

However, the protein function is determined by the three-dimensional shape the amino acid chain assumes via a folding process. The folding of the peptide, from its fully extended and biologically inactive from to the fully active and folded form, is one of the most miraculous and fascinating, but least understood, processes in modern biophysics. In the folded form, there are specific and characteristic motifs that are repeated in different proteins. These are, for example, structures where the peptide is wrapped into a tight, hydrogen-bonded, right-handed helical shape, known as the α-helix. Other shapes, such as turns, may be stabilized via covalent linkages along the peptide chain. Interactions of charged side chain residues can stabilize peptide structures, as can the formation of hydrophobic cores or hydrophilic interactions with the solvent. The resulting folded structures are referred to as the *secondary structures* or conformations of peptides. Well-known secondary structures include the α-helical, coil, β-pleated sheet, and β-turn, and random structures, some of which are shown in Fig. 8.1. Many of these secondary structures have been identified and characterized by other spectroscopic and solid phase structural methods, such as single-crystal x-ray diffraction.

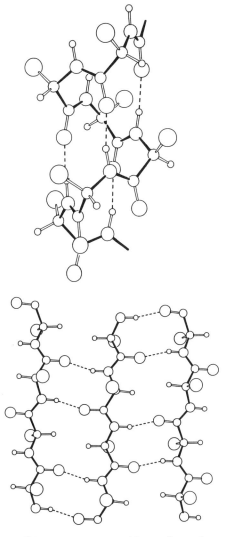

FIGURE 8.1. Structures of two common peptide conformations: α-helical (top) and β-pleated sheet (bottom). (Top: Adapted from Linus Pauling, *The Nature of the Chemical Bond, Third Edition.* Copyright © 1960 by Cornell University. Used by permission of the publisher, Cornell University Press. Bottom: From Alan Fersht, *Enzyme Structure and Mechanism,* Copyright © 1977, W. H. Freeman and Company).

When attempting to collect and analyze vibrational data of molecules with molecular masses on the order of 100,000, and about 10,000 atoms (with some 30,000 degrees of vibrational freedom!), it is clear that different procedures than the ones discussed in the first seven chapters of this book need to be developed. The data collection itself is a difficult task, and only with lasers, CCDs, and interferometric methods can the experimental problems be referred to as "routine." For the interpretation, some model compounds needed to be studied first, and one logical choice was the investigation of the vibrations of the peptide linkage (III) that connects consecutive amino acid residues. It was found that the vibrations of this linkage can be used as a qualitative indicator of the conformation of the peptide, and thus they have been studied in detail for simple model systems.

8.1.1 Vibrations of the Peptide Linkage

Efforts to understand the vibrations of peptides date back over three decades, when a simple model for the peptide linkage, N-methylacetamide (NMAA) (IV), was first investigated in detail.

$$CH_3-\underset{\underset{O}{\|}}{C}-\underset{\overset{H}{|}}{N}-CH_3$$

(IV)

The vibrational analysis of this molecule revealed some characteristic group

TABLE 8.1. Observed Frequencies [cm^{-1}] and Vibrational Assignment[a] for the Amide Vibrations

Name	Frequency	Approximate Description
Amide A	3250–3300	N—H stretch
Amide I	1630–1700	C=O stretch (C—N, C—N—H)
Amide II	1510–1570	N—H deformation (C—N)
Amide III	1230–1330	N—H/C—H deformation[b]
Amide IV	630–750	O=C—N deformation
Amide V	700–750	N—H out-of-plane deformation
Amide VI	~600	C=O out-of-plane deformation

[a]Minor contributions in parentheses.
[b]Note the different assignment given for the Amide III vibration here, as compared to the previous assignment of a N—H deformation/C—N stretch.

frequencies for the peptide linkage (cf. Table 8.1), which were analyzed in detail in the early normal mode calculations by Miyazawa et al. [1958].

In the early 1970s, R. Lord's group [Lord and Yu, 1970] collected Raman data on a protein, lysozyme, in aqueous solution, and on a mixture of amino acids with the same composition as lysozyme. The two spectra agreed reasonably well except at the frequencies where the normal vibrations of the amide moiety were known to occur. Thus it became clear that the amide vibrations could be observed even in a large protein.

The characteristic vibrations of the peptide group are now reasonably well understood and are referred to as the amide I–VII and amide A and B vibrations. Their frequencies and original assignments are listed in Table 8.1. Later, it was found [Thomas, 1975; Lord, 1977] that the frequencies of some of the peptide vibrations vary for different peptide conformations, and that these frequency variations can be correlated with secondary structure, such as β-sheets and α-helices. In the following paragraphs, the amide vibrations more commonly used in conformational studies will be introduced.

The amide A vibration at approximately $3235\,cm^{-1}$ is nearly exclusively a N—H stretching motion. It is strong in infrared absorption, and its frequency and intensity depend strongly on the hydrogen bonding of the proton and, to some degree, on the secondary structure. However, for conformational analyses, this band is less useful than others for two reasons. First, it is useful in nonaqueous solutions only, since the water peak masks the frequency region where the amide A vibration occurs. Second, the amide A band occurs as a Fermi resonance doublet with a vibration usually referred to as the amide B vibration. The other member of this doublet appears to be an overtone of the amide I vibration. Thus observation and interpretation of the amide A mode are difficult.

For the remainder of this discussion, the amide I–III vibrations of the peptide linkage are of most interest, because they have been used most extensively and are best understood. In NMAA, the amide I vibration consists mainly of the C=O stretching vibration and is observed at $1653\,cm^{-1}$. In peptides and proteins, this vibration exhibits strong infrared absorptions and medium Raman intensities. Its frequency is very sensitive to the peptide secondary structure and varies by more than $50\,cm^{-1}$ between the well-established secondary structures. Models to reproduce the large frequency shifts observed between the different secondary structures, and the large splitting between identical amide I vibrations in well-ordered polymers, will be discussed in Section 8.1.2.

Upon deuteration of the N—H hydrogen in NMAA, the amide I mode shifts to $1642\,cm^{-1}$ and is referred to as the amide I' vibration. This shift of the carbon–oxygen stretching mode upon deuteration of a neighboring N—H deformation coordinate indicates a significant contribution of this deformation. This mixing results in a notable shift of the dipole moment derivative away from the direction of the carbon–oxygen bond (*vide infra*). Good empirical force fields are now available, which properly predict the mixing of the C=O and N—H stretching coordinates.

In infrared absorption, the amide I vibration can be observed only in nonaqueous media, since water has a very broad and intense absorption (the v_2 mode) at about $1620 \, \text{cm}^{-1}$. When D_2O is used for a solvent, the labile N—H proton exchanges, and the amide I' vibration is observed instead of the amide I mode. However, this amide I' peak is easily accessible in D_2O, since the corresponding v_2 mode occurs at much lower frequency ($\sim 1210 \, \text{cm}^{-1}$ in liquid D_2O, cf. Table 7.1). Both amide I and I' vibrations are accessible via the Raman effect in both H_2O and D_2O, since water peaks are weak in Raman scattering.

The rate of the hydrogen–deuterium exchange depends on the solvent accessibility of the amide proton, and its half-life can range from milliseconds in small open-chained peptides to more than days for folded globular proteins. In fact, the intensity dependence of some of the amide hydrogen vibrational modes with time can be used as a kinetic probe.

The amide II vibration is observed at $1567 \, \text{cm}^{-1}$ and shifts to $1472 \, \text{cm}^{-1}$ upon deuteriation. This band was found [for a review, see Bandekar, 1991] to be largely due to the N—H in-plane deformation, with a much smaller contribution from the C—N stretching coordinate. It exhibits some conformational sensitivity but has not been used as extensively for conformational studies as have other amide modes.

The amide III vibration at about $1300 \, \text{cm}^{-1}$ was assigned by Miyazawa et al. [1958] in N-methylacetamide to consist of the N—H deformation and, to a lesser extent, of the stretching of the C—N bond. This assignment has been confirmed for this particular molecule by normal coordinate calculaltions by a number of research groups. However, it appears now that this mixing pattern found in the amide III region is typical only for the model compound itself, namely, N-methylacetamide. Real peptides incorporate methine hydrogens on the adjacent α-carbon (see structure V) and have methine deformation vibrations that are nearly degenerate with the (unperturbed) N—H deformation. These methine vibrations were discussed in detail for alanine in Section 7.5.

The interactions between the methine and the N—H deformations were discovered originally in a detailed vibrational study of deuteriated alanyl peptides [Oboodi et al., 1984]. A band at about $1270 \, \text{cm}^{-1}$, which previously was assigned to the "amide III" vibration, was found to disappear when the amide proton was substituted by a deuterium atom in L-Ala–L-Ala (V).

$$NH_3^+ - \underset{\underset{CH_3}{|}}{\overset{\overset{H}{|}}{C}} - \underset{\underset{O}{\parallel}}{\overset{}{C}} - \underset{}{\overset{\overset{H}{|}}{N}} - \underset{\underset{CH_3}{|}}{\overset{\overset{\overset{*}{H}}{|}}{C}} - CO_2^-$$

(V)

In terms of the hydrogen/deuterium isotopic shifts discussed in earlier chapters, this behavior is not unexpected. However, Oboodi et al. reported that the peak at 1270 cm^{-1} also disappears when the methine hydrogen marked with an asterisk in structure (V) is substituted by a deuterium. This observation suggests that both the N—H and the adjacent C—H deformation contribute to the 1270 cm^{-1} peak.

The exact nature of the amide III vibration was established via a detailed analysis of L-Ala–L-Ala and five selectively deuteriated isotopomers, the vibrational spectra of which are summarized in Table 8.2. The spectrum of L-Ala-d_1–L-Ala-d_1 (VI)

$$\text{NH}_3^+ - \underset{\underset{\text{CH}_3}{|}}{\overset{\overset{\text{D}}{|}}{\text{C}}} - \underset{\underset{\text{O}}{\|}}{\text{C}} - \text{N} - \underset{\underset{\text{CH}_3}{|}}{\overset{\overset{\text{D}}{|}}{\text{C}}} - \text{CO}_2^-$$

(VI)

exhibits only one broad band in the 1200–1350 cm^{-1} spectral region. This band, at 1336 cm^{-1}, disappears when the N—H group is deuteriated. Thus it can be assigned unequivocally to the N—H deformation motion. Since there is no other vibrational transition nearby, we refer to this band as the unperturbed amide III mode. Normal coordinate calculations (*vide infra*) confirm that this mode is composed mainly of the N—H in-plane deformation motion.

In L-Ala-d_1–L-Ala in D$_2$O (VII)

$$\text{ND}_3^+ - \underset{\underset{\text{CH}_3}{|}}{\overset{\overset{\text{D}}{|}}{\text{C}}} - \underset{\underset{\text{O}}{\|}}{\text{C}} - \text{N} - \underset{\underset{\text{CH}_3}{|}}{\overset{\overset{\text{H}}{|}}{\text{C}}} - \text{CO}_2^-$$

(VII)

the two methine deformation modes of the C-terminal alanyl residue are observed at 1279 and 1330 cm^{-1}. These frequencies are similar to those found for alanine itself. The character of these vibrations, namely, two orthogonal C—H vibrations, was established via normal mode analysis. These vibrations will be henceforth referred to as the C_C—H^I and the C_C—H^{II} modes, respectively, where the subscript C refers to the C-terminal alanyl residue.

If peptide (VII) is dissolved in H$_2$O, instead of D$_2$O, it will exchange the amide as well as the amine protons, and species (VIII) is obtained:

$$\text{NH}_3^+ - \underset{\underset{\text{CH}_3}{|}}{\overset{\overset{\text{D}}{|}}{\text{C}}} - \underset{\underset{\text{O}}{\|}}{\text{C}} - \text{N} - \underset{\underset{\text{CH}_3}{|}}{\overset{\overset{\text{H}}{|}}{\text{C}}} - \text{CO}_2^-$$

(VIII)

TABLE 8.2. Observed and Calculated Frequencies [cm^{-1}] for L-Ala–L-Ala and Various Isotopomers

Mode Description	Ala-d_1–Ala-d_1 D$_2$O Observed	Calculated	Ala-d_1–Ala-d_1 H$_2$O (VI) Observed	Calculated	Ala-d_1–Ala D$_2$O (VII) Observed	Calculated	Ala-d_1–Ala H$_2$O (VIII) Observed	Calculated	Ala-Ala D$_2$O (V) Observed	Calculated	Ala-Ala H$_2$O Observed	Calculated
Amide I	1663.0	1678.4	1680.0	1682.4	1664.0	1678.4	1680.0	1682.5	1665.0	1681.5	1680.0	1685.6
CO$_2^-$ asymmetric stretch	1584.0	1585.8	1584.0	1591.1	1590.0	1588.3	1584.0	1592.5	1592.0	1588.3	1584.0	1592.7
Amide II	1478.4	1450.3	1570.0	1576.0	1479.0	1451.0	1570.0	1578.1	1483.0	1455.4	1570.0	1578.4
CO$_2^-$ symmetric stretch	1406.0	1410.7	1406.0	1411.1	1408.0	1412.4	1407.0	1412.9	1407.0	1412.4	1407.0	1413.0
Amide III	—	—	1336.0	1314.9	—	—	1346.0	1333.5	—	—	1345.0[a]	1355.1
Amide III	—	—	—	—	—	—	1311.0	1312.2	—	—	1325.0[a]	1330.7
Amide III	—	—	—	—	—	—	—	—	—	—	1281.0[a]	1307.8
C$_N$—H deformation	—	—	—	—	—	—	—	—	1355.0	1343.2	—	—
C$_N$—H deformation	—	—	—	—	—	—	—	—	1305.0	1284.0	1302.0[a]	1284.0
C$_C$—H deformation	—	—	—	—	1330.0	1329.8	—	—	1329.0	1329.7	—	—
C$_C$—H deformation	—	—	—	—	1279.0	1278.4	1274.0	1273.8	1276.0	1278.3	1266.0[a]	1273.3

[a] Frequencies found via a band decomposition [cf. Oboodi et al., 1984].

The Raman spectra [Oboodi et al., 1984] and infrared data [Diem et al., 1992] of this molecule in the 1250–1350 cm^{-1} region show three vibrations at 1274, 1311, and 1346 cm^{-1}. The lowest one of them agrees in frequency with the C_C—H^I mode in (VII) and is assigned accordingly. The other two peaks are observed at 1346 and 1311 cm^{-1}. Neither of these frequencies agrees with the frequencies of the N—H deformation (1336 cm^{-1}) and the C_C—H^{II} (1330 cm^{-1}) modes, and one concludes that the two observed vibrations result from an interaction of the N—H and the C_C—H^{II} deformation coordinates. Like Fermi resonance (which takes place between a fundamental and an overtone), this interaction splits the vibrational states apart and mixes their wavefunctions. The two resulting coupled vibrations appear as a polarized (1311 cm^{-1}) and a depolarized (1346 cm^{-1}) peak in the Raman spectra. The polarization properties of these bands suggest that the two deformation modes combine to produce a symmetric and an antisymmetric combination state.

In L-Ala–L-Ala in D$_2$O (IX),

$$ND_3^+ - \underset{\underset{CH_3}{|}}{\overset{\overset{H}{|}}{C}} - \underset{\underset{O}{\|}}{C} - \underset{\overset{D}{|}}{N} - \underset{\underset{CH_3}{|}}{\overset{\overset{H}{|}}{C}} - CO_2^-$$

(IX)

the four C—H deformation modes are observed at 1276, 1305, 1329, and 1355 cm^{-1}, which were assigned to the C—H deformation of the C-terminal alanine 1330 (C_C—H^{II}) and 1279 cm^{-1} (C_C—H^I), and of the N-terminal alanyl residue 1355 (C_N—H^{II}) and 1305 cm^{-1} (C_N—H^I). The presence of the N—D group, with a vibrational frequency that differs greatly from those of the C—H deformation vibrations, prevents any coupling, and the four C—H deformations are observed virtually unperturbed.

For L-Ala–L-Ala in H$_2$O (V), that is, the entirely undeuterated species, the spectrum in this region is drastically different. This region was assigned to consist of two nearly uncoupled modes, the C_C—H^I deformation at 1266 cm^{-1} and the C_N—H^I deformation at 1302 cm^{-1}, and three highly coupled modes at 1281, 1325, and 1340 cm^{-1}. The latter three peaks all should be referred to as the amide III bands, since they contain various proportions of the N—H, the C_C—H^{II}, and the C_N—H^{II} deformation coordinates. Accordingly, they have been designated the AmIII^1, AmIII^2, and AmIII^3 modes.

The AmIII^1 mode at 1281 cm^{-1} is the most delocalized vibration and has contributions from the N—H in-plane deformation, the C—N stretch, and the two methine deformations. AmIII^3 also has a large contribution from the N—H in-plane deformation, coupled mostly to the C_N—H^{II} motion. AmIII^2 consists mostly of the C_C—H^{II} deformation, coupled to the C_N—H^{II} deformation with a negligible contribution from the N—H deformation. The displacement vectors of these three modes are displayed in Fig. 8.2. They depict the

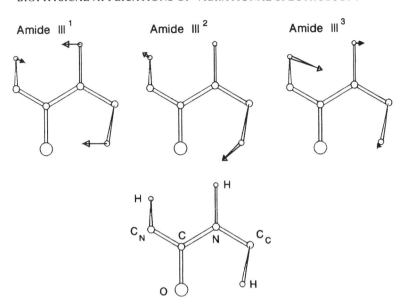

FIGURE 8.2. Atomic displacement vectors of amide III vibrations in L-alanyl–L-alanine. C_C, C-terminal alanine; C_N, N-terminal alanine.

magnitude of all atomic displacement vectors relative to each other; however, in all vibrational computations, the overall amplitude of the displacements is unknown. For normal modes of vibration, for which only one major displacement coordinate is shown, the phase of this coordinate is indeterminate as well; that is, the motion of the atom may be drawn as an increasing or decreasing internal coordinate. A complete listing of the displacement vectors has been published [Roberts, 1991].

Thus the difference between the model system, N-methylacetamide, and real peptides in the amide III vibration is the ability of the N—H vibration to couple with C—H deformations in the latter samples. In the former, there are no methine hydrogens, but only methyl groups. The closest group frequencies of the methyl groups would be the symmetric deformation mode at approximately 1375 cm^{-1}, which is too far removed from the N—H deformation for coupling to occur. Thus the C—N stretching coordinate contributes to the amide III vibration in N-methylacetamide.

This contention was further corroborated by an inspection of the displacement vectors in L-Ala-d_1–L-Ala-d_1 in H$_2$O (**VI**). This molecule resembles N-methylacetamide in that there is no methine hydrogen adjacent to the N—H group. In this molecule, the amide III vibration, indeed, consists mostly of the N—H in-plane deformation and some contribution from the C—N stretching coordinate. This is in agreement with the original findings of Miyazawa et al. [1958] for N-methylacetamide: with both methine hydrogens exchanged for deuterium, the surroundings of the peptide linkage preclude any coupling between N—H and C—H deformation coordinates. Thus the vibration is, indeed, best described as a mixture of the N—H deformation and the C—N stretching motion.

The normal mode calculations of the alanyl dipeptides referred to above yielded a detailed view of the coupling patterns and the displacement vectors of these model peptides. However, as described in Chapter 7 in the discussion of alanine, the degree of vibrational coupling is one aspect where normal coordinate calculations can be entirely misleading. Often, a good frequency fit is employed as the criterion for the quality of a force field, but calculations based on the frequency fit alone may not represent the vibrational motion at all. Thus detailed isotopic data need to be analyzed for the exact behavior of the vibrations under consideration, and subsequently, force constants need to be perturbed to reproduce *both* frequencies and the character of a vibration. Thus all six deuteriated isotopomers referred to above were included in the normal mode calculations. Calculated and observed vibrational frequencies for L-alanyl-L-alanine and five isotopomers in the amide I to amide III regions are tabulated in Table 8.2 to show the extent of agreement between observed and calculated frequencies.

The description of the amide III vibration presented here is able to account for the large frequency shifts the amide III vibration experiences when the conformation, or secondary structure, of the peptide is varied (cf. Section 8.1.2). A similar mixing of various deformation coordinates had been established before the above studies by Krimm's research group [Moore and Krimm, 1976; Bandekar and Krimm, 1988; Krimm, 1983]. Their work is based on detailed normal coordinate analyses of many different peptides. The calculated potential energy distribution in these studies often contains vibrations in the 1250–1350 cm^{-1} region, which are mixtures of N—H and C—H deformation coordinates. However, the experimental and computational results on the alanine dipeptides discussed above exactly describe the displacement vectors in the amide III region for molecules sufficiently simple to be treated by classical vibrational calculations.

The amide IV, V, and VI vibrations, observed between 600 and 730 cm^{-1}, are mostly bending and torsional modes of the C—O and C—N groups and are heavily mixed. Finally, the amide VII vibration is a low frequency (205 cm^{-1}) out-of-plane bending and torsional motion of the peptide group. The amide IV, V, and VI vibrations are used less frequently for conformational analyses due to apparent mixing with other skeletal modes, and the ambiguity in identifying these vibrations in very large molecules.

8.1.2 The Conformational Sensitivity of the Amide I and III Modes

The amide I spectral region has been used by biophysical chemists as a qualitative indicator of a peptide's solution conformation. In the past, these studies were performed in aqueous solutions (D_2O) using the amide I' mode and in nonaqueous solution using both the amide I and amide I' vibrations. Recent advances in Fourier transform infrared spectroscopy have made it possible to collect the amide I vibrations of a few very soluble peptides from a solution in H_2O.

Most of the theoretical efforts to interpret the conformational sensitivity of the amide I (or I') vibration have come from the laboratory of Krimm and co-workers [e.g., see Krimm, 1983]. Incidentally, the theoretical formalism employed in this work is nearly the same as that used by researchers in vibrational optical activity for the calculations of infrared CD intensities, to be discussed in Chapter 9. Compared to the complexity of the amide III vibrational region discussed above, the amide I mode itself is described in a relatively straightforward manner. It consists mostly of the C=O stretching coordinate, with further contributions from the C—N stretching and the N—H deformation coordinates. Its transition dipole derivative is about 0.25 debye, and the transition moment is tilted from the C=O bond direction away from the nitrogen atom by about 20°. The carbonyl group is involved in hydrogen bonding in most secondary structures of peptides. Therefore the carbonyl groups have well-defined geometries in most of the common secondary structural motifs of peptides.

The relatively strong dipole transition moment and the well-defined geometries of the carbonyl groups are responsible for strong dipolar coupling interactions among the C=O groups. This coupling manifests itself in a number of ways completely analogous to the effects of Fermi resonance: the vibrational wavefunctions of all (degenerate or near degenerate) carbonyl groups mix and form delocalized excited states. (The equivalent situation in electronic absorption spectroscopy is referred to as exciton states.)

For n interacting carbonyl transitions, n coupled vibrational excited states will be created. The splitting between these coupled states is given by the eigenvalues of the dipolar interaction energy matrix:

$$V_{ij} = \frac{\mu_i \cdot \mu_j}{|T_{ij}|^3} - \frac{3(\mu_i \cdot T_{ij})(\mu_j \cdot T_{ij})}{|T_{ij}|^5} \qquad (8.1.1)$$

where T_{ij} is the distance vector between dipole μ_i and μ_j.

Depending on the geometry of the interacting groups and the coupling energies, any one or more of the n coupled excited states can become the predominant term in the absorption process. The net effect is that, depending on the secondary structure, quite different absorption wavelengths are observed. The influence of the dipolar coupling on the vibrational frequencies of the amide I vibration can be evaluated in two ways. One of them, utilized by this author, is a straightforward application of Eq. 8.1.1 and will be discussed further in Chapter 9.

The other method to incorporate the dipole–dipole interaction into vibrational computations was pioneered by Krimm and co-workers [Moore and Krimm, 1976]. In their work, the dipolar interactions between selected dipole moments are treated as a set of additional force constants to modify the vibrational force field and, consequently, the frequencies and intensities. The dipolar interactions are required in this model to reproduce the splitting observed between the amide I vibrational modes; in the absence of the transition coupling, the splitting would amount to only a few wavenumbers. The residual splitting in the absence of dipolar interactions is due to what is referred to as kinetic energy coupling, which occurs even in mechanical systems such as coupled pendula or identical masses connected by identical springs [cf. Wilson et al., 1955, Section 2.7]. The transition dipole coupling increases the frequency splitting of the magnitudes of vibrational shifts typically observed between various secondary structures.

Since the coupling mixes the vibrational wavefunctions of the participating monomeric wavefunctions in a manner discussed in Section 3.4 for Fermi resonance, the transition intensities are affected as well. In particular, inphase and out-of-phase states appear due to the coupling, and the intensity distribution between these states depends on the geometry. Thus it may appear that the amide I bands shift in frequency between one and another secondary structures. A more accurate description is that the interacting transitions are split by $\pm 30\,\mathrm{cm}^{-1}$ in all secondary structures, and that various components of the coupled states carry most of the observed intensity. Thus it appears that the transition frequency for α-helical peptides ($\sim 1655\,\mathrm{cm}^{-1}$) is lowered to about $1630\,\mathrm{cm}^{-1}$ for β-pleated sheet structures. The arguments presented here were originally put forth over two decades ago by Krim and co-workers and have been confirmed independently in the course of infrared CD intensity calculations, to be discussed in Chapter 9.

Moore and Krimm [1976], Krimm [1983], and Krimm and Bandekar [1980] have described the methodology for the computational efforts to reproduce amide I vibrational frequencies for various peptide secondary structures. A detailed scheme of interactions between the amide linkage and the surrounding ones was introduced by these authors. This coupling occurs within a peptide strand for α-helices and β-turns but may extend over neighboring strands in sheet structures. The reader is referred to the original literature mentioned above for computational details.

For polyamino acids and peptides in well-established secondary structures, one observes specific amide I frequencies, which are reproduced in Table 8.3

TABLE 8.3. Frequencies [cm^{-1}] of the Amide I and Amide III Vibrations for Different Secondary Structures

Structure	Amide I	Amide III
α-Helix	1645–1660	1265–1300
β-Sheet	1665–1680	1230–1240
β-Turns	1640–1690	1290–1330
Unordered	1660–1670	1240–1260

for a number of peptide conformations. The values given in this table have to be used with care, since vibrational frequencies are always dependent on the exact molecular structure. Thus side group effects, solid versus solution phase, and solvent interactions will all affect the vibrational frequencies of the amide I vibration, as will, of course, the secondary structure. In addition, in proteins and naturally occurring peptides, mixtures of various secondary structures coexist: often, helical segments are connected by "unordered" peptide chains, or segments in β-pleated sheets are connected by β-turn and other turn structures. In such situations, the observed absorption spectra are broad superpositions of the various specific amide I vibrations. In fact, typical proteins may exhibit just one broad absorption with a halfwidth of over $50\,cm^{-1}$ and no immediately perceivable structure on it. Methods have been developed over the past few years to deconvolute such a broad band into composite bands due to the different structural motifs in the peptide or protein. Such a deconvolution of a broad spectral peak into bands due to the substructures is a somewhat ambiguous and indeterminate task and needs to be performed very carefully. Similar deconvolution problems exist, for example, in circular dichroism (CD) spectroscopy.

Biochemists often are interested in the percent composition of a peptide or protein in terms of α-helical, β-sheets, turns, and unordered structures; moreover, the changes in these structural compositions upon chemical reactions, denaturation, site-specific mutagenesis, and so on are very important and sometimes can even be correlated to loss of biological activity. Thus the determination of the percent composition in a protein really is a significant problem in spectroscopy.

The strategy in the deconvolution process is to artificially reduce the halfwidth of the component bands. Several procedures to deconvolute such broad peaks into narrower peaks, or groups of peaks, have been utilized. Most commonly used are a second derivative technique and Fourier self-deconvolution. In the second derivative method, the second derivative of the observed absorption spectrum is computed numerically. This can readily be accomplished using a modified Savitsky–Golay [1964] algorithm with a smoothing second derivative "window" function. The second derivative of a positive (Gaussian or Lorenzian) band will produce a negative center peak with positive low- and high-frequency side peaks. The center peak will appear at the same peak position as the original spectrum but will be narrower by about a factor of 2.7.

Fourier self-deconvolution is carried out on the spectrum in Fourier space by multiplying the interferogram by an exponentially increasing function within the apodization window. This will amplify higher Fourier frequency components, which results, upon Fourier transformation, in peaks with reduced halfwidth but with negative side bands. A reduction of the halfwidth by about a factor of 1.5 can be achieved.

Upon artificially reducing the bandwidth, the broad, unstructured amide I absorption usually exhibits resolved peaks, sometimes at frequencies where the absorptions of pure structural motifs are observed. Model studies on proteins with well-known secondary structure using these techniques have revealed important peaks in the second derivative or the self-deconvoluted spectra, which cannot be observed in the normal absorption spectra. Hemoglobin, for example, contains about 80% α-helical peptide motifs and no β-sheets. Its second derivative spectrum shows one strong peak at $1652\,\text{cm}^{-1}$ in the amide I' region, which is due to the α-helical fraction. Two further peaks, at 1638 and $1675\,\text{cm}^{-1}$, were found in the second derivative spectrum and coincide well with previously assigned frequencies of turns, and with frequencies calculated by Krimm and co-workers for β-turns [Susi and Byler, 1983].

Numerous other peptides have been studied via these two approaches, and percent compositions found from infrared data usually are in reasonable agreement with those derived from CD or x-ray experiments. Although this author is hesitant to advocate the "assignment" of a peptide secondary structure on the basis of observed amide I frequencies alone, a careful experimental approach can reveal an enormous amount of structural features from molecules as complex as peptides and proteins.

Most quantitative analysis on peptide structure has been performed using the amide I vibration. The observed frequencies of the amide III vibration, which depend also on the secondary structure, have been utilized as a qualitative probe for the solution conformation of peptides and proteins, particularly in Raman spectroscopy. Lord [1977] proposed a quantitative correlation between the conformational angle ψ and the frequency of the amide III vibration; however, for most purposes the correlation between the amide III frequencies and the secondary structure remained purely qualitative. The qualitative character of this correlation was partially due to the poor understanding of the nature of the amide III vibration at this time. Nevertheless, a number of attempts to utilize the amide III frequency, together with the amide I frequency, to monitor peptide solution conformation have been published [for a review, see Carey, 1982].

In the author's laboratory, computational efforts were undertaken to investigate the cause for amide III frequency shifts for the alanyl dipeptides (Structures V–IX) discussed earlier in this chapter. A small change in the molecular structure of L-alanine–L-alanine can be invoked by substituting a D-alanine residue for an L-alanine residue, which forces the molecule into a slightly different conformation. Indeed, one observes small frequency shifts in D-Ala–L-Ala versus L-Ala–L-Ala in the amide III region, which can be interpreted in terms of small differences in the coupling geometry, and which can be reproduced computationally. The vibrational coupling between adjacent C—H and N—H deformation coordinates, discussed earlier, explains the observed conformational sensitivity much better than any of the previous descriptions of the amide III vibration.

8.2 RESONANCE RAMAN SPECTROSCOPY OF PEPTIDES AND PROTEINS

The spectral details described in the previous section were due to infrared absorption or (nonresonant) Raman emissions of the amide backbone of peptides, observed in either the amide I or amide III vibrations. Certain side chain vibrations, such as the frequency of S—S linkages, may also be useful probes for peptide conformation at times, but most efforts have concentrated on the vibrations of the backbone. They can be monitored either off-resonance, as discussed above, or via resonance Raman studies of the backbone. In addition, resonance Raman offers the possibility to monitor the prosthetic groups of proteins. Both these methods will be introduced in the following chapter.

8.2.1 UV Resonance Raman Studies of the Peptide Linkage

Excitation of the $\pi^* \leftarrow n$ or $\pi^* \leftarrow \pi$ peptide electronic transitions and resonance enhancement of the Raman scattering of the backbone vibrations can be achieved by exciting with laser wavelengths below 200 nm. Laser pulses that far in the ultraviolet region are usually created by using a H_2 shifter. This device creates coherent laser output, via the stimulated Raman effect, of a pump laser pulse (Stokes) shifted by the H_2 vibrational energy.

With 200 nm excitation, the relative Raman intensities are quite different from those observed with visible excitation. In peptide models, such as N-methylacetamide, the most predominant peaks are the amide II vibration at about $1575\,cm^{-1}$ and the amide III vibration at $1317\,cm^{-1}$. In proteins, the most enhanced vibration is the amide II mode as well, when excitation wavelengths below 200 nm are used.

This band also shows strong hypochromism, which is an effect commonly observed for biological and other macromolecules. In absorption spectroscopy, the term hypochromism is used to describe the effect that a well-ordered polymer with n chromophores has much less absorption strength than one would expect from n independent chromophores. This effect is due to interactions of the individual chromophores and can be used as a probe of stacking. The Raman hypochromism is due to the fact that the corresponding electronic absorption, which enhances the amide II mode, exhibits absorption hypochromism.

Thus the term Raman hypochromism is used if an ordered polymer exhibits less intensity in a given mode than the same polymer in an unordered state. The amide II hypochromism can be observed between α-helical and unordered peptide structures and offers a convenient method to monitor the helical versus unordered peptide composition. It was first observed for certain peptides that undergo a pH-induced transition from α-helical to unordered states. In these, a linear decrease of the amide II scattering intensity with the helical content was found [Spiro et al., 1987].

8.2.2 Resonance Raman Spectroscopy of Prosthetic Groups

The protein backbone does not absorb light at wavelengths above about 225 nm. Certain aromatic amino acids have absorptions between 220 and 350 nm, but all amino acid residues are transparent above 350 nm. Thus proteins without any other prosthetic groups, or chromophores, are colorless. However, many naturally occurring proteins, such as hemoglobin, myoglobin, and cytochrome c, are intensely colored due to the presence of a chromophore that often consists of a macrocyclic ring system incorporating a transition metal. These metal centers are often the site of the protein's enzymatic or transport activity: the iron atoms are the active centers in the redox proteins of the cytochrome family, and oxygen binds to the iron in hemoglobin, which is a transport protein. Resonance Raman spectroscopy has been the most powerful spectroscopic tool to study the structure, the structural changes accompanying chemical reactions, and the dynamics of these reaction centers. Because of the resonance enhancement, very dilute solutions can be studied, and only selected vibrations (mostly those of the macrocycle and of the metal–ligand bond) are visible in the Raman spectra, whereas those of the peptide backbone and the solvent are entirely invisible. At the writing of this chapter, the original reports of resonance Raman enhancement of the heme (iron-porphyrin) group were published a mere two decades ago; however, in these 20 years, many hundreds of resonance Raman studies have shed light on the structural and dynamic properties of heme-protein and related compounds, and the biochemical and biophysical knowledge gained is immense.

8.2.2.1 Heme Group Resonance Raman Studies The heme moiety consists of a heterocyclic porphyrin shown in Fig. 8.3, which incorporates an iron(II) or iron(III) ion. The four nitrogen atoms occupy four sites of the six-coordinate

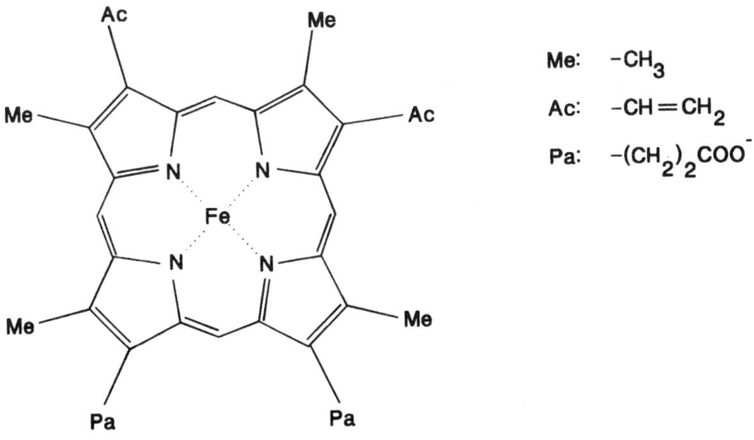

FIGURE 8.3. Structure of the porphyrin prosthetic group.

iron, with the fifth one usually occupied by a ligand from the protein, and the sixth one by a solvent molecule, or O_2 in the case of hemoglobin.

Fe(II) and Fe(III) have six and five electrons in the $3d$ orbitals, respectively. In an octahedral ligand field, the $3d$ orbitals of Fe^{2+} and Fe^{3+} are split into two groups of orbitals: the $3d_{xy}$, $3d_{xz}$, and $3d_{yz}$ orbitals referred to by their symmetry in an O_h point group as T_{2g}, and the $3d_{x^2-y^2}$ and $3d_{z^2}$ orbitals, referred to as the E_g levels. Depending on the strength of the ligand field, the splitting between the T_{2g} and E_g can be larger or about the same as the spin pairing energy; thus both Fe^{2+} and Fe^{3+} can exist in high spin and low spin configurations. Although the ligand field that the iron experiences in a protein is not octahedral, one describes the resulting energy states of the iron atom in terms of the octahedral configurations.

The visible/near UV absorption spectrum of heme groups is dominated by a strong band with an absorption coefficient of about 100,000 at about 400 nm. This band is known as the Soret band and is an in-phase mixing of two inplane $\pi^* \leftarrow \pi$ transitions of the porphyrin group. Its position and intensity depend somewhat on the oxidation and spin states of the iron atom.

In addition, there are two weaker absorptions, known as the α- and the β-bands, at about 550 and 510 nm, respectively. The α-band is the out-of-phase combination of the same two $\pi^* \leftarrow \pi$ transitions that give rise to the Soret band of the porphyrin, whereas the β-band is due to vibronic interactions involving the α-band and vibrational quanta of about $1300\,cm^{-1}$. Excitation of the Raman spectra into the Soret band leads to large enhancement of totally symmetric vibrations of the heme unit, which are observed at 1360 and $674\,cm^{-1}$. The enhancement is attributed to the A-type mechanism discussed in Section 5.6. The enhanced vibrations are polarized and, with excitation into the Soret band, none of the "inverse polarization" (*vide infra*) common in 500–525 nm excitation is observed.

Excitation into the α- and β-bands also produces intense resonance enhancement, shown in Fig. 8.4. In contrast to Soret band excitation, many more vibrational transitions are resonance enhanced, and the resulting Raman bands are depolarized or inversely polarized (*vide infra*). They result from a B-type mechanism, which involves the mixing of electronic states via the vibrational transition. The most enhanced vibrations are those due to in-plane C—C and C—N stretching modes of the porphyrin and in-plane C—H deformations. These modes are responsible for the vibronic mixing of the Soret and the α-transitions.

Inverse polarization is a phenomenon where the depolarization ratio ρ is infinity. A depolarization value of infinity corresponds to the rotation of the plane of polarization during the Raman scattering process; that is, the parallel component is not observed at all, whereas the scattered component polarized perpendicularly to the laser is very strong. A depolarization ratio of infinity is expected on theoretical grounds for some vibrations under D_{4h} symmetry with excitation into the α-band. In reality, the ρ values observed for these bands do not reach infinity, but values of $0.75 < \rho < \infty$ are found. This discrepancy is

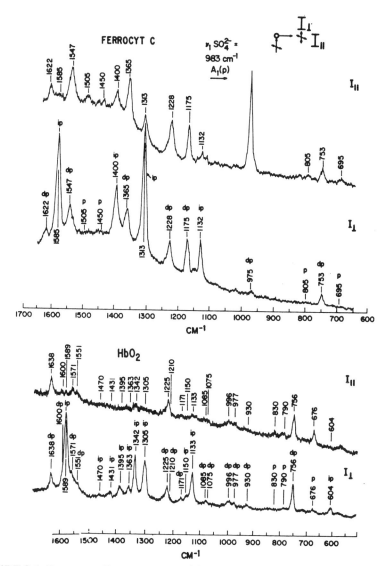

FIGURE 8.4. Resonance Raman spectra of ferrocytochrome c. Excitation wavelength: (a) 514.5 nm (β absorption band) and (b) 568.2 nm (α absorption band). [Adapted from Spiro and Strekas, *Proc. Nat. Acad. Sci. USA*, 69, 2622 (1972)].

due to the fact that the symmetry of the iron–ligand moiety is not D_{4h} but is lowered by the different R groups attached to the porphyrin and by interactions with the protein.

Typical examples for inversely polarized bands are the vibrations at 1585, 1400, 1313, and 1132 cm^{-1} in ferrocytochrome c (Fig. 8.4). A detailed analysis of the symmetry of the electronic states [for a review, see Carey, 1982] yields that the vibrations exhibiting inverse polarization belong to the A_{2g} representation, whereas vibrations belonging to B_{1g} and B_{2g} (under local D_{4h} symmetry) should be depolarized. The inverse polarization of the A_{2g} species vibrations results from the fact that the scattering tensor is antisymmetric; that is, $\alpha_{ij} = -\alpha_{ji}$.

In the mid-1970s, the sensitivity of a number of porphyrin vibrations toward the oxidation and spin states of the central iron atom was established. The frequency shifts of some of these vibrations are summarized in Table 8.4. Band I, for example, serves as a marker for the iron oxidation state (1360 versus 1375 cm^{-1} for Fe(II) and Fe(III), respectively) and is not much affected by the spin state (low or high spin) of the iron. On the other hand, band IV is observed around 1585 cm^{-1} for low spin and 1555–1570 cm^{-1} for high spin iron. These differences have been explained in a number of models: the oxidation state of the iron affects the back donation of iron electrons into the π-system of the porphyrin and thereby changes the force constants between double bonded atoms. The spin state affects mostly the size of the iron ion; low spin Fe(II) and Fe(III) are sufficiently small to fit between the four N atoms in the porphyrin ring, whereas the high spin species are too big to fit and are located above the plane of the four N atoms.

8.2.2.2 Heme Group Dynamic Studies Resonance Raman spectroscopy offers the unique possibility to study not only static structural features but dynamic processes as well. A number of studies have been carried out on carrier proteins, such as hemoglobin or myoglobin with either O_2 or CO bound to the heme iron. The carboxy-myoglobin (MbCO), for example, can be photodissociated by excitation with 532 nm pulses. With pulse lengths of 30 ps, complete photodissociation into Mb and CO occurs, and the same pulse producing the photolysis of the sample is used to record the Raman spectra. Within the 30 ps pulse length, a resonance Raman spectrum was obtained that resembles that of deoxy-Mb. From time-resolved Raman and absorption experiments, the following dynamic scheme could be derived. Within fractions of a picosecond, relaxation of the Fe core to a high spin Fe(II) occurs, followed by a resizing of the porphyrin core within 200 ps. In the photodissociation of carboxy-hemoglobin (HbCO), the appearance of the spectra indicating an

TABLE 8.4. Vibrational Frequencies of Certain Porphyrin Bands as a Function of Iron Oxidation and Spin State

Spin State	Band I	Band II	Band IV	Band V
Low spin Fe(II)	1358–1370	1490–1498	1584	1625
High spin Fe(II)	1358	1472	1555	1606
Low spin Fe(III)	1375	1505	1585	1640
High spin Fe(III)	1374	1495	1570	1630

expanded porphyrin core is as fast as it is in MbCO, within less than 20 ps. However, the core resizes about a thousand times more slowly. These differences in core relaxation have been correlated with the differences in the protein pocket in which the heme group is located.

8.2.2.3 Other Prosthetic Groups One other success story of resonance Raman and time-resolved resonance Raman spectroscopies is the work on the visual pigment rhodopsin and its analogue, bacteriorhodopsin. The former is found in the retina of all mammals and many other living beings and is responsible for the primary process of vision, namely, the capture of a visible photon and its conversion to chemical energy. The second of the proteins above occurs in the purple membranes of an algae which is found in very saline water. This algae uses bacteriorhodopsin as an antenna for visible light and utilizes the energy captured to transport protons through the bacterium's membrane.

Both of these proteins contain retinal as their prosthetic groups. Retinal is a polyene, the structure of which is shown in Fig. 8.5. The mechanism of the conversion of light to chemical energy is believed to be very similar for the two proteins, particularly in the early steps. Since the purple membrane pigment is easier to work with than the visual cells in the retina, most models studies have been carried out for bacteriorhodopsin (bR). For bR, the following photochemical cycle is believed to be responsible for its activity:

$$bR_{570} \underset{}{\overset{h\nu}{\rightleftharpoons}} K_{590} \rightarrow L_{550} \xrightarrow{-H^+} M_{412} \xrightarrow{+H^+} O_{640}$$

Here, the subscripts denote the wavelengths of maximum absorption. Unfortunately, slightly different nomenclatures for the intermediates and a varying number of intermediates have been used by different workers in the field.

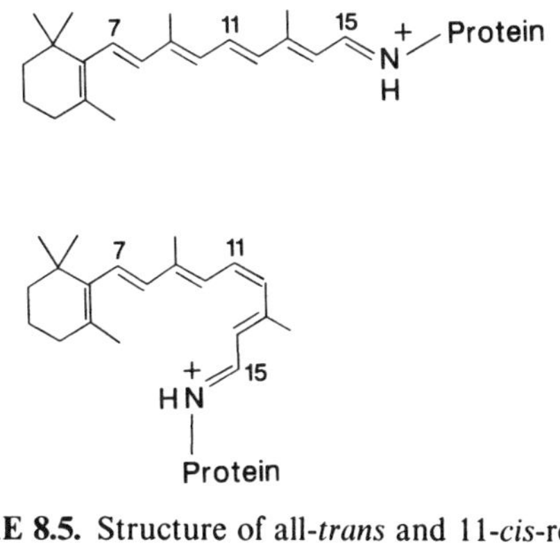

FIGURE 8.5. Structure of all-*trans* and 11-*cis*-retinal.

However, all theories seem to agree that, originally, the retinal in bR_{570} exists in the all *trans* form. After absorption of a photon in the first step, a distortion about the double bond 13 occurs, which leads to a complete isomerization to 13-*cis*-retinal shown in Fig. 8.5. Only the first step in this cycle is light driven and is believed to occur within 40 ps, with a number of even faster reactions (on the 1 ps time scale) that lead to the formation of K_{590}.

A number of relatively slow reactions follow. During the transformation of the L_{550} to M_{412}, which occurs within 40 μs, the protonated Schiff base, with which the retinal is bound to the (side) amino group (a lysine residue of the protein), is deprotonated. This proton is the one transported through the membrane. Thus the photochemical reaction establishes a proton gradient across the membrane, which in turn is used to synthesize ATP. After abstracting a proton from the medium, bR is recreated within a few milliseconds to complete the cycle.

Time-resolved resonance Raman spectroscopy has been used extensively to study this photochemical cycle, in particular, the early steps. For this purpose, specially prepared membranes containing bR are suspended in water and subject to (one or two) laser pulses as described in Chapter 5.

In one-pulse experiments, with a pulse width of 10 ps, it was found that the Raman spectra of bR_{570} change somewhat with increasing laser power, which was attributed to the fast creation of a mixture of a photoexcitation product and the original species within the length of the laser pulse. This product could be identified further by pump-probe experiments, with delay times of between 0 and 40 ps. It was found via the analysis of a number of marker bands that the first product was most likely an intermediate, twisted stage between all *trans*-retinal and 13-*cis*-retinal. The marker bands had been defined via detailed deuterium substitutions along the olefinic backbone of retinal.

As experimental techniques are being refined, more details of these reaction pathways, and their dynamics, will undoubtedly be found, and other techniques, such as time-resolved infrared and UV visible absorption spectroscopies, have been adapted to study such fast processes. However, to date, time-resolved and resonance Raman techniques have proved invaluable in the determination of the processes involved in many biochemical reactions. In addition to the prosthetic groups introduced so far, a large number of different reaction centers of biological molecules have been studied, among them chlorophyll, iron-sulfur centers, and the active site in vitamin B.

8.3 NUCLEIC ACIDS: DNA AND RNA

In the nuclei of eukaryotic cells, macromolecules are found hat carry the genetic code of the organism to which the cell belongs. The genetic code is the chemically stored information transmitted by a cell, during cell division, to the new generation of cells. This information is stored via the sequence of four different "bases" in polymers known as ribonucleic acids (RNA) or deoxyribonucleic acids (DNA). There is quite a structural difference between these two nucleic acids, and they exhibit a very different chemistry and biochemistry. In this chapter, mostly the spectroscopy of DNA will be elaborated on.

DNA is a polymer of hundreds of thousands of monomers known as nucleotides. A nucleotide consists of a ribose (a five-membered sugar) and an ionized phosphate portion, shown in Fig. 8.6. In DNA, the 2′ carbon atom of the ribose has a hydrogen, instead of the OH group shown, which explains the nomenclature of *deoxy*ribonucleic acid. Attached to the ribose unit is one of the four "bases," which are aromatic, heterocyclic purine or pyrimidine derivatives shown in Fig. 8.6. In DNA, the bases denoted by the symbols A (adenine), C (cytosine), G (guanine), and T (thymine) are found, whereas in RNA, a U (uracil) base takes the place of T in DNA.

The nucleotide monomers, also known as mononucleotides, are connected via phosphodiester linkages to form a structure depicted in Fig. 8.6. The

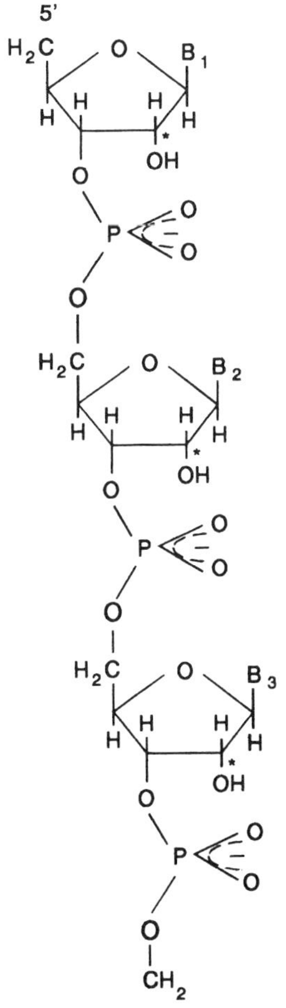

FIGURE 8.6. Generalized structure of nucleic acids. B_1 to B_3 denote one of four bases. The position marked with asterisk (*) bears an H atom, instead of an —OH group, in RNA.

strands formed by polymerization of the monomers may interact with each other through hydrogen bonding, as shown in Fig. 8.7, to form helical, double stranded structues. It was found that in natural, double stranded DNA, the base C on one strand is always opposed by, and hydrogen bonded to, a G base on the other strand. Similarly, A and T (or U in RNA) form base pairs between opposing strands. Thus the chemical information is stored in duplicate, since the complementary character of the bases on either strand contains basically the same information. The sequence of a DNA fragment is given from the 5' to the 3' end (cf. Fig. 8.6).

The basic and most common form of DNA in the crystalline, fibrous, and solution phases is the so-called B-form, where two strands of DNA are wrapped into a right-handed helix. The directon of the two strands in the double helix is opposite. In the B-form, there are 10 base pairs per turn of the helix, and the bases are stacked perpendicularly to the helix axis, about 3.3 Å

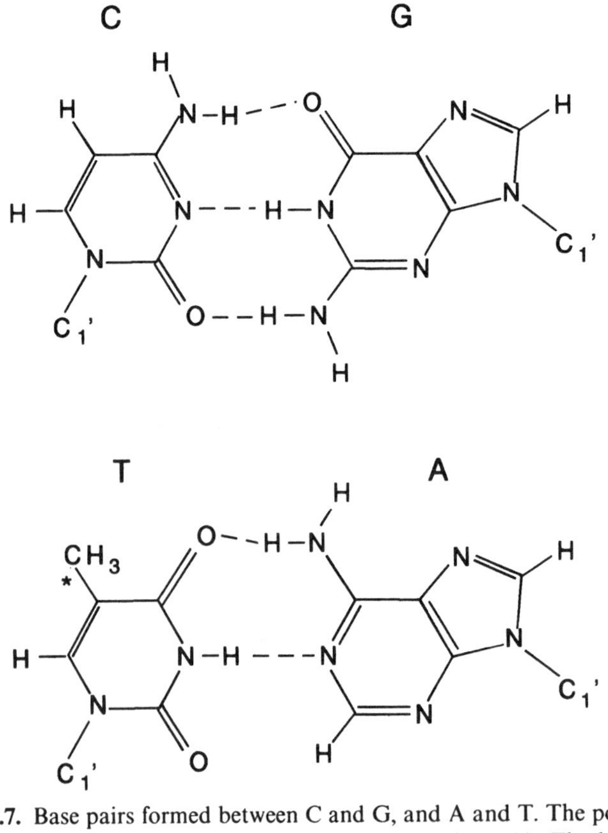

FIGURE 8.7. Base pairs formed between C and G, and A and T. The position marked with asterisk (*) bears an H atom, instead of a —CH_3, in RNA. The base with an H atom is known as U.

apart. Another common right-handed helix is the A-form, typically assumed by RNA molecules and by DNA under dehydrating conditions. Particularly fascinating is a form of double stranded DNA where the helicity is reversed to a left-handed form, known as Z-DNA. This structure is formed spontaneously in solution at high salt conditions or very low concentrations of certain polyamines, such as spermidine. *In vivo*, the macromolecular polymers are found wrapped around protein matrices to form chromosomes. In analogy to peptides and proteins, this is referred to as tertiary structure.

The solution structure, that is, the shape a DNA or a model oligonucleotide assumes in solution, is one major area of investigation in modern biophysical research. In particular, the different secondary structures, which are manifested in various different helical forms of DNA, and its tertiary structure are being investigated in order to understand the mechanism of certain physiological processes involving DNA. One of the most fascinating of these processes is the recognition of certain base sequences by enzymes with phenomenal specificity. At this point, the question arises whether or not a given sequence assumes a static or dynamic structure ever so slightly different from the rest of the molecule to permit recognition. These are the kinds of questions one wishes to answer by studying spectroscopic characteristics of DNA.

Efforts to investigate the solution structures of nucleic acids, using vibrational spectroscopy, date back to the early 1970s, when laser Raman studies on these molecules became practical. The observed Raman spectra can logically be divided into spectral features due to the ribose, the phosphate, and the bases.

8.3.1 Phosphodiester Vibrations

In DNA, the link between the various ribose units is a phoshodiester unit, shown in Fig. 8.6. There are two different phosphorus–oxygen bonds around the P atom, namely, the O—P stretching vibrations of the actual phosphodiester group C—O—P—O—C, and the free oxygen atoms, which bear a delocalized charge of -1, and whose bond order is about 1.5 (cf. Fig. 8.6). The Raman spectrum exhibits the symmetric PO_2^- stretching mode at about 1090 cm^{-1}, whereas the corresponding PO_2^- antisymmetric stretching vibration is observed in the infrared absorption spectrum at about 1220 cm^{-1}.

The phosphodiester vibrations can also be described to be a symmetric and antisymmetric combination of the P—O stretching coordinates. An intense Raman line at about 800 cm^{-1} is believed to be due to the O—P—O diester symmetrc stretching mode, and an infrared band at only slightly higher wavenumber (~ 830 cm^{-1}) is believed to be due to the O—P—O diester antisymmetric stretching mode [Prescott et al., 1984]. In addition, two C—O stretching vibrations are observed at 1017 and 1058 cm^{-1}. These vibrations, in particular the phosphodiester symmetric stretching vibration, are very sensitive to the conformation of DNA (*vide infra*).

8.3.2 Ribose Vibrations

The vibrations of the ribose or deoxyribose moiety are very difficult to assign, due to the enormous amount of mixing of C—C and C—O stretching coordinates within the five-membered ring, and the mixing with the vibrations of the neighboring C—O vibrations. In the Raman spectra, (deoxy)ribose modes are observed at 830, 895, 917, 975, 1144, 1448, and 1462 cm^{-1}. Again, some of these modes are very sensitive to the backbone conformation. This is due to the fact that the pucker of the five-membered ring varies as the conformation of the nucleic acid changes. Particularly important are the conformations about the 2' and 3' carbon atoms. These can be in the endo and exo forms. In the former, the 2'-*endo* form, for example, the 2' carbon atom is *endo* (puckered out of the sugar plane and on the same side of the plane as the 5' carbon). In the 2'-*exo* form, the 2' carbon is puckered out of the plane and on the opposie side of the plane than the 5' carbon atom. The lowest frequency band listed above for the ribose vibration is believed to be sensitive to the existence of the B-form with the ribose C2' in *endo* configuration. When DNA undergoes a transition to the left-handed (Z) form, the intensity of this peak decreases and a band at 627 cm^{-1}, due to the 3'-*endo* conformation, is observed.

8.3.3 Base Vibrations

Most Raman lines in the spectra of native DNA may be attributed to the vibrations of the bases, which have been analyzed in detail by Tsuboi et al. [1973]. The ring modes are often strong in the Raman effect because of the π-character of many of the bonds, but the infrared spectra also show strong vibrations, mainly due to the C=O vibration at about 1650 cm^{-1}. Although one should expect the vibrations of the ring to be fairly well-defined group frequencies, and therefore not very sensitive to the conformation of the nucleic acid, one finds that certain ring modes, particularly those that also involve the glycosidic bond to the ribose, are very sensitive to the conformation. A number of the marker bands are listed in Table 8.5.

8.3.4 Conformational Studies on Model Nucleotides

In order to simplify the problem that exists with naturally occurring DNA, with its enormous size and the variability in sequence, many model studies have been carried out on systems such as poly(dG-dC)·poly(dG-dC) or poly(dG)·poly(dC), where the nomenclature employed here indicates that all nucleotides are composed of *deoxy*ribose units. The former of the two species contains an alternating d(··GCGC··) sequence in each strand, whereas the latter has d(··CCCC··) in one strand and d(··GGGG··) in the other. The dot between the strands implies a hydrogen bonded, double helical structure.

TABLE 8.5. Frequencies of Conformation Specific
Marker Bands in Model Polynucleotides[a]

Z Form	B-Form	Assignment[b]
627		G, 3'-endo
	835	BK, 2'-endo
1355	1362	G
1424	1421	G, BK
	1530	G

[a]From Thomas and Peticolas [1984].
[b]G, guanine; BK, backbone.

In such model systems, the characteristic vibrations due to the A, B, and Z conformations could be established by recording the Raman spectra under carefully controlled conditions and monitoring the structure via other techniques, such as circular dichroism. The major conformational marker bands are listed in Table 8.5. Most noticeable is the intensity change of a strong vibration at 681 cm^{-1} in B-form nucleotides, which originates from a guanine ring vibration coupled with the vibration of the ribose when it is in the C2'-endo conformation. This band all but disappears when the polymer switches to the Z-conformation [Thomas and Peticolas, 1984].

In addition, the stacking of the bases can be followed conveniently in Raman spectroscopy via hypochromism, which was introduced before in Section 8.2.1. Hypochromism is very easily observed for nucleic acids in the UV spectral region and is due to stacking interactions. Due to the change in the extinction upon stacking, Raman hypochromism is observed. A vibration at 1245 cm^{-1}, assigned to a cytosine ring mode, becomes much less intense when G and C containing polymers undergo a transition from a disordered state (e.g., at elevated temperature) to the base-stacked double helical state.

Infrared spectroscopy has been used for conformational studies to a lesser extent, due to the difficulties in collecting infrared absorption data from aqueous solutions. However, the C=O stretching vibrations of the nucleotide bases, which are arranged in a fixed geometry in a double helical DNA or DNA model, interact strongly by a dipolar coupling mechanism identical to the one responsible for the coupling of the amide I vibrations. This coupling produces conformational sensitivity in the carbonyl stretching region, which has not yet been fully exploited. In Chapter 9, in the discussion of the infrared CD features of model nucleic acids, this aspect will be elaborated on in detail.

The discussion presented so far on the vibrational spectroscopy of nucleic acids demonstrates that this field has made enormous progress over the past two decades, and qualitative correlations between certain band frequencies and the conformation of the polymer are available. However, a good and useful theoretical treatment, as is available for peptides, has not yet been worked out for nucleic acids. This is partially due to the enormous complexity of the problem: whereas a peptide requires normally only two parameters (the conformational angles ψ and ϕ for each amino acid), each nucleotide monomer requires six conformational angles for the backbone alone, in addition to the ribose conformational degrees of freedom, and those affecting the base orientation.

8.4 LIPIDS

A third important class of biological compounds are lipids, which are the molecular constituents of biological membranes. Membranes are assemblies of molecules, such as phospholipids (*vide infra*), which are held in their positions by noncovalent forces, such as hydrophilic and hydrophobic interactions. Located on the surface of membranes, or partially inserted into the membrane or even spanning it completely, are a number of proteins responsible for transport or other catalytic functions. The locations of the protein components on the membrane surface, or in the membrane, are random, and the mechanisms of how a protein binds to the membrane or inserts itself through the membrane are not completely understood at this time.

However, a fairly good description of the phospholipid component of the membrane does exist. The phospholipid molecule itself consists of a polar head group and non polar fatty acid side groups. The fatty acids, typically consisting of C_{14} to C_{24} chains, are linked to two oxygen atoms of glycerol via ester linkages. The third oxygen of the glycerol carries a phosphoether group, as shown in structure (X):

$$
\begin{array}{c}
\underset{\text{Fatty acids}}{\underbrace{\begin{array}{c} C_nH_{2n+1}\overset{O}{\underset{\|}{C}}-O-CH_2 \\ C_mH_{2m+1}\underset{\|}{\overset{}{C}}-O-CH \\ O \end{array}}} \quad \underset{\text{Glycerol}}{\underbrace{\begin{array}{c} \\ \\ CH_2-O- \end{array}}} \quad \underset{\text{Phosphate}}{\underbrace{\begin{array}{c} O^- \\ | \\ P-O- \\ \| \\ O \end{array}}} \quad \underset{\text{Alcohol}}{\underbrace{(CH_2)_2-\overset{+}{N}(CH_3)_3}}
\end{array}
$$

(X)

There are a number of different phospholipids, differing in the particular alcohol attached to the phosphate group. The alcohol group shown in structure (X) is known as choline, and the entire structure is called phosphatidylcholine.

Phospholipids are organized in structures known as *lipid bilayers* shown in Fig. 8.8. In these structures, which are between 40 and 80 Å thick, two layers of lipids are arranged such that the zwitterionic and polar head groups point

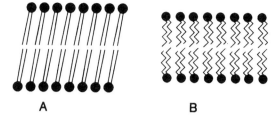

FIGURE 8.8. Schematic of the structure of lipid bilayer membranes: (A) gel phase (all *trans* polyethylene chain) and (B) liquid crystalline phase (partial *gauche* polyethylene chain).

toward the aqueous phase, whereas the fatty acid tails form a hydrophobic region that points toward the fatty acids of the second layer. Often, lamellar structures are formed, in which concentric, spherical bilayers are arranged like onion skins, separated by aqueous regions. Such lamellar structures can be several micrometers in diameter.

The aliphatic chains of each fatty acid residue can be in an all-*trans* conformation. This structure is referred to as the gel phase. If the aliphatic chain contains *gauche* C—C groups, a liquid crystalline phase of the phospholipid is obtained.

In contrast to peptides, proteins, and nucleic acids, natural membranes are inhomogeneous mixtures of many compounds held together by noncovalent forces. Consequently, x-ray crystallography cannot be used as a structural tool. Vibrational spectroscopy is an extremely powerful and noninvasive technique to monitor certain structural properties of membranes, such as the conformational changes accompanying the gel–liquid crystal phase transition. Among vibrational techniques, Raman studies were used originally to study membranes, but with modern FT and reflectance techniques, infrared techniques have been applied as well.

It was found that the vibrational frequencies and intensities of the hydrocarbon chain vary between the gel and liquid crystalline forms, and that the vibrations of the lipid head groups are not affected by this phase transition. The most useful vibration to probe the all-*trans* to the part-*gauche* transition of a polyethylene chain was found to be the C—H stretching modes of the methylene groups. In the gel state, two sharp and intense lines are observed in the Raman spectra at 2847 and 2883 cm^{-1}, corresponding to the methylene symmetric and antisymmetric stretching modes [Levine, 1985]. The terminal methyl group exhibits the symmetric and antisymmetric stretching modes at 2936 and 2962 cm^{-1}, respectively. As expected by the ratio of methylene to methyl groups in the fatty acid chain, the methyl vibrations are weak compared to those of the methylene groups.

In the liquid crystalline phase, the linewidth of the band at 2883 cm^{-1} increases and its intensity decreases, while the band at 2935 cm^{-1} becomes more intense. Thus the ratio of Raman intensities at 2935 and 2883 cm^{-1}, I_{2935}/I_{2883}, serves as a marker of the transition between gel and liquid crystalline phases. For phosphatidylcholine with various chain lengths, dispersed in aqueous solution to produce lamella, different gel–liquid crystal transition temperatures can be observed. These transition temperatures vary from 23.8°C for C_{14} chains to 73.8°C for C_{22} chains, with very sharp, cooperative phase transitions.

Other vibrations, notably the skeletal optical mode at 1128 and 1064 cm^{-1} (where the term *optical mode* implies vibrations perpendicular to the chain, whereas the term *acoustical mode* implies a vibration in the direction of the chain), also show the gel–liquid crystal transition. Furthermore, infrared studies on completely deuteriated fatty acid side chains showed that the carbon–deuterium stretching vibrations can be analyzed similarly. In a multicomponent system, for example, which may contain dispersions of lipids and membrane proteins, the use of deuteriated fatty acids allows an analysis of the lipid structure to be carried out even in the presence of peptides and proteins [e.g., see Carey, 1982, Chapter 8, and references therein].

REFERENCES

J. Bandekar, *Biochim. Biophys. Acta, 1120*, 123 (1991).

J. Bandekar and S. Krimm, *Biopolymers, 27*, 909 (1988).

P. R. Carey, *Biochemical Applications of Raman and Resonance Raman Spectroscopies*, Academic Press, New York, 1982.

M. Diem, O. Lee, and G. M. Roberts, *J. Phys. Chem., 96(3)*, 548 (1992).

S. Krimm, *Biopolymers, 22*, 217 (1983).

S. Krimm and J. Bandekar, *Biopolymers, 19*, 1 (1980).

I. W. Levine, in *Chemical, Biological and Industrial Applications of Infrared Spectroscopy*, J. R. Durig, Ed., Wiley, New York, 1985, p. 173.

R. C. Lord and N. T. Yu, *J. Mol. Biol., 50*, 509 (1970).

R. C. Lord, *Appl. Spectrosc., 31*, 187 (1977).

T. Miyazawa, T. Shimanouchi, and S. Mizushima, *J. Chem. Phys., 29*, 611 (1958).

W. H. Moore and S. Krimm, *Biopolymers, 15*, 2439 (1976).

M. R. Oboodi, C. Alva, and M. Diem, *J. Phys. Chem., 88*, 501 (1984).

B. Prescott, W. Steinmetz, and G. J. Thomas, *Biopolymers, 23*, 235 (1984).

G. M. Roberts, Ph.D. Dissertation, City University of New York, 1991.

A. Savitsky and M. J. E. Golay, *Anal. Chem., 36*, 1627 (1964).

T. G. Spiro, R. P. Rava, S. P. A. Fodor, S. Dasgupta, and R. A. Copeland, in *Time Resolved Vibrational Spectroscopy*, G. H. Atkinson, Ed., Gordon & Breach Science Publishers, New York, 1987.

H. Susi and D. M. Byler, *Biochem. Biophys. Res. Commun., 115*, 392 (1983).

G. J. Thomas, Jr., in *Vibrational Spectra and Structure*, Vol. 3, J. R. Durig, Ed., Elsevier Science Publishers, Amsterdam, The Netherlands, 1975, p. 239.

G. A. Thomas and W. L. Peticolas, *Biochemistry, 23*, 3202 (1984).

M. Tsuboi, S. Takahashi, and I. Harada, in *Physico-chemical Properties of Nucleic Acids*, Vol. 2, J. Duchesne, Ed., Academic Press, New York, 1973, p. 91.

E. B. Wilson, J. C. Decius, and P. C. Cross, *Molecular Vibrations: The Theory of Infrared and Raman Vibrational Spectra*, McGraw-Hill, New York, 1955.

9

VIBRATIONAL OPTICAL ACTIVITY

Vibrational optical activity (VOA) is a collective term applied to two new spectroscopic techniques discovered during the early 1970s. The two techniques, which had been predicted much earlier, are infrared or vibrational circular dichroism (VCD) and Raman optical activity (ROA). Both these techniques are manifestations of natural optical activity, which has been observed in electronic transitions via two techniques, either the rotation of the plane of polarization of light or the differential absorption of left and right circularly polarized light upon passing through the sample. These two effects are known as optical rotatory dispersion (ORD) and circular dichroism (CD), respectively, and are often introduced during the discussion of stereochemistry in introductory courses in organic chemistry. CD and ORD are mathematically related to each other and measure the same phenomenon, referred to as optical activity, which is a property exhibited by enantiomerically pure, chiral molecules. Chirality, or handedness, describes the relationship between two molecules that are related to each other like a left hand is related to a right hand, namely, by being mirror images of each other. A molecule is said to be chiral when it is nonsuperimposable on its mirror image.

Chiroptical techniques, such as CD, are among the most fascinating spectroscopic phenomena, since they can distinguish molecules that are identical except for their configuration, that is, their handedness. Chiroptical methods use circularly polarized light (cf. Section 2.8, Fig. 2.6) to distinguish the two forms of molecules. Since circularly polarized light itself has handedness (it can be either left or right circularly polarized), different interactions of the two forms of light occur with chiral molecules and give rise to the observable effects enumerated above.

In VOA, the effects of the chirality on vibrational, and not electronic, transitions is observed. Thus both ROA and VCD are truly forms of vibrational spectroscopy. It was therefore felt that this material should be included in this text, which is intended to present an overview in developments in modern vibrational spectroscopy.

VCD is conceptually a straightforward experiment in which the differential absorption of left and right circularly polarized infrared radiation by a vibrational transition of a chiral molecule is observed. As such, it is a direct extension of the principles of electronic CD toward a different spectral range, namely, that involving vibrational transitions.

ROA, like Raman spectroscopy, is a scattering, rather than an absorption, phenomenon. It is observed by exciting the sample alternatively with left and right circularly polarized laser radiation and measuring the differential Raman scattering cross section as an intensity differential. Both these effects will be discussed separately in this chapter.

9.1 INFRARED (VIBRATIONAL) CIRCULAR DICHROISM

9.1.1 Phenomenological Description and Basic Equations

Conceptually, one may view VCD as an extension of the principles of electronic circular dichroism (CD), normally observed in the ultraviolet spectral region, into the domain of vibrational transitions in the infrared spectral region.

In infrared VCD, the experimental result is the differential absorption between left and right circularly polarized infrared radiation, defined as

$$\Delta A = A_L - A_R \qquad (9.1.1)$$

Here, as in the remainder of this chapter, capital subscripts R and L are used to denote right and left circularly polarized radiation. For a given transition, the differential absorption ΔA can be related to the quantum mechanical observable known as the *rotatory strength* R_{01} by converting from ΔA units to $\Delta \varepsilon$ units, using the Lambert–Beer law (Eq. 5.1.1) and integrating over the VCD peak:

$$R_{01} = \int (\Delta \varepsilon / \tilde{\nu}) \, d\tilde{\nu} \qquad (9.1.2)$$

Similarly, the dipole strength D_{01} is obtained by converting from absorbance units A to units of ε and integrating (see Eq. 3.10.1)

$$D_{01} = \int (\varepsilon / \tilde{\nu}) \, d\tilde{\nu} \qquad (3.10.1)$$

The ratio of the integrated VCD and absorption peaks can be defined as

$$\Delta A/A = 4R_{01}/D_{01} \qquad (9.1.3)$$

and is typically on the order of 5×10^{-4} to 5×10^{-5}.

The *rotatory strength* is given in terms of the molecular transition moments by

$$R_{01} = \text{Im}[\langle 0|\mu|1\rangle \cdot \langle 1|m|0\rangle] \qquad (9.1.4)$$

in analogy to the dipole strength of a transition defined in Chapter 2:

$$D_{01} = \langle 0|\mu|1\rangle^2 \qquad (2.4.2)$$

In these last equations, $|0\rangle$ and $|1\rangle$ denote vibrational ground and excited state wavefunctions, and μ and m are the electric and magnetic dipole operators, respectively, defined in Eqs. 2.3.6 and 9.1.5:

$$\mathbf{m} = \sum_i (e_i/2m_i)(\mathbf{r}_i \times \mathbf{p}_i) \qquad (9.1.5)$$

where e and m are the charges and masses of the particles, $\mathbf{r}$ and $\mathbf{p}$ are the position and momentum of particle i, and the summation is over all particles.

As in all manifestations of natural optical activity, the desired observables —that is, ΔA or R_{01}—arise through the interference of electric and magnetic dipole transition moments. The sensitivity of chiroptical techniques toward the handedness of the molecule results directly from the form of the magnetic moment operator: it contains the vector product of momentum and position vectors, the result of which is another vector. The sign of this vector is determined by the handedness of the coordinate system, and it is well known that a vector product will change sign upon converting from a left- to a right-handed coordinate system. By changing the configuration of the molecule from one to the other enantiomer, the sign of the magnetic transition moment changes. Similarly, by keeping the configuration of the molecule fixed and changing from left to right circularly polarized light, we reverse the sign of the magnetic transition moment. The equations presented here are, of course, completely analogous to the expressions governing electronic CD in the visible/ultraviolet spectral range.

Since VCD is a very small effect, its observation was and still is rather difficult. However, through the diligent efforts of a few research groups in the field, VCD has now become a generally applicable spectroscopic technique.

9.1.2 Observation of VCD

VCD has been observed via dispersive or Fourier transform (FT) instrumentation with equal success. Since the VCD effect is so small (the differential absorbance ΔA in VCD is only on the order of 5×10^{-4} to 5×10^{-5} of the infrared absorption A), it requires sophisticated modulation techniques for its observation. Thus VCD is observed by switching the incident infrared light beam at high frequency between left and right circular polarization via a photoelastic modulator (PEM, *vide infra*). The desired VCD information is then contained in an AC signal at the frequency of the PEM.

Thus the central piece of all chiroptical instruments is the device that produces alternating left and right circularly polarized light. This can be a photoelastic modulator or an electro-optic modulator (cf. Section 9.4). A PEM uses the phenomenon of stress birefringence to introduce a phase difference between two orthogonally polarized components of light incident on the modulator. The PEM crystal consists of a uniaxial piece of material that is transparent in the spectral range of interest and is aligned with its unique axis (the z-axis) along the propagation direction of the light. Under the influence of a mechanical stress or strain, applied perpendicularly to the propagation direction of the light and along the crystal's x- or y-axis, the refractive indices along these axes, n_x and n_y, become unequal, causing light waves polarized along the x- and y-directions to travel with different velocities through the crystal. At the exit face of the crystal, circularly polarized light is produced if the retardation between the two orthogonal components of light is $\lambda/4$.

The alternating strain/stress is produced by squeezing the modulator crystal between two piezoelectric drivers. An AC voltage is applied to the piezoelectric crystals, and the amplitude of this voltage determines the stress/strain and therewith the retardation. Commercial modulators are available for the infrared region, using CaF_2 above 1200 cm^{-1} and ZnSe down to about 600 cm^{-1} as stress optical materials.

In the measurement of VCD, the actual differential signal, at constant circular dichroism of the sample, will vary with the light level transmitted at a given wavelength. Thus two intensities must be monitored and their ratio determined in order to obtain ΔA. Denoting the signal at the modulator frequency as $I_{AC}(\tilde{v})$, and the overall transmission of the sample as $I_{DC}(\tilde{v})$, it can be shown from an analysis of the radiant energy at the detector that

$$I_{AC}(\tilde{v})/I_{DC}(\tilde{v}) = \tanh(1.15\,\Delta A)\sin\alpha \qquad (9.1.6)$$

Here, α denotes the retardation between the two linearly polarized components of light used to produce circularly polarized light (*vide supra*). Equation 9.1.6 demonstrates that the ratio $I_{AC}(\tilde{v})/I_{DC}(\tilde{v})$ is proportional to ΔA. The term $\sin\alpha$ itself varies sinusoidally with time, since the retardation α may be written as

$$\alpha = \alpha_0 \sin(\omega_M t) \qquad (9.1.7)$$

where ω_M is the oscillation frequency of the modulator. Using a number of simplifications, an approximation for the observed signal can be obtained:

$$I_{AC}(\tilde{v})/I_{DC}(\tilde{v}) = (1.15\,\Delta A)J_1(\alpha_0)\sin(\omega_M t) \qquad (9.1.8)$$

where α_0 is the amplitude of the retardation, and J_1 is the first-order Bessel function. Inspection of Eq. 9.1.8 yields that for the observation of ΔA, a signal in phase with ω_M must be monitored via a lock-in amplifier and continuously divided by the DC signal. Details of the derivation have been reviewed in a number of publications [Lee, 1992].

A typical dispersive, double modulating VCD spectrometer is shown in Fig. 9.1. Light from a very hot infrared source is focused via gold coated optical components into a monochromator optimized for a relatively narrow region in the mid-infrared. The light is chopped mechanically at low frequency and converted to alternating left and right circularly polarized light at the PEM. After passing the sample, the light is detected via an infrared detector operating at liquid nitrogen temperature. Signal analysis is carried out using two lock-in amplifiers in tandem, first demodulating the signal at the modulator frequency ω_M and subsequently at the chopper frequency to yield I_{AC}. I_{DC} is measured separately using a third lock-in amplifier, and the ratio of I_{AC}/I_{DC} is computed by a dedicated personal computer [Diem et al., 1988; Lee and Diem, 1992]. FT–VCD instruments with similar sensitivity have been constructed as well.

9.2 APPLICATIONS OF VCD: SMALL MOLECULES

Both techniques of vibrational optical activity had relatively slow starts, due to a number of reasons. First, both ROA and VCD are experimentally very difficult techniques, and only after a number of instrumental breakthroughs are

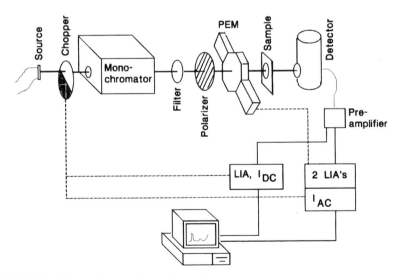

FIGURE 9.1. Schematic of a dispersive VCD instrument. Heavy line, light path; solid line, signal path; dashed lines, reference signals, PEM, photoelastic modulator; LIA, lock-in amplifier.

the effects observed routinely. Second, the interpretation of the observed spectra is difficult. In VCD, there was no generally applicable theory available for the interpretation of the first VCD results, and a number of empirical correlations and approximate computational procedures were used with relatively low success. In addition, since the molecules are by necessity chiral, even vibrational assignment and calculations are difficult. Finally, in a new technique, such as VCD, many molecules need to be investigated until the real power of the method becomes apparent. The molecules studied now by VCD, for which an enormous amount of molecular structure has been obtained, could not be studied originally when VCD was first introduced due to a number of reasons, such as wavelength limitations and the lack of sensitivity to permit studies in suitable solvents. However, during the past 10 years or so, the situation has changed drastically, and VCD is now an established method that excels in two areas—biological molecules and very small molecules, particularly those where the chirality is due to isotopic effects. The VCD of small molecules will be reviewed briefly in this section, and its application to biological systems in Section 9.3.

After its original discovery, VCD was reported in a number of chiral molecules as neat liquids or as solutions. In Fig. 9.2 are shown the VCD spectra in the mid-infrared region of α-pinene, an organic terpene derivative This spectrum exemplifies the VCD effort of the early years: mostly neat liquids were studied, which were available in enantiomerically pure form. Since α-pinene is a bicyclic molecule, its vibrations in the mid-infrared are very complex, and although strong signals were observed, their interpretation has

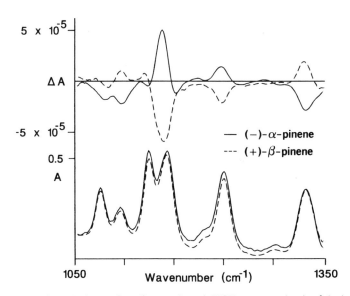

FIGURE 9.2. Infrared absorption (bottom) and VCD spectra (top) of (+) and (−) α-pinene.

not yet been performed. However, the spectra of α-pinene have emerged as the calibration standard for both ROA and VCD and are reported here to demonstrate the appearance of a VCD spectrum.

For the original VCD results, qualitative interpretative methods (such as the "perturbed degenerate mode model" and the "coupled oscillator model") and simple computational approaches (such as the "fixed partial charge model", cf. Section 3.10) were developed and utilized, with limited success. It was found that these models could be used successfully for isolated cases only and were not generally applicable.

Thus the emphasis has shifted away from molecules such as pinene, 3-methyl cyclohexanone, or other species with one or two chiral carbon centers, such as derivatives of lactic and tartaric acid, to molecular systems with fewer degrees of vibrational freedom and with less conformational variation. Simultaneously, more generally applicable theoretical models, such as the "localized molecular orbital" approach [Nafie and Walnut, 1977; Freedman and Nafie, 1984] (cf. Section 3.10), and computational methods based on it, have been developed. These methods have been applied recently to molecular systems that are small, rigid, devoid of elements past chlorine in the periodic chart (for computational reasons), and, of course, chiral. The molecules of recent interest are those shown in structure (I) below:

$$\begin{array}{c} X \\ / \backslash \\ D-C-C-H \\ / \quad \backslash \\ H \quad D \end{array}$$

(I)

When X is a CH_2 group, the molecule is *trans*-dideuteriocyclopropane. Other molecules of this series, which have been reported recently, are the ones where X is a $-CH_2-CH_2-$ (*trans*-dideuteriocyclobutane) or an oxygen or a sulfur atom (oxirane and thiirane, respectively). These molecules are dissymmetric and have two chiral centers. The chirality is due to isotopic substitution only, and in the case of the deuteriated cyclopropane and cyclobutane, the absence of a nearby electronic transition virtually prohibits chiroptical studies except in vibrational transitions.

Finally, asymmetric species of similar general structure (II) have been studied as well:

$$\begin{array}{c} X \\ / \backslash \\ CH_3-C-C-H \\ / \quad \backslash \\ H \quad H \end{array}$$

(II)

When X is an oxygen atom, molecule (**II**) is known as propylene oxide. Solution and gas phase VCD and absorption spectra for many of these molecules above have been reported and have pushed forward the frontier of experiment (e.g., high-resolution rotational–vibrational VCD) and computational aspects.

These molecules have served to develop and test *ab initio* methods to compute the force fields, vibrational intensities, and VCD intensities, and the level of agreement between observed and computed spectra is now outstanding. The reader is referred to the original literature and some review articles for many of the details of these studies [cyclopropane: Cianciosi et al., 1990; cyclobutane: Annamalai et al., 1985; oxirane: Pickard et al., 1992; thiirane: Polavarapu et al., 1991].

Thus, for small molecules, VCD has had tremendous influence on the field of vibrational spectroscopy, by forcing new computational methods to be developed to compute molecular vibrations and infrared absorption intensities. VCD also has expanded chiroptical methods by allowing studies to be carried out for molecules without nearby electronic transitions. This is not possible in visible/ultraviolet CD spectroscopy, which depends on the accessibility of an electronic transition. Furthermore, it has opened up a field of chiroptical studies for molecules where the chirality is due to isotopic substitution only.

9.3 APPLICATIONS OF VCD: BIOMOLECULES

One of the most fascinating aspects of VCD is its sensitivity toward the molecular conformation. So far, the discussion has emphasized the ability of chiroptical techniques to distinguish the configuration (i.e., the chirality) of molecules. However, chiroptical techniques such as CD and VCD are also sensitive to the structural changes a molecule may undergo by internal rotation about single bonds. These conformational changes occur without breakage and reformation of chemical bonds. A typical conformational transition occurring in macromolecules is the one between an α-helical and a coil conformation in peptides, which was discussed in the previous chapter. Such conformational changes alter the dihedral angles between certain probe groups, such as the amide I vibration (cf. Chapter 8) in peptides. VCD intensity can be produced by the dipolar coupling of (virtually achiral) vibrational transitions, which are in a fixed, dissymmetric geometric pattern, such as a helix. This mechanism for the creation of VCD intensity is thought to be one of the dominant mechanisms giving rise to the conformational sensitivity of VCD in biomolecules to be discussed next.

9.3.1 VCD of Peptides and Homo-oligoamino Acids

The conformational sensitivity of VCD in the amide I region of peptides is believed to be due to the dipolar coupling of the C=O stretching vibrations of the peptide linkages. The coupling of these transitions produces distinct VCD "couplets" (adjacent VCD bands of positive and negative intensities), which have permitted in some cases a quantitative determination of dihedral angles between the coupled oscillators and thus of the molecular solution conformation.

Application of VCD to biomolecular systems has mostly utilized the 6 μm spectral region, in which the C=O stretching vibration occurs, although the amide II and III regions have been studied to a lesser extent. The ensuing discussion will concentrate on results from the amide I region. The first direct VCD results indicating the sensitivity of VCD toward peptide secondary structure came from the laboratory of Keiderling [Yasui and Keiderling, 1986]. Polyamino acids, such as poly-L-Tyr or poly-L-Lys, for which the secondary structure is well known and can be varied as a function of solvent acidity, were studied via VCD, and distinctly different results for the established conformations were observed. In poly-L-Tyr, for example, the α-helical conformation exhibits a distinct, sharp, and near-conservative positive/negative couplet in the amide I region. In the context of this discussion, *conservative* couplet implies equal positive and negative VCD intensities, and a description of a positive/negative couplet always implies low to high wavenumber. The α-helical conformation is assumed by poly-L-Tyr in acidified DMSO solution (80%:20% by volume of DMSO and trifluoroacetic acid). In neutral DMSO, a couplet with opposite sign is observed. Infrared absorption and VCD spectra of the two forms of poly-L-Tyr are shown in Fig. 9.3.

In order to carry out a computational interpretation of these results, the exciton model was borrowed from the theory of CD spectroscopy [Tinoco, 1963]. In this model, one assumes that there are a large number of degenerate transitions, such as the C=O groups in a peptide helix, arranged in a fixed geometry. If one vibrational quantum is absorbed by these degenerate oscillators, the resulting vibrationally excited state is best described by a sum over all possible one-quantum excitations. This implies that the excitation is no longer localized on one of the oscillators but is delocalized over the entire array of identical oscillators. This delocalized excitation is referred to as an "exciton." The dipolar coupling between the transitions lifts their degeneracy; consequently, one observes as many discrete exciton energy levels as there are interacting dipoles.

In the "degenerate extended coupled oscillator" (DECO) description of the optical activity of n interacting dipoles, the rotational strength R, and hence the VCD intensities, is given by

$$R_k = -(\pi \tilde{\nu}_0/c) \sum_{i=1}^{n} \sum_{j>i}^{n} C_{ik} C_{jk} [\mathbf{T}_{ij} \cdot \boldsymbol{\mu}_i \times \boldsymbol{\mu}_j] \qquad (9.3.1)$$

where c is the velocity of light, and the C_{ij} are the eigenvector components of

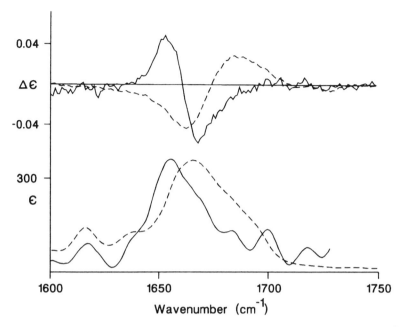

FIGURE 9.3. Observed infrared absorption (bottom) and VCD spectra (top) of poly-L-Tyr. Solid traces, α-helical (acidic DMSO); dashed traces, extended helical (neutral DMSO).

the (dipole–dipole) interaction matrix,

$$V_{ij} = \frac{\mu_i \cdot \mu_j}{|T_{ij}|^3} - \frac{3(\mu_i \cdot T_{ij})(\mu_j \cdot T_{ij})}{|T_{ij}|^5} \tag{9.3.2}$$

T_{ij} is the distance vector between dipoles μ_i and μ_j, $\tilde{\nu}_0$ is the center frequency (in wavenumbers) of the unperturbed transition, and the subscript k refers to the kth exciton component $(1 < k < n)$. The infrared absorption intensities can be obtained from the dipole strengths D, defined by

$$D_k = \sum_{i=1}^{n} \sum_{j=1}^{n} C_{ik} C_{jk} (\mu_i \cdot \mu_j) \tag{9.3.3}$$

The computations of dipole and rotational strengths for the polymers are carried out using Cartesian coordinates of the C and O atoms of the various carbonyl groups. Furthermore, the transition frequency $\tilde{\nu}_0$ of an unperturbed carbonyl transition is needed and its monomeric dipole strength D, which is proportional to μ^2. Once the geometry of the dipole transition moments is defined, the interaction energies of all dipoles with each other can be calculated according to Eq. 9.3.2, and the interaction matrix is diagonalized numerically. The eigenvalues of V_{ij} are the frequency displacements for each exciton component from $\tilde{v}_0$, and the eigenvector components are used, according to Eqs. 9.3.2 and 9.3.3, to compute the rotational and dipole strengths of the exciton components. The atomic coordinates of the carbonyl groups can be derived from crystallographic data, or from any of a number of molecular graphics programs.

Observed VCD and calculated results, using the exciton approach, are shown in Figs. 9.3 and 9.4 for the case of α-helical poly-L-Tyr. The agreement between experimental and theoretical data is within a factor of 2, with the experimental data larger than the computed ones. This discrepancy was attributed to the underestimation of the splitting between the exciton components and, consequently, cancellation of positive and negative computed VCD intensity [Birke et al., 1992].

In neutral DMSO solution, the peptide assumes a conformation that has been referred to as the "random coil." However, previous vibrational studies had suggested that this "random" conformation is, in fact, a very elongated left-handed helix known as the extended helix [Sengupta and Krimm, 1987]. The VCD features calculated for such an extended helix reproduced the observed data well, as shown in Fig. 9.4.

Thus the α-helical and the extended helical structures are well established in VCD, and there exists a simple method for the interpretation of the data. The VCD features of the β-pleated sheet structure appear reasonably well established too, although its interpretation is much more difficult. Since the data are monosignate, the exciton model is not appropriate, because it always predicts

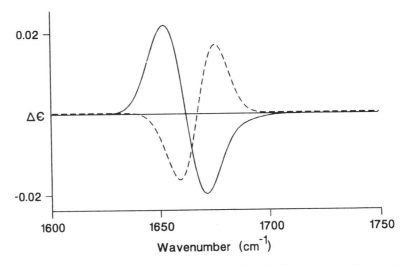

FIGURE 9.4. Calculated VCD spectra of poly-L-Tyr. Solid trace, α-helical; dashed trace, extended helical (neutral DMSO).

conservative couplets. Nafie and co-workers explained such monosignate VCD in terms of a model with nearly collinear (and antiparallel) electric and magnetic dipole transition moments [Paterlini et al., 1986].

Small peptides have been studied by VCD as well. Here, a series of alanine polymers studied in the author's laboratory will be discussed. These studies follow logically the detailed efforts on the vibrational assignments of alanine and alanyl peptides, discussed in Sections 7.5 and 8.1.1. In (L-Ala)$_3$, there are only two peptide linkages. Consequently, according to Eq. 9.3.1, one expects a positive/negative or negative/positive VCD couplet, depending on the geometry, and an absorption spectrum consisting of two peaks. This is what was observed experimentally in the VCD studies on the tripeptide L-Ala–L-Ala–L-Ala, which exhibits a distinct, near-conservative VCD spectrum in the amide I' region, shown in Fig. 9.5. The infrared absorption shows two overlapping peaks in the amide I' region, at 1650 and 1675 cm^{-1}. These spectral data were interpreted via a coupled oscillator model, and calculations were carried out to find a molecular conformation in Φ and Ψ space around the center alanine residue which best reproduces the observed spectral data.

This analysis of the VCD results on L-Ala–L-Ala–L-Ala suggested a left-handed, twisted structure with Φ and Ψ angles (120° and −25°, respectively) that is not a common conformation. This structure is believed to be stabilized by a zwitterionic interaction, because of the proximity of the —ND$_3^+$ and the

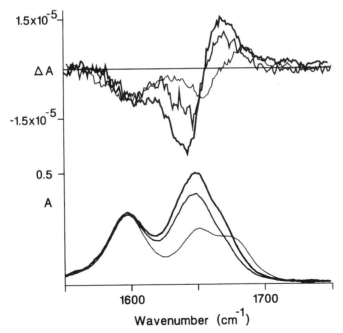

FIGURE 9.5. Observed infrared absorption (bottom) and VCD spectra (top) of (L-Ala)$_3$ (thin trace), (L-Ala)$_4$ (medium trace), and (L-Ala)$_5$ (heavy trace).

—CO_2^- groups in this structure, and because of the fact that the VCD features disappear in solutions of low and high pH values, which do not support a zwitterion [Lee et al., 1989].

Larger (L-Ala)$_n$ oligomers, with $n = 4$, 5, and 6, in aqueous solution, exhibit solution phase VCD and absorption spectra, shown in Fig. 9.5, which are different from those of (L-Ala)$_3$. These spectra are characterized by an increase in the ratio of intensities of the amide I' peak (1648 cm^{-1}) to the carboxylate antisymmetric stretching peak (1598 cm^{-1}), and an increase in the VCD amplitude, as n increases.

The observed absorption spectra do not exhibit resolved amide I' components, as in the case of (L-Ala)$_3$, but show a broad absorption envelope. Compared to (L-Ala)$_3$, there is a drastic increase in the VCD intensity, a sharpening of the VCD features, and a shift toward lower wavenumbers of the VCD zero crossing. These observed results indicate a different solution structure for the trimer and the higher oligomers, although the overall spectra appear similar at first glance.

VCD intensity calculations have demonstrated that a left-handed extended helix ($\Phi = -110°$, $\Psi = 120°$) fits the observed data best for tetra-, penta-, and hexa-alanine, whereas the trimer was reproduced best by a structure with $\Phi = 120°$ and $\Psi = -25°$. The differences in the structures of the trimer and the higher homologues appear to be caused the differences in the stabilization of the oligomers: the trimer is stabilized by zwitterionic interactions, which lose importance in the larger oligomers and are overshadowed by hydrogen bonding and other effects.

The tripeptide L-Pro–L-Pro–L-Pro in aqueous solution exhibits VCD spectra similar to those observed for L-Ala–L-Ala–L-Ala, although the couplet is more negatively biased [Dukor and Keiderling, 1993]. No analysis of these results in terms of a solution conformation based on the coupled oscillator model was carried out; however, there is clear evidence that in aqueous and nonaqueous solutions, distinct solution conformation is reached at the level of the tripeptide and persists similarly up to dodecamers. Whether or not the (L-Pro)$_3$ assumes the same solution conformation as does (L-Ala)$_3$ is not clear at this point, and one needs to keep in mind that the proline residues are significantly less flexible than alanine. However, their VCD spectra are certainly similar.

The VCD results on peptides demonstrate that VCD is a much more powerful solution conformational probe than IR spectroscopy alone. Presently, the methods delineated in Chapter 8 are used frequently to determine the percent composition of peptides, based on an analysis of the amide I absorption peak. VCD exhibits much more distinct signatures for the established secondary structures and appears to be sensitive to the occurrence of turns as well. Thus VCD should be used more frequently in the analysis of the secondary structure of peptides and proteins to be discussed in the next section.

One further aspect needs to be pointed out about the structural sensitivity of VCD. The results on small peptides discussed above point toward an interesting difference between the solution structures deduced by VCD and NMR spectroscopies. Since VCD is a form of vibrational spectroscopy, it probes peptide structures on a very fast time scale. In fact, the time scale of VCD (in the picosecond regime) is faster than the torsional motions that a peptide may undergo between similar conformations. Thus VCD samples an ensemble of slightly different structures, and the VCD signal is an average of these conformations. NMR, on the other hand, samples the conformation at a time scale slower than the torsional motions, and at times may not detect a solution conformation at all. Thus one may find statements in the literature indicating that small peptides do not exhibit any solution conformation at all, since techniques for monitoring solution conformation of small peptides were often unable to detect a preferred structure.

9.3.2 VCD of Proteins

The first VCD spectra of proteins were published by Keiderling's group [Pancoska et al., 1989]. The sensitivity of the instrumentation was sufficient at this point to observe these data from solutions in D_2O. A large number of proteins have been studied since then, with all the work being done at Keiderling's laboratory. These proteins have included myoglobin, hemoglobin, chymotrypsin, papain, lysozyme, and ribonuclease. A number of major conclusions can be drawn from these studies. First, the secondary structures apparent from a cursory comparison of UV–CD and VCD spectra may be very different for some proteins. In particular, myoglobin, cytochrome, lysozyme, and chymotrypsin all exhibit UV–CD spectra that one would classify as mostly α-helical, but their VCD spectra are distinctly nonhelical, with the exception of myoglobin.

Second, a quantitative interpretation of the amide I' VCD region is possible. Using factor analysis methodology, reliable percentages of α-helical, β-sheet, and other contributions have been obtained, using prototypical peptide VCD spectra as basis sets. The factor analytical data appear to agree well with solid phase x-ray structures, and to a somewhat lesser extent with UV–CD data. Thus it appears that the instrumental and computational methods are available to utilize VCD as a novel spectroscopic technique for structural studies on peptides and proteins.

In Chapter 8, methods were discussed to elucidate the percent compositions of peptides and proteins via an analysis of the frequencies and infrared intensities. It should be clear at this point that the use of VCD intensities to deduce the same information is a much more sensitive method, since the different secondary structures have much more distinct patterns in VCD than in infrared absorption spectroscopy. Although this has not yet been widely recognized, it is hoped that VCD will become a more widely applied technique in the study of protein structure.

9.3.3 VCD of Canonical Nucleic Acids Models

In DNA and RNA, the dipole moments of the bases' carbonyl groups (cf. Fig. 8.6) couple and give rise to distinct, conformation-dependent VCD signals between 1550 and 1750 cm^{-1}. Since these modes are localized on groups that are achiral, the VCD is nearly entirely due to the dissymmetric coupling of these transitions according to Eq. 9.3.1, which is sensitive to the geometry between the groups.

In analogy to the discussion of the VCD of various peptide structural motifs, the VCD of the common DNA structures will be introduced first. Single stranded ribonucleic acids, such as poly(rU) and poly(rG), exist in aqueous solution at low temperature as base-stacked, helical polymers. They exhibit a slightly positively biased, positive/negative VCD couplet in the carbonyl stretching vibration of the bases. This VCD signal is attributed mostly to coupling of the C=O stretching vibrations. Poly(rA), which does not contain any carbonyl groups, exhibits a similar, but somewhat smaller, VCD signal at lower frequency. This signal was attributed to other double bond stretching vibrations, which interact with each other as do the C=O stretching vibrations. Thus VCD of nucleic acids may be observed in the absence of hydrogen bonding between complementary strands, and sufficient order exists even in single stranded polymers for dipolar coupling to occur and to produce exciton-type VCD features [Annamalai and Keiderling, 1987].

Poly(rC)·poly(rG), which forms a solution conformation of the *A-family*, exhibits a positive/negative couplet with an additional small low frequency feature (cf. Fig. 9.6). This signal is similar to the VCD observed for the deoxy analogue, poly(dC)·poly(dG).

The VCD of a double stranded DNA model, poly(dG-dC)·poly(dG-dC) in the B-form, is shown in Fig. 9.7. The B-form may be described by negative VCD at 1700 cm^{-1}, a positive peak at 1682 cm^{-1}, and a negative shoulder at 1660 cm^{-1}. The corresponding absorption peaks are at 1690, 1655, and 1625 cm^{-1}. The VCD of poly(dG)·poly(dC) in the B-form (Fig. 9.6) is similar to that of poly(dG-dC)·poly(dG-dC).

The spectra of B-form poly(dA-dT)·poly(dA-dT) and poly(dA)·poly(dT) show more complex VCD patterns than the G-C polymers. In poly(dA)·poly(dT), the highest frequency absorption peak exhibits no VCD intensity, and a large negative/positive couplet (1675/1650 cm^{-1}) is followed by a positive/negative couplet (1640/1625 cm^{-1}). In poly(dA-dT)·poly(dA-dT), there is an additional negative/positive VCD couplet at about 1700 cm^{-1} under the absorption peak, which shows no signal in poly(dA)·poly(dT) [Zhong et al., 1990].

High ionic strength solutions (e.g., aqueous solutions containing high concentrations of salt) are known to induce a phase transition to a left-handed helical form in ...CGCG... sequences. VCD detects this phase transition by a reversal of the VCD couplets. The VCD spectrum of the Z-form of poly(dG-

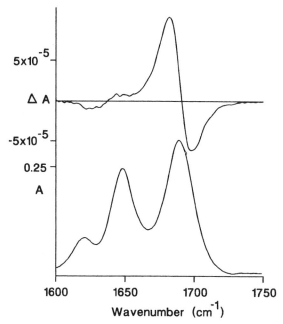

FIGURE 9.6. Observed infrared absorption (bottom) and VCD spectra (top) of poly(rG)·poly(rC).

dC)·poly(dG-dC) in 3 M NaCl exhibits a positive peak at 1677 cm^{-1} followed by a broad negative peak with minima at 1658 and 1643 cm^{-1}. The corresponding IR absorption peaks occur at approximately 1690 and 1650 cm^{-1} and the VCD spectral features of the Z-form appear positively biased [Gulotta et al., 1989].

The observed Z-form spectra depend on the choice of the counterion used to induce the B → Z transition, with Mg^{2+} causing larger VCD intensities and shifts to lower frequency. It is not clear, at this point, whether or not the spectral differences between the Na$^+$- and Mg^{2+}-induced Z-forms are due to slightly different structures or to different interactions of the metal ions with the carbonyl groups.

Triple stranded RNA, incorporating Hoogstein base pairs, can be formed by poly(rA)·poly(rU)·poly(rU) at low temperature. Yang and Keiderling [1993] reported for these polymers an overall VCD signal similar to that observed for double stranded poly(dA)·poly(dT), namely, a large negative/positive–positive/ negative couplet with an additional, positive low frequency band. The spectrum of the triple helix was also observed when poly(rA)·poly(rU) was slowly heated, and unwinding of two double helices into a triple and a single strand polymer occurred. This study demonstrated nicely the conformational sensitivity of VCD at the phase transition.

9.3.4 VCD Intensity Calculations for d(CG)$_5$ · d(CG)$_5$

VCD intensity calculations, using the DECO formalism (cf. Eqs. 9.3.1–9.3.3) were carried out for an alternating, double stranded polymer, d(CG)$_5$ · d(CG)$_5$, which can be considered to be a small segment of poly(dG-dC)·poly(dG-dC). Canonical B-form coordinates for the carbonyl geometries were used, along with a dipole moment derivative derived from the integrated band intensity of a monomeric unit. Without any adjustable parameters in the computations, the agreement between observed and computed spectral features was outstanding as shown in Fig. 9.7.

Using the geometries of G-C base pairs in an alternating polymer, the computed VCD spectra for even smaller (G-C) polymers were investigated as well. The simplest model system used was that of two base pairs:

$$
\begin{array}{cc}
\text{dG—dC} & \text{dC—dG} \\
\vdots\vdots & \vdots\vdots \\
\text{dC—dG} & \text{dG—dC} \\
\text{d(CG)} & \text{d(GC)}
\end{array}
$$

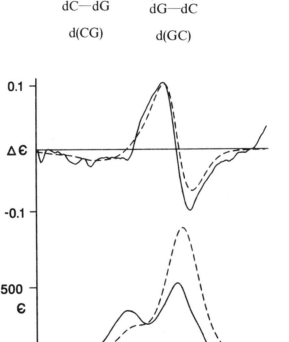

FIGURE 9.7. Observed infrared absorption (bottom) and VCD spectra (top) of poly(dG-dC)·poly(dG-dC): observed (solid trace) and calculated (dashed trace).

In these two segments, the orientations of the four carbonyl groups differ drastically, and consequently, entirely different VCD patterns are expected. In double stranded 5'd(CG)3', the four carbonyl groups are virtually all parallel, albeit in different planes. Thus the induced chirality is low, but the coupling is large, as manifested by the large splitting in the IR absorption spectra. In double stranded d(GC), on the other hand, the second set of carbonyl groups is twisted nearly 90° with respect to the first (lower) set of dipoles. This arrangement results in large induced chirality with a sign pattern similar to the one observed in poly(dG-dC)·poly(dG-dC) and d(CG)$_5$ [Zhong et al., 1990]. The arrangement pattern of dipoles found in d(CG) and d(GC) is prototypical for B-type double helices consisting of CG bases: there will be, between consecutive base pairs, either the "near parallel" alignment of dipoles or the "near 90° twist." These calculations were confirmed experimentally in the study of a number of small oligomers, to be discussed next.

9.3.5 VCD of Small Oligomers

VCD spectra have been reported for the self-complementary tetranucleotides, 5'd(CGCG)3', 5'd(GCGC)3', 5'd(CCGG)3', and 5'd(GGCC)3'. This work demonstrated that different conformational information may be obtained from solution phase VCD, electronic CD, and Raman spectroscopy.

Figure 9.8 shows the observed IR absorption and VCD spectra of CGCG and GCGC. Under these conditions, the nucleotides are thought to exist in a right-handed double stranded form, although the exact conformation is not well established (*vide infra*). The observed VCD spectrum of CGCG exhibits a negative/positive couplet that resembles that of B-form poly(dG-dC)·poly(dG-dC). Calculated VCD features for the A-form geometries agree somewhat better with the experimental data than the B-form structures.

The observed VCD spectra of CCGG and GGCC are shown in Fig. 9.9. Among the four tetramers studied, CCGG exhibits the simplest VCD spectrum, consisting basically of a simple, slightly positively biased couplet. Although previous studies agreed that CCGG is least B-like, the VCD spectrum resembles that observed for B-family poly(dG-dC)·poly(dG-dC). GGCC, on the other hand, shows the most complex VCD patterns (cf. Fig. 9.9). The VCD spectrum exhibits a negative/positive couplet (1688/1672 cm^{-1}), which is followed by a minimum at 1665 cm^{-1} and a strong positive peak at 1650 cm^{-1}. The couplets are nearly conservative. Of all tetramers studied, GGCC has the least overall agreement between observed and any of the calculated spectra [Birke et al., 1993].

The four self-complementary tetra-deoxynucleotides reported here have been studied before using different experimental methods. Raman data on all four tetramers were reported by Thomas and Peticolas [1984]. In this Raman study, the structures of the tetramers were elucidated by a comparison of their Raman spectra to reference spectra of poly(dG-dC)·poly(dG-dC) in the stan-

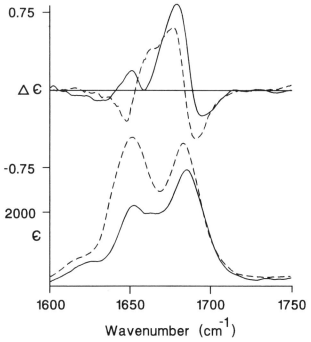

FIGURE 9.8. Observed infrared absorption (bottom) and VCD spectra (top) of 5'd(CGCG)3' (solid trace) and 5'd(GCGC)3' (dashed trace).

dard B- and Z-forms. These Raman results indicate that CGCG, GCGC, and GGCC are in a right-handed conformation similar to that of poly(dG-dC)·poly(dG-dC), and that only CCGG is markedly different from a standard B-form. CCGG, however, did not show Raman marker bands typical of the A-form.

The VCD data on all four tetramers are sufficiently different from those of $d(CG)_5$ or poly(dG-dC)·poly(dG-dC) to suggest a significant perturbation of the canonical B-form conformation. In addition, the VCD spectra of the four tetramers differ markedly among themselves, whereas the CD spectra were similar within the alternating and the nonalternating series. In general, the VCD features are best reproduced by a mixture of A- and B-form geometries. However, that does not necessarily imply that the tetramers exist in an A-form structure, but that the VCD is better reproduced for the less tightly wound geometry encountered in the canonical B-form. These changes in structure could be static or dynamic: VCD cannot differentiate whether or not there is a fast dynamic equilibration between A-, B-, or any other geometries, or whether the true geometry is static and corresponds to structural parameters in-between those of the extreme A- and B-form geometries.

The differences between the VCD spectra of the four tetramers can be explained relatively easily. The most dominant interactions giving rise to VCD intensities are those between adjacent base pairs in a double strand. Since base

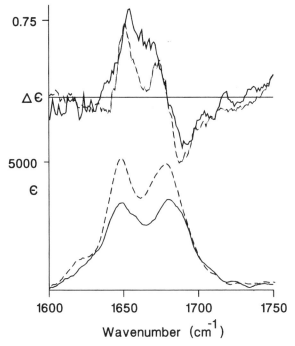

FIGURE 9.9. Observed infrared absorption (bottom) and VCD spectra (top) of 5'd(CCGG)3' (solid trace) and 5'd(GGCC)3' (dashed trace).

fraying will occur to some extent at the ends of the tetramers, the VCD spectra are basically determined by the center two base pairs. The orientation of the four carbonyl groups in the center two base pairs is either such that all four carbonyl groups are nearly parallel or are twisted by nearly 90°. Due to the handedness of the twist in the second case, a large optical activity is induced by the dipolar coupling of the transitions. Thus it is not surprising that the observed and computed VCD features of 5'd(CGCG)3' and 5d(GCGC)3' are different. Similar arguments can be made for the nonalternating tetramers.

Some discrepancies between Raman and VCD derived conformations must be sought in terms of the different regions of the molecules sampled by the two techniques, rather than their time scales. Most Raman interpretations are based on the vibrations of a certain moiety of the oligomer or polymer. When a polymer undergoes a B → Z transition, every other (deoxy-) ribose group will change from the C2'-*endo*-furanose to the *exo* conformation. However, the presence of the marker band due to the C2'-*endo*-furanose in the tetramers does not imply necessarily that the *overall* conformation is unchanged. It merely implies that the furanose conformation is unchanged and that a slight unwinding or opening of the duplex could have happened undetected.

Concluding the discussion of VCD results on DNA and DNA models, it should be emphasized that VCD reveals different aspects of the solution conformation than other spectroscopic techniques. In particular, it is important to visualize the differences between other vibrational techniques and VCD as discussed above. Finally, as pointed out in the discussion of peptide conformational analyses using vibrational spectroscopy and VCD, the additional information revealed in VCD via through-space coupling of transitions makes this technique a powerful newcomer to the available solution conformational probes.

9.4 RAMAN OPTICAL ACTIVITY

9.4.1 Phenomenological Description

ROA is the differential scattering of circularly polarized photons from chiral molecules. However, since Raman scattered light has some of the polarization properties of the incident radiation, the photons scattered during an ROA experiment maintain some of the polarization characteristics of the incident light. Consequently, there are a number of slightly different ROA experiments, depending on whether just the intensity differentials or the intensity and polarization of the scattered photons are investigated. To simplify the discussion in the context of this book, only the simplest case will be treated here, where only the intensity differential of the scattered radiation is of interest.

In this experiment (termed ICP for *incident circular polarization*), the circular intensity differential Δ is observed, which is given by

$$\Delta = \frac{I_R - I_L}{I_R + I_L} \tag{9.4.1}$$

Here, I denotes Raman intensities scattered for left and right circularly polarized light.

The denominator in Eq. 9.4.1 is just (twice) the ordinary Raman scattering and, for incident circularly polarized light, is either $6\beta^2$ or $45\alpha^2 + 7\beta^2$, depending on whether the polarization analyzer is parallel or perpendicular to the scattering plane (the YZ-plane in Fig. 5.5; cf. Section 5.5.1). The numerator of Eq. 9.4.1, namely, the differential scattering, depends on the appropriate averages of the *rotatory polarizability* tensor terms.

Just as the regular polarizability terms were defined in terms of a sum over all intermediate states of two dipole transition moments (cf. Eq. 5.3.5), the rotatory polarizability is defined as

$$G_{\alpha\beta} = \frac{-2}{\hbar} \sum_m \frac{\omega}{\omega^2 - \omega^2} \operatorname{Im}\{\langle \psi_{ev}|\mu_\alpha|\psi_{e'v'}\rangle\langle \psi_{e'v'}|m_\beta|\psi_{ev'}\rangle\} \tag{9.4.2}$$

where the meaning of the symbols is the same as in Eq. 5.3.5, and m is the magnetic dipole operator, defined in Eq. 9.1.5. Thus it can be seen that ROA, like all optical activity, is due to the interaction of electric and magnetic transition moments. In contrast to absorption measurements, where the derivative $\partial(\mu \cdot m)/\partial Q$ is measured directly, the inelastic light scattering experiment proceeds through the intermediate, virtual states, which is expressed in Eq. 9.4.2.

The expression in the numerator contains orientationally averaged cross terms between α and G. Polavarapu [1990] defined the necessary averages as follows:

$$2\gamma^2 = (\alpha_{xx} - \alpha_{yy})(G_{xx} - G_{yy}) + (\alpha_{xx} - \alpha_{zz})(G_{xx} - G_{zz}) + (\alpha_{yy} - \alpha_{zz})(G_{yy} - G_{zz})$$
$$+ 6[\alpha_{xy}G_{xy} + \alpha_{xz}G_{xz} + \alpha_{yz}G_{yz}] \quad (9.4.3)$$

in analogy to Eq. 5.5.24. In Eq. 9.4.3, the polarizability and rotatory polarizability terms listed are actually the derivatives with respect to the normal coordinates, as discussed before in the case of Raman spectroscopy. If the polarizability tensor terms themselves are used in Eq. 9.4.3, the equivalent of Rayleigh optical activity would be described.

In addition to rotatory polarizability terms (also known as electric dipole–magnetic dipole polarizability), the interference between electric dipole and electric quadrupole transition moments also gives rise to circular differential scattering. The electric dipole–electric quadrupole polarizability $A_{\alpha\beta\gamma}$ is given by an expression similar to that of the regular polarizability, except that the second transition moment is substituted by terms of the form

$$\langle \psi_{e'v'} | \Theta_{\beta\gamma} | \psi_{ev'} \rangle \quad (9.4.4)$$

where Θ is the electric quadrupole transition operator, defined by

$$\Theta_{\alpha\beta} = \frac{1}{2} \sum e_i (3 r_{i\alpha} r_{i\beta} - r_i^2 \delta_{\alpha\beta}) \quad (9.4.5)$$

where r is the position vector of particle i, and δ is the Kronecker delta symbol. The appropriate form of the electric dipole–electric quadrupole polarizability terms, averaged over all possible orientations, is somewhat more complicated than for regular polarizability (Eq. 5.5.24) and electric dipole–magnetic dipole polarizability (Eq. 9.4.3). Let it suffice here that this corresponding average is denoted by δ^2, the exact form of which has been published [Polavarapu, 1990].

Using the anisotropies of the αG and αA, and the isotropic parts $\overline{\alpha G}$, the "polarized" intensity differential is given by

$$(I_R - I_L)_X = (2/c)[45\overline{\alpha G} + 7\gamma^2 + \delta^2] \quad (9.4.6)$$

whereas the depolarized intensity differential is

$$(I_R - I_L)_Z = (12/c)[7\gamma^2 - \tfrac{1}{3}\delta^2] \quad (9.4.7)$$

The analyzer in the scattered beam can be set at an angle, termed the "magic angle" by Hecht and Barron [1990], where the contributions of the quadrupole polarizability cancel. For observation of this particular experiment, the analyzer is set at an angle of 35.3° with respect to the X-axis. In this case, the Raman intensity is given by $30\alpha^2 + \frac{20}{3}\beta^2$, and the corresponding differential intensities by

$$(I_R - I_L)_M = (20/3c)[9\overline{\alpha G} + 2\gamma^2] \quad (9.4.8)$$

Typical differential scattering intensities are about 10^{-3} to 10^{-4} of the Raman intensities. Although this effect is somewhat larger than VCD discussed before, its observation is significantly more complicated than that of VCD, and much more susceptible to artifacts.

9.4.2 Observation of ROA

The classical ROA experiment (as opposed to the one where the polarization of the scattered photons is measured) is carried out using a standard Raman instrument to which light modulation optics has been added. Since Raman scattering is a weak effect, multichannel detection of the scattered light is required to achieve reasonable data acquisition times. All modern multichannel detectors, either diode arrays or CCDs (cf. Section 6.3.2), are relatively slow in their readout procedure. Thus it is advantageous to modulate the incident laser relatively slowly between left and right circular polarization states and read the array detector only once during each modulation cycle. The slow modulation rate (typically less than 1 Hz) precludes the use of photoelastic modulators, discussed in Section 9.1.2. Furthermore, it is advantageous to switch directly from one circular polarization state to the other in ROA, unlike in VCD, where the sinusoidal modulation of the PEM produces intermediate (i.e., circular, elliptical, linear, elliptical and circular) polarization states during a modulation half-cycle. Therefore electro-optic modulators (EOM) are commonly used in ROA. In an EOM, the birefringence needed to produce circularly polarized light is induced by a high voltage applied across a uniaxial crystal, such as KD_2PO_4, along the propagation direction of the laser. By reversing the direction of the electric field applied to the crystal, both left and right circularly polarized light can be produced. A detailed account of the design of state-of-the art ROA instrumentation was recently published [Hecht et al., 1992].

ROA dates back to the early 1970s, when scanning monochromators, photomultipliers, and dual channel photon counters were used for the observation of ROA. Although some effort was expended to utilize diode array detectors for the observation of multichannel ROA, it was not until the end of the 1980s, when CCDs became readily available, that ROA became a truly useful chiroptical technique. Although diode array spectrometers could produce ROA features of neat liquids in a few hours of data acquisition time, they basically failed for aqueous solutions, and therefore no ROA of biological molecules was reported until 1989. However, with CCD-based ROA instruments, a number of publications have appeared for peptides and proteins, to be discussed in Section 9.6.

9.5 APPLICATIONS OF ROA: SMALL MOLECULES

More than 90% of experimental work and a similar amount of theoretical efforts of ROA have come from the laboratory of Barron [1982]. The first observation of ROA by Barron [for a review, see Barron, 1989] nearly two decades ago were for small cyclic molecules, such as α-pinene, or asymmetric species, such as α-phenylethylamine. In particular, the cyclic molecules with a rigid, bicyclic structure were found to exhibit a rich ROA spectrum, although the difficulties in the interpretation of the vibrational spectra, aside from qualitative, group-frequency-based arguments, precluded a detailed interpretation on the origin of ROA. Some molecules containing locally symmetric moieties such as methyl groups showed the near-degenerate vibrations of E symmetry to be split into positive/negative couplets due to the lifting of the local symmetry, and chiral perturbation of these vibrations. The antisymmetric methyl deformation modes of α-phenylethylamine, at about $1460\,cm^{-1}$, were found early to be split into a couplet as described in the case of VCD. However, the different amount of vibrational coupling of these modes with the aromatic ring deformations, and other nearby vibrations, precluded these vibrations to be employed as a probe for local stereochemistry. Incidentally, similar intensity patterns in near-degenerate, antisymmetric methyl stretching vibrations were observed in VCD.

A number of qualitative interpretations of species such as 3-methylcyclohexanone, and comparisons of related species, such as pinane and α- and β-pinene, and a few other molecules were published; however, in the early years of ROA, limitations on the experimental and theoretical fronts kept the progress relatively slow.

However, enormous progress was made over the past five years or so, on both the experimental (*vide supra*) and theoretical fronts. The formalism for *ab initio* computation of ROA intensities is now possible, mostly due to the pioneering efforts in the laboratories of Barron [Barron et al., 1992] and Polavarapu [1990]. As pointed out in Chapter 5, methods to compute Raman intensities were developed, to a large extent, during the efforts to develop the formalism for ROA computations. For both, an elegant method was developed to compute the polarizability or rotatory polarizability via a static electric field perturbation, and not as a sum over all excited states. The derivatives of the polarizability tensor with respect to the normal coordinates can be performed analytically, whereas a numerical derivative needs to be evaluated for the rotatory polarizability, due to the form of the magnetic moment operator.

The small molecules investigated and reported on recently in ROA studies are basically the same listed among the small molecule studies for VCD. Good agreement between the observed and computed ROA spectra for oxirane and thiirane have been reported, and it appears that the computations are reliable. With the ever increasing computing power available, it is anticipated that the ROA of molecules much larger than the present limit of about 10 atoms can be interpreted quantitatively in the near future.

9.6 APPLICATIONS OF ROA: BIOLOGICAL MOLECULES

The first reported ROA of a biological molecule is that of the resonance enhanced ROA spectrum of cytochrome c [Barron, 1975], at a time when the observation of ROA was still in its infancy. It was found that resonance ROA, in general, may be a very difficult experiment to carry out, due to the circular dichroism accompanying the visible absorption of a chiral molecule. However, in the case of cytochrome c, it turned out that at a wavelength of 514.5 nm, the resonance enhancement is strong (cf. Section 8.2.2.1), but the CD is virtually zero. Thus resonance ROA and magnetic resonance ROA were obtained with minimal interference from the electronic CD of the sample.

Nonresonance ROA of biological molecules as aqueous solutions was not possible until the late 1980s, when the first CCD-based ROA spectrometers were built. It is now possible to collect ROA spectra of (relatively concentrated) aqueous solutions in a few hours of data acquisition time. The ROA spectra of a number of amino acids, small peptides, proteins, carbohydrates, and a few other species have been reported.

ROA of alanine, both in the zwitterionic form in neutral water and in 6 N HCl and 6 N NaOH, has been reported by Barron et al. [1992]. These studies revealed an enormous change in the Raman as well as the ROA spectra of the three forms of alanine, present under these conditions, which are $H_3N^+-CH(CH_3)-CO_2^-$, $H_3N^+-CH(CH_3)-CO_2H$, and $H_2N-CH(CH_3)-CO_2^-$, respectively. *Ab initio* ROA and Raman intensities were reported along with the experimental studies. Alanyl–alanine and the corresponding tripeptide were reported Barron et al. [1992]. Interestingly, the ROA spectra of the di-, tri-, and tetra-alanine are very similar. Thus it appears that in ROA there is no long-range coupling mechanism that produces direct conformational sensitivity, at least not in the amide I signals, which are negative/positive couplets for most peptides reported so far.

Whenever there is a direct comparison possible between observed VCD and ROA spectra, one finds that there is no correlation between the two forms of vibrational optical activity. In VCD, for example, the amide I signal is the most predominant feature in the spectra of most peptides and proteins. In ROA, on the other hand, di- and tripeptides as well as the few proteins reported so far show most intensity in the amide III region. However, sufficient data are not yet available to ascertain whether or not the amide I and amide III vibrations are conformationally sensitive in ROA.

REFERENCES

A. Annamalai and T. A. Keiderling, *J. Am. Chem. Soc. 109*, 3125 (1987).

A. Annamalai, T. A. Keiderling, and J. S. Chickos, *J. Am. Chem. Soc., 107*, 2285 (1985).

L. D. Barron, *Nature (London), 257*, 372 (1975).

L. D. Barron, *Molecular Light Scattering and Optical Activity*, Cambridge University Press, Cambridge, UK, 1982.

L. D. Barron, in *Vibrational Spectra and Structure*, Vol. 17B, H. D. Bist, J. R. Durig, and J. F. Sullivan, Eds., Elsevier, Amsterdam, The Netherlands, 1989, p. 343.

L. D. Barron, A. R. Gargaro, L. Hecht, and P. L. Polaravapu, *Spectrochim. Acta, 48A*, 261 (1992).

L. D. Barron, A. R. Gargaro, L. Hecht, P. L. Polaravapu, and H. Sugeta, *Spectrochim. Acta, 48A*, 1051 (1992).

L. D. Barron, Z. Q. Wen, and L. Hecht, *J. Am. Chem. Soc., 114*, 784 (1992).

S. S. Birke, I. Agbaje, and M. Diem, *Biochemistry 31*, 450 (1992).

S. S. Birke, M. Moses, M. Gulotta, B. Kagarlovsky, D. Jano, and M. Diem, *Biophysical Journal, 65*, 1262 (1993).

S. J. Cianciosi, N. Ragunathan, T. B. Freedman, L. A. Nafie, and J. E. Baldwin, *J. Am. Chem. Soc., 112*, 8204 (1990).

M. Diem, G. M. Roberts, O. Lee, and A. Barlow, *Appl. Spectrosc., 42*, 20 (1988).

R. K. Dukor and T. A. Keiderling, *Biopolymers, 31*, 1747 (1991).

T. B. Freedman and L. A. Nafie, *J. Phys. Chem., 88*, 496 (1984).

M. Gulotta, D. J. Goss, and M. Diem, *Biopolymers, 28*, 2047 (1989).

L. Hecht and L. D. Barron, *Appl. Spectrosc., 44*, 483 (1990).

L. Hecht, L. D. Barron, A. R. Gargaro, Z. Q. Wen, and W. Hug, *J. Raman Spectrosc., 23*, 401 (1992).

O. Lee, Ph.D. Dissertation, City University of New York, 1992.

O. Lee and M. Diem, *Anal. Instrument., 20*, 23 (1992).

O. Lee, G. M. Roberts, and M. Diem, *Biopolymers, 29*, 1759 (1989).

L. A Nafie and T. H. Walnut, *Chem. Phys. Lett., 49*, 441 (1977).

P. Pancoska, S. C. Yasui, and T. A. Keiderling, *Biochemistry, 28*, 5917 (1989).

G. M. Paterlini, T. B. Freedman, and L. A. Nafie, *Biopolymers, 25*, 1751 (1986).

S. T. Pickard, H. E. Smith, P. L. Polavarapu, T. M. Black, A. Rauk, and D. Yang, *J. Am. Chem. Soc., 114*, 6850 (1992).

P. L. Polavarapu, *J. Phys. Chem., 94*, 8106 (1990).

P. L. Polavarapu, S. T. Pickard, H. E. Smith, T. M. Black, A. Rauk, and D. Yang, *J. Am. Chem. Soc., 113*, 9747 (1991).

P. K. Sengupta and S. Krimm, *Biopolymers, 26*, S99 (1987).

I. Tinoco, *Radiat. Res., 20*, 133 (1963).

G. A. Thomas and W. L. Peticolas, *Biochemistry, 23*, 3202 (1984).

L. Yang and T. A. Keiderling, *Biopolymers, 33*, 315 (1993).

S. C. Yasui and T. A. Keiderling, *Biopolymers, 25*, 5 (1986).

W. Zhong, M. Gulotta, D. J. Goss, and M. Diem, *Biochemistry, 29*, 7485 (1990).

APPENDIX I

VIBRATIONAL FREQUENCIES [cm^{-1}] FOR SOME COMMON GROUPS

Frequency Range	Group	Description of Vibration
3400	ROH	OH stretch
3330–3400	—NH$_2$	Antisymmetric stretch
3250–3300	—NH$_2$	Symmetric stretch
3300–3330	≡CH	Acetylenic C—H stretch
3050–3065	Arom-H	Aromatic C—H stretch
2960–3020	—CH$_3$	Antisymmetric stretch
2910–2930	—CH$_2$	Antisymmetric stretch
2880–2970	—CH$_3$	Symmetric stretch
2850–2860	—CH$_2$	Symmetric stretch
2560–2590	RSH	Thiole SH stretch
2350–2600	B—H	
2275–2450	P—H	
2230–2300	C≡C	CC triple bond stretch
2230–2260	C≡N	CN triple bond stretch
2100–2250	Si—H	
2100–2250	C—D	Various CD, CD$_2$, CD$_3$ group vibrations
1760	C=O	Organic acids
1700–1725	C=O	Ketones
1650–1725	C=O	Amide I
1650–1660	C=N	
1625–1700	C=O	Amide I'
1640–1660	C=C	
1595–1600	—CO$_2^-$	Antisymmetric carboxylate stretch

VIBRATIONAL FREQUENCIES FOR SOME COMMON GROUPS

Frequency Range	Group	Description of Vibration
1450–1600	N=O	NO stretch in organic nitrates
1440–1470	—CH_3	Antisymmetric deformation
1440–1470	—CH_2	Symmetric and antisymmetric deformation
1420–1430	C—OH	Carbon–oxygen stretch in organic acids
1390–1410	—CO_2^-	Symmetric carboxylate stretch
1370–1390	—CH_3	Symmetric deformation
1350	—CF_3	
1250–1340		Amide III
1250–1330	—CH	Methine deformation
1250–1265	—COH	COH deformation, organic acids
1260		Ring breathing, oxirane
1140–1300	P=O	
1120–1300	CF_2	
1100–1200	S=O	Sulfoxides, sulfonic acids
1060–1120	C—F	
1100–1300	C—O	
950–1150	C—C	
1040–1070	S=O	
1060	C=S	
1000–1030	CCH	Aromatic in-plane C—H deformation
992		Benzene ring breathing
830–930	COC	Symmetric stretch, ethers
720–730	C—Cl	
705	—CCl_2	Symmetric stretch
665	—CCl_3	Symmetric stretch
640–650	C—Br	
610	—CBr_2	Symmetric stretch
540	—CBr_3	Symmetric stretch
510–525	—S—S—	
520	C—I	

APPENDIX II

REFINED SET OF UREY–BRADLEY FORCE CONSTANTS

(a) Stretching Force Constants

Designation	Value	Molecule
$K(C-H)$	4.555	CH_3-Cl
	4.150	C*—H, alanine
	3.883	$CHCl_3$
$K(C-C)$	2.425	CH_3-CCl_3
	2.400	$CH_3-CHFClBr$
	2.400	C*—CO_2, alanine
	1.800	C*—CH_3, alanine
$K(C-N)$	6.50	C—N, peptide linkage, $1\frac{1}{2}$ bond order
	2.400	C*—N, alanine
$K(C=O)$	10.0	R—C=O(—OH), organic acids
	8.40	C=O, peptide linkage
$K(C-O)$	7.45	—CO_2^- anion,
$K(C-O)$	6.90	R—C—OH(=O), organic acids
$K(C-F)$	3.95	HCFCl—R
$K(C-Cl)$	2.155	CH_3Cl
	2.00	HCFClBr
	1.86	—CCl_3
$K(C-Br)$	1.56	HCFClBr
$K(N-H)$	5.100	NH_3^+, alanine
$K(O-H)$	6.70	RC(=O)O—H, organic acids

(b) Deformation Force Constants				
Diagonal		Off-diagonal		
Designation	Value	Designation	Value	Molecule
H(H—C—H)	0.450	F(H—C—H)	0.1	CH_3Cl
H(H—C—C)	0.290	F(H—C—C)	0.655	CCl_3—CH_3
H(H—C*—C)	0.390	F(H—C*—C)	0.78	Alanine
H(H—C—F)	0.10	F(H—C—F)	0.10	HCFCl—R
H(H—C—Cl)	0.350	F(H—C—Cl)	0.60	CH_3Cl
H(C—C—C)	0.500	F(C—C—C)	0.500	Alanine
H(C—C—N)	0.720	F(C—C—N)	0.70	Alanine
H(C—C—O)	0.300	F(C—C—O)	1.00	Alanine
H(C—C—F)	0.420	F(C—C—F)	1.10	HCFCl—R
H(C—C—Cl)	0.610	F(C—C—Cl)	0.60	CCl_3—CH_3
H(N—C—O)	0.45	F(N—C—O)	1.5	Peptide
H(O—C—O)	0.130	F(O—C—O)	3.1	—CO_2^-
H(Cl—C—Cl)	0.15	F(Cl—C—Cl)	0.7	CCl_3—CH_3

(c) Out-of-Plane Deformation (Wagging)		
Designation	Value	Molecule
W(C—O)	0.500	Peptide linkage
W(N—H)	0.185	Peptide linkage
$W(CO_2)$	0.429	—CO_2^-

APPENDIX III

CHARACTER TABLES FOR CHEMICALLY IMPORTANT SYMMETRY GROUPS

THE NONAXIAL GROUPS

C_1	E
A	1

C_s	E	σ_h		
A'	1	1	x, y, R_z	x^2, y^2, z^2, xy
A''	1	-1	z, R_x, R_y	yz, xz

C_i	E	i		
A_g	1	1	R_x, R_y, R_z	$x^2, y^2, z^2, xy, xz, yz$
A_u	1	-1	x, y, z	

THE C_n GROUPS

C_2	E	C_2		
A	1	1	z, R_z	x^2, y^2, z^2, xy
B	1	-1	x, y, R_x, R_y	yz, xz

266

CHARACTER TABLES

C_3	E	C_3	C_3^2			$\varepsilon = \exp(2\pi i/3)$
A	1	1	1		z, R_z	x^2+y^2, z^2
E	$\begin{cases}1\\1\end{cases}$	$\begin{matrix}\varepsilon\\ \varepsilon^*\end{matrix}$	$\begin{matrix}\varepsilon^*\\ \varepsilon\end{matrix}\Big\}$		$(x,y)(R_x, R_y)$	$(x^2-y^2, xy)(yz, xz)$

C_4	E	C_4	C_2	C_4^3		
A	1	1	1	1	z, R_z	x^2+y^2, z^2
B	1	-1	1	-1		x^2-y^2, xy
E	$\begin{cases}1\\1\end{cases}$	$\begin{matrix}i\\-1\end{matrix}$	$\begin{matrix}-1\\-1\end{matrix}$	$\begin{matrix}-i\\i\end{matrix}\Big\}$	$(x,y)(R_x, R_y)$	(yz, xz)

C_5	E	C_5	C_5^2	C_5^3	C_5^4			$\varepsilon = \exp(2\pi i/5)$
A	1	1	1	1	1		z, R_z	x^2-y^2, z^2
E_1	$\begin{cases}1\\1\end{cases}$	$\begin{matrix}\varepsilon\\\varepsilon^*\end{matrix}$	$\begin{matrix}\varepsilon^2\\\varepsilon^{2*}\end{matrix}$	$\begin{matrix}\varepsilon^{2*}\\\varepsilon^2\end{matrix}$	$\begin{matrix}\varepsilon^*\\\varepsilon\end{matrix}\Big\}$		$(x,y)(R_x, R_y)$	(yz, xz)
E_2	$\begin{cases}1\\1\end{cases}$	$\begin{matrix}\varepsilon^2\\\varepsilon^{2*}\end{matrix}$	$\begin{matrix}\varepsilon^*\\\varepsilon\end{matrix}$	$\begin{matrix}\varepsilon\\\varepsilon^*\end{matrix}$	$\begin{matrix}\varepsilon^{2*}\\\varepsilon^2\end{matrix}\Big\}$			(x^2-y^2, xy)

C_6	E	C_6	C_3	C_2	C_3^2	C_6^5			$\varepsilon = \exp(2\pi i/6)$
A	1	1	1	1	1	1		z, R_z	x^2+y^2, z^2
B	1	-1	1	-1	1	-1			
E_1	$\begin{cases}1\\1\end{cases}$	$\begin{matrix}\varepsilon\\\varepsilon^*\end{matrix}$	$\begin{matrix}-\varepsilon^*\\-\varepsilon\end{matrix}$	$\begin{matrix}-1\\-1\end{matrix}$	$\begin{matrix}-\varepsilon\\-\varepsilon^*\end{matrix}$	$\begin{matrix}\varepsilon^*\\\varepsilon\end{matrix}\Big\}$		(x,y) (R_x, R_y)	(xz, yz)
E_2	$\begin{cases}1\\1\end{cases}$	$\begin{matrix}-\varepsilon^*\\-\varepsilon\end{matrix}$	$\begin{matrix}-\varepsilon\\-\varepsilon^*\end{matrix}$	$\begin{matrix}1\\1\end{matrix}$	$\begin{matrix}-\varepsilon^*\\-\varepsilon\end{matrix}$	$\begin{matrix}-\varepsilon\\-\varepsilon^*\end{matrix}\Big\}$			(x^2-y^2, xy)

C_7	E	C_7	C_7^2	C_7^3	C_7^4	C_7^5	C_7^6			$\varepsilon = \exp(2\pi i/7)$
A	1	1	1	1	1	1	1		z, R_z	x^2-y^2, z^2
E_1	$\begin{cases}1\\1\end{cases}$	$\begin{matrix}\varepsilon\\\varepsilon^*\end{matrix}$	$\begin{matrix}\varepsilon^2\\\varepsilon^{2*}\end{matrix}$	$\begin{matrix}\varepsilon^3\\\varepsilon^{3*}\end{matrix}$	$\begin{matrix}\varepsilon^{3*}\\\varepsilon^3\end{matrix}$	$\begin{matrix}\varepsilon^{2*}\\\varepsilon^2\end{matrix}$	$\begin{matrix}\varepsilon^*\\\varepsilon\end{matrix}\Big\}$		(x,y) (R_x, R_y)	(xz, yz)
E_2	$\begin{cases}1\\1\end{cases}$	$\begin{matrix}\varepsilon^2\\\varepsilon^{2*}\end{matrix}$	$\begin{matrix}\varepsilon^{3*}\\\varepsilon^3\end{matrix}$	$\begin{matrix}\varepsilon^*\\\varepsilon\end{matrix}$	$\begin{matrix}\varepsilon\\\varepsilon^*\end{matrix}$	$\begin{matrix}\varepsilon^3\\\varepsilon^{3*}\end{matrix}$	$\begin{matrix}\varepsilon^{2*}\\\varepsilon^2\end{matrix}\Big\}$			(x^2-y^2, xy)
E_3	$\begin{cases}1\\1\end{cases}$	$\begin{matrix}\varepsilon^3\\\varepsilon^{3*}\end{matrix}$	$\begin{matrix}\varepsilon^*\\\varepsilon\end{matrix}$	$\begin{matrix}\varepsilon^2\\\varepsilon^{2*}\end{matrix}$	$\begin{matrix}\varepsilon^{2*}\\\varepsilon^2\end{matrix}$	$\begin{matrix}\varepsilon\\\varepsilon^*\end{matrix}$	$\begin{matrix}\varepsilon^{3*}\\\varepsilon^3\end{matrix}\Big\}$			

C_8	E	C_3	C_4	C_2	C_7^3	C_8^3	C_8^5	C_8^7		$\varepsilon = \exp(2\pi i/8)$
A	1	1	1	1	1	1	1	1	z, R_z	$x^2 - y^2, z^2$
B	1	-1	1	1	1	-1	-1	-1		
E_1	$\begin{cases} 1 \\ 1 \end{cases}$	$\begin{matrix}\varepsilon \\ \varepsilon^*\end{matrix}$	$\begin{matrix}i \\ -i\end{matrix}$	$\begin{matrix}-1 \\ -1\end{matrix}$	$\begin{matrix}-i \\ i\end{matrix}$	$\begin{matrix}-\varepsilon^* \\ -\varepsilon\end{matrix}$	$\begin{matrix}-\varepsilon \\ -\varepsilon^*\end{matrix}$	$\begin{matrix}\varepsilon^* \\ \varepsilon\end{matrix}$	$\begin{matrix}(x, y) \\ (R_x, R_y)\end{matrix}$	(xz, yz)
E_2	$\begin{cases} 1 \\ 1 \end{cases}$	$\begin{matrix}i \\ -i\end{matrix}$	$\begin{matrix}-1 \\ -1\end{matrix}$	$\begin{matrix}1 \\ 1\end{matrix}$	$\begin{matrix}-1 \\ -1\end{matrix}$	$\begin{matrix}-i \\ i\end{matrix}$	$\begin{matrix}i \\ -i\end{matrix}$	$\begin{matrix}-i \\ i\end{matrix}$		$(x^2 - y^2, xy)$
E_3	$\begin{cases} 1 \\ 1 \end{cases}$	$\begin{matrix}-\varepsilon \\ -\varepsilon^*\end{matrix}$	$\begin{matrix}i \\ -i\end{matrix}$	$\begin{matrix}-1 \\ -1\end{matrix}$	$\begin{matrix}-i \\ i\end{matrix}$	$\begin{matrix}\varepsilon^* \\ \varepsilon\end{matrix}$	$\begin{matrix}\varepsilon \\ \varepsilon^*\end{matrix}$	$\begin{matrix}-\varepsilon^* \\ -\varepsilon\end{matrix}$		

THE D_n GROUPS

D_2	E	$C_2(z)$	$C_2(y)$	$C_2(x)$		
A	1	1	1	1		x^2, y^2, z^2
B_1	1	1	-1	-1	z, R_z	xy
B_2	1	-1	1	-1	y, R_y	xz
B_3	1	-1	-1	1	x, R_x	yz

D_3	E	$2C_3$	$3C_2$		
A_1	1	1	1		$x^2 - y^2, z^2$
A_2	1	1	-1	z, R_z	
E	2	-1	0	$(x, y)(R_x, R_y)$	$(x^2 - y^2, xy)(xz, yz)$

D_4	E	$2C_4$	$C_2(=C_4^2)$	$2C_2'$	$2C_2''$		
A_1	1	1	1	1	1		$x^2 - y^2, z^2$
A_2	1	1	1	-1	-1	z, R_z	
B_1	1	-1	1	1	-1		$x^2 - y^2$
B_2	1	-1	1	-1	1		xy
E	2	0	-2	0	0	$(x, y)(R_x, R_y)$	(xz, yz)

D_5	E	$2C_5$	$2C_5^2$	$5C_2$		
A_1	1	1	1	1		$x^2 - y^2, z^2$
A_2	1	1	1	-1	z, R_z	
E_1	2	$2\cos 72°$	$2\cos 144°$	0	$(x, y)(R_x, R_y)$	(xz, yz)
E_2	2	$2\cos 144°$	$2\cos 72°$	0		$(x^2 - y^2, xy)$

CHARACTER TABLES

D_6	E	$2C_6$	$2C_3$	C_2	$3C_2'$	$3C_2''$		
A_1	1	1	1	1	1	1		$x^2 - y^2, z^2$
A_2	1	1	1	1	−1	−1	z, R_z	
B_1	1	−1	1	−1	1	−1		
B_2	1	−1	1	−1	−1	1		
E_1	2	1	−1	−2	0	0	$(x, y)(R_x, R_y)$	(xz, yz)
E_2	2	−1	−1	2	0	0		$(x^2 - y^2, xy)$

THE C_{nv} GROUPS

C_{2v}	E	C_2	$\sigma_v(xz)$	$\sigma_v'(yz)$		
A_1	1	1	1	1	z	x^2, y^2, z^2
A_2	1	1	−1	−1	R_z	xy
B_1	1	−1	1	−1	x, R_y	xz
B_2	1	−1	−1	1	y, R_x	yz

C_{3v}	E	$2C_3$	$3\sigma_v$		
A_1	1	1	1	z	$x^2 - y^2, z^2$
A_2	1	1	−1	R_z	
E	2	−1	0	$(x, y)(R_x, R_y)$	$(x^2 - y^2, xy)(xz, yz)$

C_{4v}	E	$2C_4$	C_2	$2\sigma_v$	$2\sigma_d$		
A_1	1	1	1	1	1	z	$x^2 - y^2, z^2$
A_2	1	1	1	−1	−1	R_z	
B_1	1	−1	1	1	−1		$x^2 - y^2$
B_2	1	−1	1	−1	1		xy
E	2	0	−2	0	0	$(x, y)(R_x, R_y)$	(xz, yz)

C_{5v}	E	$2C_5$	$2C_5^2$	$5\sigma_v$		
A_1	1	1	1	1	z	$x^2 + y^2, z^2$
A_2	1	1	1	−1	R_z	
E_1	2	$2\cos 72°$	$2\cos 144°$	0	$(x, y)(R_x, R_y)$	(xz, yz)
E_2	2	$2\cos 144°$	$2\cos 72°$	0		$(x^2 - y^2, xy)$

C_{6v}	E	$2C_6$	$2C_3$	C_2	$3\sigma_v$	$3\sigma_d$		
A_1	1	1	1	1	1	1	z	$x^2 + y^2, z^2$
A_2	1	1	1	1	−1	−1	R_z	
B_1	1	−1	1	−1	1	−1		
B_2	1	−1	1	−1	−1	1		
E_1	2	1	−1	−2	0	0	$(x, y)(R_x, R_y)$	(xz, yz)
E_2	2	−1	−1	2	0	0		$(x^2 - y^2, xy)$

THE C_{nh} GROUPS

C_{2h}	E	C_2	i	σ_h		
A_g	1	1	1	1	R_z	x^2, y^2, z^2, xy
B_g	1	-1	1	-1	R_x, R_y	xz, yz
A_u	1	1	-1	-1	z	
B_u	1	-1	-1	1	x, y	

C_{3h}	E	C_3	C_3^2	σ_h	S_3	S_3^5			$\varepsilon = \exp(2\pi i/3)$
A'	1	1	1	1	1	1	R_z	x^2+y^2, z^2	
E'	$\begin{cases}1\\1\end{cases}$	ε ε^*	ε^* ε	1 1	ε ε^*	ε^* ε	(x, y)	(x^2-y^2, xy)	
A''	1	1	1	-1	-1	-1	z		
E''	$\begin{cases}1\\1\end{cases}$	ε ε^*	ε^* ε	-1 -1	$-\varepsilon$ $-\varepsilon^*$	$-\varepsilon^*$ $-\varepsilon$	(R_x, R_y)	(xz, yz)	

C_{4h}	E	C_4	C_2	C_4^3	i	S_4^3	σ_h	S_4		
A_g	1	1	1	1	1	1	1	1	R_z	x^2+y^2, z^2
B_g	1	-1	1	-1	1	-1	1	-1		x^2-y^2, xy
E_g	$\begin{cases}1\\1\end{cases}$	i $-i$	-1 -1	$-i$ i	1 1	i $-i$	-1 -1	$-i$ i	(R_x, R_y)	(xz, yz)
A_u	1	1	1	1	-1	-1	-1	-1	z	
B_u	1	-1	1	-1	-1	1	-1	1		
E_u	$\begin{cases}1\\1\end{cases}$	i $-i$	-1 -1	$-i$ i	-1 -1	$-i$ i	1 1	i $-i$	(x, y)	

C_{5h}	E	C_5	C_5^2	C_5^3	C_5^4	σ_h	S_5	S_5^7	S_5^3	S_5^9			$\varepsilon = \exp(2\pi i/5)$
A'	1	1	1	1	1	1	1	1	1	1	R_z	x^2+y^2, z^2	
E_1'	$\begin{cases}1\\1\end{cases}$	ε ε^*	ε^2 ε^{2*}	ε^{2*} ε^2	ε^* ε	1 1	ε ε^*	ε^2 ε^{2*}	ε^{2*} ε^2	ε^* ε	(x, y)		
E_2'	$\begin{cases}1\\1\end{cases}$	ε^2 ε^{2*}	ε^* ε	ε ε^*	ε^{2*} ε^2	1 1	ε^2 ε^{2*}	ε^* ε	ε ε^*	ε^{2*} ε^2		(x^2-y^2, xy)	
A''	1	1	1	1	1	-1	-1	-1	-1	-1	z		
E_1''	$\begin{cases}1\\1\end{cases}$	ε ε^*	ε^2 ε^{2*}	ε^{2*} ε^2	ε^* ε	-1 -1	$-\varepsilon$ $-\varepsilon^*$	$-\varepsilon^2$ $-\varepsilon^{2*}$	$-\varepsilon^{2*}$ $-\varepsilon^2$	$-\varepsilon^*$ $-\varepsilon$	(R_x, R_y)	(xz, yz)	
E_2''	$\begin{cases}1\\1\end{cases}$	ε^2 ε^{2*}	ε^* ε	ε ε^*	ε^{2*} ε^2	-1 -1	$-\varepsilon^2$ $-\varepsilon^{2*}$	$-\varepsilon^*$ $-\varepsilon$	$-\varepsilon$ $-\varepsilon^*$	$-\varepsilon^{2*}$ $-\varepsilon^2$			

CHARACTER TABLES   271

C_{6h}	E	C_6	C_3	C_2	C_3^2	C_6^5	i	S_3^5	S_6^5	σ_h	S_6	S_3		$\varepsilon=\exp(2\pi i/6)$
A_g	1	1	1	1	1	1	1	1	1	1	1	1	R_z	x^2+y^2, z^2
B_g	1	-1	1	-1	1	-1	1	-1	1	-1	1	-1		
E_{1g}	1	ε	$-\varepsilon^*$	-1	$-\varepsilon$	ε^*	1	ε	$-\varepsilon^*$	-1	$-\varepsilon$	ε^*	(R_x, R_y)	(xz, yz)
	1	ε^*	$-\varepsilon$	-1	$-\varepsilon^*$	ε	1	ε^*	$-\varepsilon$	-1	$-\varepsilon^*$	ε		
E_{2g}	1	$-\varepsilon^*$	$-\varepsilon$	1	$-\varepsilon^*$	$-\varepsilon$	1	$-\varepsilon^*$	$-\varepsilon$	1	$-\varepsilon^*$	$-\varepsilon$		(x^2-y^2, xy)
	1	$-\varepsilon$	$-\varepsilon^*$	1	$-\varepsilon$	$-\varepsilon^*$	1	$-\varepsilon$	$-\varepsilon^*$	1	$-\varepsilon$	$-\varepsilon^*$		
A_u	1	1	1	1	1	1	-1	-1	-1	-1	-1	-1	z	
B_u	1	-1	1	-1	1	-1	-1	1	-1	1	-1	1		
E_{1u}	1	ε	$-\varepsilon^*$	-1	$-\varepsilon$	ε^*	-1	$-\varepsilon$	ε^*	1	ε	$-\varepsilon^*$	(x, y)	
	1	ε^*	$-\varepsilon$	-1	$-\varepsilon^*$	ε	-1	$-\varepsilon^*$	ε	1	ε^*	$-\varepsilon$		
E_{2u}	1	$-\varepsilon^*$	$-\varepsilon$	1	$-\varepsilon^*$	$-\varepsilon$	-1	ε^*	ε	-1	ε^*	ε		
	1	$-\varepsilon$	$-\varepsilon^*$	1	$-\varepsilon$	$-\varepsilon^*$	-1	ε	ε^*	-1	ε	ε^*		

THE D_{nh} GROUPS

D_{2h}	E	$C_2(z)$	$C_2(y)$	$C_2(x)$	i	$\sigma(xy)$	$\sigma(xz)$	$\sigma(yz)$		
A_g	1	1	1	1	1	1	1	1		x^2, y^2, z^2
B_{1g}	1	1	-1	-1	1	1	-1	-1	R_z	xy
B_{2g}	1	-1	1	-1	1	-1	1	-1	R_y	xz
B_{3g}	1	-1	-1	1	1	-1	-1	1	R_x	yz
A_u	1	1	1	1	-1	-1	-1	-1		
B_{1u}	1	1	-1	-1	-1	-1	1	1	z	
B_{2u}	1	-1	1	-1	-1	1	-1	1	y	
B_{3u}	1	-1	-1	1	-1	1	1	-1	x	

D_{3h}	E	$2C_3$	$3C_2$	σ_h	$2S_3$	$3\sigma_v$		
A_1'	1	1	1	1	1	1		x^2+y^2, z^2
A_2'	1	1	-1	1	1	-1	R_z	
E'	2	-1	0	2	-1	0	(x, y)	(x^2-y^2, xy)
A_1''	1	1	1	-1	-1	-1		
A_2''	1	1	-1	-1	-1	1	z	
E''	2	-1	0	-2	1	0	(R_x, R_y)	(xz, yz)

D_{4h}	E	$2C_4$	C_2	$2C_2'$	$2C_2''$	i	$2S_4$	σ_h	$2\sigma_v$	$2\sigma_d$		
A_{1g}	1	1	1	1	1	1	1	1	1	1		x^2+y^2, z^2
A_{2g}	1	1	1	-1	-1	1	1	1	-1	-1	R_z	
B_{1g}	1	-1	1	1	-1	1	-1	1	1	-1		x^2-y^2
B_{2g}	1	-1	1	-1	1	1	-1	1	-1	1		xy
E_g	2	0	-2	0	0	2	0	-2	0	0	(R_x, R_y)	(xz, yz)
A_{1u}	1	1	1	1	1	-1	-1	-1	-1	-1		
A_{2u}	1	1	1	-1	-1	-1	-1	-1	1	1	z	
B_{1u}	1	-1	1	1	-1	-1	1	-1	-1	1		
B_{2u}	1	-1	1	-1	1	-1	1	-1	1	-1		
E_u	2	0	-2	0	0	-2	0	2	0	0	(x, y)	

D_{5h}	E	$2C_5$	$2C_5^2$	$5C_2$	σ_h	$2S_5$	$2S_5^3$	$5\sigma_v$		
A_1'	1	1	1	1	1	1	1	1		x^2+y^2, z^2
A_2'	1	1	1	-1	1	1	1	-1	R_z	
E_1'	2	$2\cos 72°$	$2\cos 144°$	0	2	$2\cos 72°$	$2\cos 144°$	0	(x, y)	
E_2'	2	$2\cos 144°$	$2\cos 72°$	0	2	$2\cos 144°$	$2\cos 72°$	0		(x^2-y^2, xy)
A_1''	1	1	1	1	-1	-1	-1	-1		
A_2''	1	1	1	-1	-1	-1	-1	1	z	
E_1''	2	$2\cos 72°$	$2\cos 144°$	0	-2	$-2\cos 72°$	$-2\cos 144°$	0	(R_x, R_y)	(xz, yz)
E_2''	2	$2\cos 144°$	$2\cos 72°$	0	-2	$-2\cos 144°$	$-2\cos 72°$	0		

D_{6h}	E	$2C_6$	$2C_3$	C_2	$3C_2'$	$3C_2''$	i	$2S_3$	$2S_6$	σ_h	$3\sigma_d$	$3\sigma_v$		
A_{1g}	1	1	1	1	1	1	1	1	1	1	1	1		x^2+y^2, z^2
A_{2g}	1	1	1	1	-1	-1	1	1	1	1	-1	-1	R_z	
B_{1g}	1	-1	1	-1	1	-1	1	-1	1	-1	1	-1		
B_{2g}	1	-1	1	-1	-1	1	1	-1	1	-1	-1	1		
E_{1g}	2	1	-1	-2	0	0	2	1	-1	-2	0	0	(R_x, R_y)	(xz, yz)
E_{2g}	2	-1	-1	2	0	0	2	-1	-1	2	0	0		(x^2-y^2, xy)
A_{1u}	1	1	1	1	1	1	-1	-1	-1	-1	-1	-1		
A_{2u}	1	1	1	1	-1	-1	-1	-1	-1	-1	1	1	z	
B_{1u}	1	-1	1	-1	1	-1	-1	1	-1	1	-1	1		
B_{2u}	1	-1	1	-1	-1	1	-1	1	-1	1	1	-1		
E_{1u}	2	1	-1	-2	0	0	-2	-1	1	2	0	0	(x, y)	
E_{2u}	2	-1	-1	2	0	0	-2	1	1	-2	0	0		

CHARACTER TABLES

D_{8h}	E	$2C_8$	$2C_8^3$	$2C_4$	C_2	$4C_2'$	$4C_2''$	i	$2S_8$	$2S_8^3$	$2S_4$	σ_h	$4\sigma_d$	$4\sigma_v$		
A_{1g}	1	1	1	1	1	1	1	1	1	1	1	1	1	1		x^2+y^2, z^2
A_{2g}	1	1	1	1	1	-1	-1	1	1	1	1	1	-1	-1	R_z	
B_{1g}	1	-1	-1	1	1	1	-1	1	-1	-1	1	1	1	-1		
B_{2g}	1	-1	-1	1	1	-1	1	1	-1	-1	1	1	-1	1		
E_{1g}	2	$\sqrt{2}$	$-\sqrt{2}$	0	-2	0	0	2	$\sqrt{2}$	$-\sqrt{2}$	0	-2	0	0	(R_x, R_y)	(xz, yz)
E_{2g}	2	0	0	-2	2	0	0	2	0	0	-2	2	0	0		(x^2-y^2, xy)
E_{3g}	2	$-\sqrt{2}$	$\sqrt{2}$	0	-2	0	0	2	$-\sqrt{2}$	$\sqrt{2}$	0	-2	0	0		
A_{1u}	1	1	1	1	1	1	1	-1	-1	-1	-1	-1	-1	-1		
A_{2u}	1	1	1	1	1	-1	-1	-1	-1	-1	-1	-1	1	1	z	
B_{1u}	1	-1	-1	1	1	1	-1	-1	1	1	-1	-1	-1	1		
B_{2u}	1	-1	-1	1	1	-1	1	-1	1	1	-1	-1	1	-1		
E_{1u}	2	$\sqrt{2}$	$-\sqrt{2}$	0	-2	0	0	-2	$-\sqrt{2}$	$\sqrt{2}$	0	2	0	0	(x, y)	
E_{2u}	2	0	0	-2	2	0	0	-2	0	0	2	-2	0	0		
E_{3u}	2	$-\sqrt{2}$	$\sqrt{2}$	0	-2	0	0	-2	$\sqrt{2}$	$-\sqrt{2}$	0	2	0	0		

THE D_{nd} GROUPS

D_{2d}	E	$2S_4$	C_2	$2C_2'$	$2\sigma_d$		
A_1	1	1	1	1	1		x^2+y^2, z^2
A_2	1	1	1	-1	-1	R_z	
B_1	1	-1	1	1	-1		x^2-y^2
B_2	1	-1	1	-1	1	z	xy
E	2	0	-2	0	0	$(x, y); (R_x, R_y)$	(xz, yz)

D_{3d}	E	$2C$	$3C_2$	i	$2S_6$	$3\sigma_d$		
A_{1g}	1	1	1	1	1	1		x^2+y^2, z^2
A_{2g}	1	1	-1	1	1	-1	R_z	
E_g	2	-1	0	2	-1	0	(R_x, R_y)	$(x^2-y^2, xy),$ (xz, yz)
A_{1u}	1	1	1	-1	-1	-1		
A_{2u}	1	1	-1	-1	-1	1	z	
E_u	2	-1	0	-2	1	0	(x, y)	

D_{4d}	E	$2S_8$	$2C_4$	$2S_8^3$	C_2	$4C_2'$	$4\sigma_d$		
A_1	1	1	1	1	1	1	1		x^2+y^2, z^2
A_2	1	1	1	1	1	-1	-1	R_z	
B_1	1	-1	1	-1	1	1	-1		
B_2	1	-1	1	-1	1	-1	1	z	
E_1	2	$\sqrt{2}$	0	$-\sqrt{2}$	-2	0	0	(x, y)	
E_2	2	0	-2	0	2	0	0		(x^2-y^2, xy)
E_3	2	$-\sqrt{2}$	0	$\sqrt{2}$	-2	0	0	(R_x, R_y)	(xz, yz)

D_{5d}	E	$2C_5$	$2C_5^2$	$5C_2$	i	$2S_{10}^3$	$2S_{10}$	$5\sigma_d$		
A_{1g}	1	1	1	1	1	1	1	1		x^2+y^2, z^2
A_{2g}	1	1	1	-1	1	1	1	-1	R_z	
E_{1g}	2	2 cos 72°	2 cos 144°	0	2	cos 72°	2 cos 144°	0	(R_x, R_y)	(xz, yz)
E_{2g}	2	2 cos 144°	2 cos 72°	0	2	2 cos 144°	2 cos 72°	0		(x^2-y^2, xy)
A_{1u}	1	1	1	1	-1	-1	-1	-1		
A_{2u}	1	1	1	-1	-1	-1	-1	1	z	
E_{1u}	2	2 cos 72°	2 cos 44°	-2	-2	$-2\cos 72°$	$-2\cos 14°$	0	(x, y)	
E_{2u}	2	2 cos 14°	2 cos 72°	0	-2	$-2\cos 144°$	$-2\cos 72°$	0		

D_{6d}	E	$2S_{12}$	$2C_6$	$2S_4$	$2C_3$	$2S_{12}^5$	C_2	$6C_2'$	$6\sigma_d$		
A_1	1	1	1	1	1	1	1	1	1		x^2+y^2, z^2
A_2	1	1	1	1	1	1	1	-1	-1	R_z	
B_1	1	-1	1	-1	1	-1	1	1	-1		
B_2	1	-1	1	-1	1	-1	1	-1	1	z	
E_1	2	$\sqrt{3}$	1	0	-1	$-\sqrt{3}$	-2	0	0	(x, y)	
E_2	2	1	-1	-2	-1	1	2	0	0		$(x^2-y^2, x$
E_3	2	0	-2	0	2	0	-2	0	0		
E_4	2	-1	-1	2	-1	-1	2	0	0		
E_5	2	$-\sqrt{3}$	1	0	-1	$\sqrt{3}$	-2	0	0	(R_x, R_y)	(xz, yz)

THE S_n GROUPS

S_4	E	S_4	C_2	S_4^3		
A	1	1	1	1	R_z	x^2+y^2, z^2
B	1	-1	1	-1	z	x^2-y^2, xy
E	$\begin{cases}1 \\ 1\end{cases}$	$\begin{matrix}i \\ -i\end{matrix}$	$\begin{matrix}-1 \\ -1\end{matrix}$	$\begin{matrix}-i \\ i\end{matrix}$	$(x, y); (R_x, R_y)$	(xz, yz)

S_6	E	C_3	C_3^2	i	S_6^5	S_6			$\varepsilon = \exp(2\pi i/3)$
A_g	1	1	1	1	1	1	R_z		x^2+y^2, z^2
E_g	$\begin{cases}1\\1\end{cases}$	$\begin{matrix}\varepsilon\\\varepsilon^*\end{matrix}$	$\begin{matrix}\varepsilon^*\\\varepsilon\end{matrix}$	$\begin{matrix}1\\1\end{matrix}$	$\begin{matrix}\varepsilon\\\varepsilon^*\end{matrix}$	$\begin{matrix}\varepsilon^*\\\varepsilon\end{matrix}\Big\}$	(R_x, R_y)		$(x^2-y^2, xy);$ (xz, yz)
A_u	1	1	1	-1	-1	-1	z		
E_u	$\begin{cases}1\\1\end{cases}$	$\begin{matrix}\varepsilon\\\varepsilon^*\end{matrix}$	$\begin{matrix}\varepsilon^*\\\varepsilon\end{matrix}$	$\begin{matrix}-1\\-1\end{matrix}$	$\begin{matrix}-\varepsilon\\-\varepsilon^*\end{matrix}$	$\begin{matrix}-\varepsilon^*\\-\varepsilon\end{matrix}\Big\}$	(x, y)		

S_8	E	S_8	C_4	S_8^3	C_2	S_8^5	C_4^3	S_8^7		$\varepsilon = \exp(2\pi i/8)$
A	1	1	1	1	1	1	1	1	R_z	x^2+y^2, z^2
B	1	-1	1	-1	1	-1	1	-1	z	
E_1	$\begin{cases}1\\1\end{cases}$	$\begin{matrix}\varepsilon\\\varepsilon^*\end{matrix}$	$\begin{matrix}i\\-i\end{matrix}$	$\begin{matrix}-\varepsilon^*\\-\varepsilon\end{matrix}$	$\begin{matrix}-1\\-1\end{matrix}$	$\begin{matrix}-\varepsilon\\-\varepsilon^*\end{matrix}$	$\begin{matrix}-i\\i\end{matrix}$	$\begin{matrix}\varepsilon^*\\\varepsilon\end{matrix}\Big\}$	$(x, y);$ (R_x, R_y)	
E_2	$\begin{cases}1\\1\end{cases}$	$\begin{matrix}i\\-i\end{matrix}$	$\begin{matrix}-1\\-1\end{matrix}$	$\begin{matrix}-i\\i\end{matrix}$	$\begin{matrix}1\\1\end{matrix}$	$\begin{matrix}i\\-i\end{matrix}$	$\begin{matrix}-1\\-1\end{matrix}$	$\begin{matrix}-i\\i\end{matrix}\Big\}$		(x^2-y^2, xy)
E_3	$\begin{cases}1\\1\end{cases}$	$\begin{matrix}-\varepsilon^*\\-\varepsilon\end{matrix}$	$\begin{matrix}-i\\i\end{matrix}$	$\begin{matrix}\varepsilon\\\varepsilon^*\end{matrix}$	$\begin{matrix}-1\\-1\end{matrix}$	$\begin{matrix}\varepsilon^*\\\varepsilon\end{matrix}$	$\begin{matrix}i\\-i\end{matrix}$	$\begin{matrix}-\varepsilon\\-\varepsilon^*\end{matrix}\Big\}$		(xz, yz)

THE CUBIC GROUPS

T	E	$4C_3$	$4C_3^2$	$3C_2$			$\varepsilon = \exp(2\pi i/3)$
A	1	1	1	1			$x^2+y^2+z^2$
E	$\begin{cases}1\\1\end{cases}$	$\begin{matrix}\varepsilon\\\varepsilon^*\end{matrix}$	$\begin{matrix}\varepsilon^*\\\varepsilon\end{matrix}$	$\begin{matrix}1\\1\end{matrix}$			$(2z^2-x^2-y^2,$ $x^2-y^2)$
T	3	0	0	-1	$(R_x, R_y, R_z); (x, y, z)$		(xy, xz, yz)

T_h	E	$4C_3$	$4C_3^2$	$3C_2$	i	$4S_6$	$4S_6^5$	$3\sigma_h$		$\varepsilon = \exp(2\pi i/3)$
A_g	1	1	1	1	1	1	1	1		$x^2+y^2+z^2$
A_u	1	1	1	1	-1	-1	-1	-1		
E_g	$\begin{cases}1\\1\end{cases}$	$\begin{matrix}\varepsilon\\\varepsilon^*\end{matrix}$	$\begin{matrix}\varepsilon^*\\\varepsilon\end{matrix}$	$\begin{matrix}1\\1\end{matrix}$	$\begin{matrix}1\\1\end{matrix}$	$\begin{matrix}\varepsilon\\\varepsilon^*\end{matrix}$	$\begin{matrix}\varepsilon^*\\\varepsilon\end{matrix}$	$\begin{matrix}1\\1\end{matrix}\Big\}$		$(2z^2-x^2-y^2,$ $x^2-y^2)$
E_u	$\begin{cases}1\\1\end{cases}$	$\begin{matrix}\varepsilon\\\varepsilon^*\end{matrix}$	$\begin{matrix}\varepsilon^*\\\varepsilon\end{matrix}$	$\begin{matrix}1\\1\end{matrix}$	$\begin{matrix}-1\\-1\end{matrix}$	$\begin{matrix}-\varepsilon\\-\varepsilon^*\end{matrix}$	$\begin{matrix}-\varepsilon^*\\-\varepsilon\end{matrix}$	$\begin{matrix}-1\\-1\end{matrix}\Big\}$		
T_g	3	0	0	-1	-3	0	0	-1	(R_x, R_y, R_z)	(xz, yz, xy)
T_u	3	0	0	-1	-3	0	0	1	(x, y, z)	

T_d	E	$8C_3$	$3C_2$	$6S_4$	$6\sigma_d$		
A_1	1	1	1	1	1		$x^2+y^2+z^2$
A_2	1	1	1	-1	-1		
E	2	-1	2	0	0		$(2z^2-x^2-y^2,$ $x^2-y^2)$
T_1	3	0	-1	1	-1	(R_x, R_y, R_z)	
T_2	3	0	-1	-1	1	(x, y, z)	(xy, xz, yz)

O	E	$6C_4$	$3C_2(=C_4^2)$	$8C_3$	$6C_2$		
A_1	1	1	1	1	1		$x^2+y^2+z^2$
A_2	1	-1	1	1	-1		
E	2	0	2	-1	0		$(2z^2-x^2-y^2,$ $x^2-y^2)$
T_1	3	1	-1	0	-1	$(R_x, R_y, R_z); (x, y, z)$	
T_2	3	-1	-1	0	1		(xy, xz, yz)

O_h	E	$8C_3$	$6C_2$	$6C_4$	$3C_2(=C_4^2)$	i	$6S_4$	$8S_6$	$3\sigma_h$	$6\sigma_d$		
A_{1g}	1	1	1	1	1	1	1	1	1	1		$x^2+y^2+z^2$
A_{2g}	1	1	-1	-1	1	1	-1	1	1	-1		
E_g	2	-1	0	0	2	2	0	-1	2	0		$(2z^2-x^2-y^2,$ $x^2-y^2)$
T_{1g}	3	0	-1	1	-1	3	1	0	-1	-1	(R_x, R_y, R_z)	
T_{2g}	3	0	1	-1	-1	3	-1	0	-1	1		(xz, yz, xy)
A_{1u}	1	1	1	1	1	-1	-1	-1	-1	-1		
A_{2u}	1	1	-1	-1	1	-1	1	-1	-1	1		
E_u	2	-1	0	0	2	-2	0	1	-2	0		
T_{1u}	3	0	-1	1	-1	-3	-1	0	1	1	(x, y, z)	
T_{2u}	3	0	1	-1	-1	-3	1	0	1	-1		

THE GROUPS $C_{\infty v}$ AND $D_{\infty h}$ FOR LINEAR MOLECULES

$C_{\infty v}$	E	$2C_\infty^\Phi$	$\cdots$	$\infty \sigma_v$		
$A_1 \equiv \Sigma^+$	1	1	$\cdots$	1	z	x^2+y^2, z^2
$A_2 \equiv \Sigma^-$	1	1	$\cdots$	-1	R_z	
$E_1 \equiv \Pi$	2	$2\cos\Phi$	$\cdots$	0	$(x,y); (R_x, R_y)$	(xz, yz)
$E_2 \equiv \Delta$	2	$2\cos 2\Phi$	$\cdots$	0		(x^2-y^2, xy)
$E_3 \equiv \Phi$	2	$2\cos 3\Phi$	$\cdots$	0		
$\cdots$	$\cdots$	$\cdots$	$\cdots$	$\cdots$		

$D_{\infty h}$	E	$2C_\infty^\Phi$	$\cdots$	$\infty \sigma_v$	i	$2S_\infty^\Phi$	$\cdots$	∞C_2		
Σ_g^+	1	1	$\cdots$	1	1	1	$\cdots$	1		x^2+y^2, z^2
Σ_g^-	1	1	$\cdots$	-1	1	1	$\cdots$	-1	R_z	
Π_g	2	$2\cos\Phi$	$\cdots$	0	2	$-2\cos\Phi$	$\cdots$	0	(R_x, R_y)	(xz, yz)
Δ_g	2	$2\cos 2\Phi$	$\cdots$	0	2	$2\cos 2\Phi$	$\cdots$	0		(x^2-y^2, xy)
$\cdots$	$\cdots$	$\cdots$	$\cdots$	$\cdots$	$\cdots$	$\cdots$	$\cdots$	$\cdots$		
Σ_u^+	1	1	$\cdots$	1	-1	-1	$\cdots$	-1	z	
Σ_u^-	1	1	$\cdots$	-1	-1	-1	$\cdots$	1		
Π_u	2	$2\cos\Phi$	$\cdots$	0	-2	$2\cos\Phi$	$\cdots$	0	(x,y)	
Δ_u	2	$2\cos 2\Phi$	$\cdots$	0	-2	$-2\cos 2\Phi$	$\cdots$	0		
$\cdots$	$\cdots$	$\cdots$	$\cdots$	$\cdots$	$\cdots$	$\cdots$	$\cdots$	$\cdots$		

THE ICOSAHEDRAL GROUPS*

	E	$12C_5$	$12C_5^2$	$20C_3$	$15C_2$	i	$12S_{10}$	$12S_{10}^3$	$20S_6$	15σ		
A_g	1	1	1	1	1	1	1	1	1	1		$x^2+y^2+z^2$
T_{1g}	3	$\frac{1}{2}(1+\sqrt{5})$	$\frac{1}{2}(1-\sqrt{5})$	0	-1	3	$\frac{1}{2}(1-\sqrt{5})$	$\frac{1}{2}(1+\sqrt{5})$	0	-1	(R_x, R_y, R_z)	
T_{2g}	3	$\frac{1}{2}(1-\sqrt{5})$	$\frac{1}{2}(1+\sqrt{5})$	0	-1	3	$\frac{1}{2}(1+\sqrt{5})$	$\frac{1}{2}(1-\sqrt{5})$	0	-1		
G_g	4	-1	-1	1	0	4	-1	-1	1	0		
H_g	5	0	0	-1	1	5	0	0	-1	1		$(2z^2-x^2-y^2,$ $x^2-y^2,$ $xy, yz, zx)$
A_u	1	1	1	1	1	-1	-1	-1	-1	-1		
T_{1u}	3	$\frac{1}{2}(1+\sqrt{5})$	$\frac{1}{2}(1-\sqrt{5})$	0	-1	-3	$-\frac{1}{2}(1-\sqrt{5})$	$-\frac{1}{2}(1+\sqrt{5})$	0	1	(x, y, z)	
T_{2u}	3	$\frac{1}{2}(1-\sqrt{5})$	$\frac{1}{2}(1+\sqrt{5})$	0	-1	-3	$-\frac{1}{2}(1+\sqrt{5})$	$-\frac{1}{2}(1-\sqrt{5})$	0	1		
G_u	4	-1	-1	1	0	-4	1	1	-1	0		
H_u	5	0	0	-1	1	-5	0	0	1	-1		

*For the pure rotation group I, the outlined section in the upper left is the character table; the g subscripts should, of course, be dropped and (x, y, z) assigned to the T_1 representation.

INDEX

ab initio force fields, 77
Accidental degeneracy, 58
Acetone, 164
Acetonitrile, 164
Acoustical mode, 234
Additive dispersion, 168
Adenine, 227
A-form of nucleic acid, 229, 231, 250
AgBr, as an optical material, 158
AgCl, as an optical material, 158
Alanine, 194, 260
Alanyl alanine, 210–213, 219, 260
Alanyl alanyl alanine, 247, 248
Alanyl dipeptide, *see* Alanyl alanine
Alanyl oligomer, 248
Alanyl tripeptide, *see* Alanyl alanyl alanine
α-Band, 224
α-Helix, peptide structure, 206, 246
α-Phenylethylamine, 259
α-Pinene, 241, 259
Amide group, 206
Amide vibrations, amide A,B,I–VII, 209–212, 243, 244, 248, 260
Amino acid, 194, 205
Ammonia:
 potential function, 57
 symmetry, 101
Amplitude, vibrational, 50
Angular dispersion, 147
Angular frequency, 24
Angular momentum, 9, 36

Anharmonic force constants, 11
Anharmonicity, 42, 55
Anisotropy, of tensor, 121
Anomalous dispersion, 166
Anthraquinone, 132
Anti-Stokes (Raman) scattering, 111, 114
Apodization, 218
Array detector, 168
Asymmetric molecules, 188
Asymmetric top rotor, 34, 42
Associated Legendre functions, 35
Atomic displacement vectors, 71, 86
Atomic polar tensor, 88
ATR, *see* Attenuated total reflection
Attenuated total reflection (ATR), 164, 166
A-type band, 41
Axes of inertia, 34
Axes of rotation, 92

Backscattering, 126
Bacteriorhodopsin, 225
BaF_2, as an optical material, 158
Base fraying, 255
Base pairs, in nucleic acids, 228
Base stacking, in nucleic acids, 231
Base vibration, in nucleic acids, 230
Beam waist, 167
Benzene:
 symmetry, 93
 vibration, 64, 198
Bessel function, 240

279

β-Band, 222
β-Sheet, peptide structure, 207, 246
β-Turn, peptide structure, 207
B-form of nucleic acid, 228, 231, 250
Binning (a CCD detector), 171
Black body radiator, 159
B-matrix, 67, 79
Boltzmann's law, 111
Bond dissociation energy, 22, 26
Boundary conditions, 17
Bromochlorofluoro methane (HCFClBr), 188
Bromoform ($CDBr_3$), 164
B-type band, 41

CaF_2, as an optical material, 158, 239
Carbon dioxide, 176
Carbon disulfide, 164, 176
Carbon tetrachloride (CCl_4), 141, 164, 180
Carbonyl, 178
Carboxylate, 248
Carboxylic acid function, 195
CARS, see Coherent anti-Stokes Raman spectroscopy
Cartesian displacement coordinates, 47, 79
CCD, see Charge coupled device
CCl_4, see Carbon tetrachloride
C_2Cl_4, 141
$CDBr_3$, see Bromoform
$CDCl_3$, see Chloroform
Center of inversion, 92
Centrifugal distortion, 38, 44
Centrosymmetric group, 118, 178
Character, 97
Character table, 98, 266–276
Charge coupled device (CCD), 169
$CHCl_3$, see Chloroform
Chirality, 236, 243
Chloroform ($CDCl_3$ and $CHCl_3$), 164, 183, 185
Chromophore, 117
Chymotrypsin, 249
Circular dichroism, 236
Circularly polarized light, see Polarization
Coherent anti-Stokes Raman spectroscopy (CARS), 3, 112, 134, 135
Coherent Raman effect, 134
Coil, peptide structure, 207
Collisional deactivation, 28
Collisional excitation, 28
Combination band, 58, 164, 181
Commutation of operators, 35
Configuration, of molecules, 243
Conformation of peptides, see Secondary structure

Conformational sensitivity, 215, 243
Conne advantage, 162
Conservative couplet, 244
C_2 operation, 91
Coupled oscillator, 242
Couplet, 244, 247
CsI, as an optical material, 158
CS_2, see Carbon disulfide
C-type band, 41
Cubic force constants, see Anharmonic force constants
Cyclopentene, potential function, 57
Cytochrome, 129, 132, 221, 260
Cytosine, 227
Czerny–Turner monochromator, 146

D-Ala-L-Ala, see Alanyl alanine
de Broglie's relation, 30
Deconvolution, 218
Degenerate modes, 179, 244
δ-Function, 154
Denaturation, 218
Deoxyribonucleic acid (DNA), 204, 227
Deoxyribose, 230
Depolarization ratio, 118, 123, 178
Detectors, infrared, 160
Deuteriated triglycine sulfate (DTGS), as detector material, 160
Deuteriochloroform, 185
Deuteriomethyl group, 187
Diagonalization, 68, 70, 246
Dielectric constant, 113
Dielectric susceptibility, 133
Differential absorption, 237
Diffraction grating, 125, 145
Diffuse reflection, 164, 166
Diode array detector, 169
Diphenylbutadiene, 133
Dipolar coupling, 216, 231, 243
Dipole–dipole interactions, 216
Dipole operator, 13
Dipole strength, 15, 87, 237, 246
Dipole transition moment, 115
Direction cosine matrix, 122
Discrete Fourier transform, 154
Dispersion, 147
Dispersive monochromators, 145
Dimethyl sulfate, 164
DMSO, 164
DNA, see Deoxyribonucleic acid
DTGS, see Deuteriated triglycine sulfate
Dual beam instrument, 161
Dynamic studies of heme group, 224

Eigenfunction, 9, 18
Eigenvalues, 17
  symmetric top rotor, 36
  vibrational, 23
Eigenvector matrix, 70
Einstein coefficients, 31
Electric field perturbation, 127, 259
Electric quadrupole moment, 257
Electromagnetic radiation, 28
Electro-optic modulator (EOM), 239, 258
Encodegram, 173
Endo conformation, of ribonucleic acid, 231, 255
Exo conformation, of ribonucleic acid, 231, 255
EOM, *see* Electro-optic modulator
Equilibrium coordinates, 78
Ethane, 92, 191
  derivatives, 191
  symmetry, 92
Even function, 20
Excitation profile, 116
Exciton, 216, 244, 250
Exclusion rule, 116, 118, 178
Extended helix, peptide structure, 246
Extinction coefficient, 165

Factor analysis, 249
Fast Fourier transform (FFT), 155
Fatty acid, 232
Fe(II) and Fe(III), 222
Fellgett advantage, 162
Fermi resonance, 55, 58, 102, 177, 181, 186, 189, 196, 213, 216
FFT, *see* Fast Fourier transform
5'd(CCGG)3', *see* Tetranucleotide
5'd(CGCG)3', *see* Tetranucleotide
5'd(GCGC)3', *see* Tetranucleotide
5'd(GGCC)3', *see* Tetranucleotide
f-Number, 126
Fixed partial charge model, 87, 242
Fluorescence, 28, 170, 172
F-matrix, 67, 69, 70, 80, 107
Force constant, 22, 49, 73, 80, 264–265
Force field, 72
Formaldehyde, 177
  symmetry, 94
Forward scattering, 126
Fourier pair, 152
Fourier self-deconvolution, 218
Fourier transform, 149, 154
Fourier transform infrared spectrometer, 162
Fourier transform Raman spectroscopy, 112, 172

Freon, 188

Gain equation, 135
Gating, detector, 170
Gaussian function, 154, 218
Gel phase, 233
Generalized valence force field (GVFF), 72, 83
Genetic code, 226
Germanium, detector, 161
Glass, 158
Globar, 160
G-matrix, 68
Grating, *see* Diffraction grating
Group, 94
Group frequencies, 61, 183, 262–263
Group theory, 90
Guanine, 227, 231
GVFF, *see* Generalized valence force field

Hadamard transform, 173
Hamilton operator, 8, 17, 18
Hamiltonian, *see* Hamilton operator
Harmonic analysis, 152, 154, 173
HCFClBr, *see* Bromochlorofluoro methane
Heat radiation, 159
Heisenberg uncertainty principle, 6
Heme group, 221
Hemoglobin, 129, 132, 221, 224, 249
Hermite's differential equation, 23
Hermite polynomials, 23, 26
HgCdTe, *see* Mercury–cadmium–telluride detector
High spin, 222
Holographic grating, 168
Homogeneous linear equations, 51
Homo-oligoamino acids, 243
Hoogstein base pair, 251
Hook's law, 22, 47, 49
Hot band, 55
Hyper-Raman spectroscopy, 3, 112, 134, 140
Hyper-Rayleigh effect, 134, 141
Hyperpolarizability tensor, 133, 140
Hypochromism, 220, 231

Identity element, 92, 94
Image intensifier, 170
Induced dipole moment, 113–115
Indium antimonide (InSb), 161
Indium gallium arsenide (InGaAs), 161, 172
Inelastic light scattering, 109
Infrared absorption intensity, 87
Infrared absorption spectroscopy, 109, 204
InGaAs, *see* Indium gallium arsenide

InSb, see Indium antimonide
Instrumentation:
  for infrared spectroscopy, 157
  for Raman spectroscopy, 166
Interferogram, 152
Interferometer, 145, 149
Internal coordinates, 62, 66, 74
Internal rotation, 57, 192
Inverse polarization, 222
Inverse Raman effect, 112, 134, 139
Irreducible representation, 95, 96, 98, 117, 141
Isotopic effect, 181
Isotopic shift, 185, 210
Isotopic species, see Isotopomer
Isotopic substitution, 63, 242
Isotopomer, 57, 63, 72
Isotropic part, of tensor, 121

Jacquinot advantage, 162

KBr, as an optical material, 158
Kinetic isotope effect, 57
KRS-5, as an optical material, 158, 166

L-Ala-L-Ala, see Alanyl alanine
L-Ala-L-Ala-L-Ala, see Alanyl alanyl alanine
L-Pro-L-Pro-L-Pro, see Prolyl prolyl proline
Lagrange's equation of motion, 47, 49
Lambert–Beer law, 109, 163
Laser noise, in FT Raman spectroscopy, 172
Lead sulfide, 161
Librational motion, 192
Life time, 129
Light collection optics, 148
Linear dispersion, 147, 171
Linear force constants, 190
Linear momentum, 8
Linear molecules, rotational energy, 36, 38
Linearly polarized light, see Polarization
Lipids, 232
Lipid bilayer, 232
Liquid crystalline phase, 233
Localized molecular orbital method, 88, 242
Lock-in amplifier, 240
Lorenzian band, 218
Low spin, 222
Lysozyme, 249

Macrocycle, 221
Magic angle, in Raman optical activity, 258
Mass-weighted Cartesian displacement coordinates, 47, 48, 66
MCT, see Mercury–cadmium–telluride detector
Membrane, 232

Mercury–cadmium–telluride detector (MCT) (HgCdTe), 161
Methane, 181
Methane derivatives, 179
N-methylacetamide, 208, 210, 214, 220
Methyl chloride, 186
Methyl deformation vibrations, 186
Methyl rocking vibrations, 187
Methyl stretching vibrations, 186
Methylene chloride, 187
Methylene stretching vibrations, in lipids, 233
Michelson interferometer, see Interferometer
Microchannel plate, 170
Microwave spectroscopy, see Rotational spectroscopy
Mirror reflections, see Reflection
Molar extinction coefficient, 109
Moment of inertia, 9
Momentum, 7, 238
Mononucleotide, 227
Morse function, see Morse potential
Morse potential, 10, 22
Multichannel spectrometers, 169
Multiplexing advantage, 162, 169, 174
Multiplexing disadvantage, 173
Mutual exclusion principle, see Exclusion rule
Myoglobin, 132, 224, 249

NaCl, as an optical material, 158
Naturally polarized light, see Polarization
Nernst glower, 160
Newtonian mechanics, 5
Newton's equation of motion, 47
Nichrome, 160
Nonbonded interactions, 75. See also Urey–Bradley force field
Nonlinear (Raman) effects, 133
Nonsuperimposable mirror image, 236
Nontrivial solution, secular determinant, 51
Normal coordinate, 52, 70
Normal mode, 47, 52, 113
  analysis, 65
  computation, 78
Nucleotide, 227

Oblate top, 36
Odd function, 20
Oligonucleotide, 229
Optical activity, 236
Optical delay line, 130
Optical mode, 234
Optical resolution, 147
Optical rotatory dispersion (ORD), 235

ORD, *see* Optical rotatory dispersion
Orthogonality theorem, 95
Out-of-plane deformation, 178
Overtone band, 43, 56, 61, 164, 181
Oxirane, 242, 260

Papain, 249
Parallel band, 41
Particle-in-a-box, 16
P-branch, 40
PbS, *see* Lead sulfide
PED, *see* Potential energy distribution
PEM, *see* Photoelastic modulator
Pentatomic molecules, 179, 180
Peptide conformation, 215–220, 244
Peptide group, *see* Peptide linkage
Peptide linkage, 204, 206
Percent composition, of peptides, 218, 248
Permanent dipole, in relation to rotational
 spectroscopy, 183
Perpendicular band, 41
Perturbed degenerate modes, 242
Phase factor, 50
Phase sensitive detector, 161
Phosphatidylcholine, 232, 234
Phosphodiester, 229
Phospholipid, 232
Phosphorescence, 28, 33
Photoconductive detector, 161
Photodiode, 169
Photoelastic modulator, 239
Photomultiplier, 167
Photon counting, 167
Photovoltaic detector, 161
Plane polarized light, *see* Polarization
Pockels effect, 134
Polar tensor model, 87
Polarizability, 14, 110, 112, 115
Polarizability tensor, 113–116, 123, 127
Polarization:
 of light, 30, 31, 119, 124, 237
 modulation, 239
 of Raman effect, 118
 scrambler, 124, 126
Poly(dA-dT)·poly(dA-dT), 250
Poly(dA)·poly(dT), 250
Poly(dG-dC)·poly(dG-dC), 230, 250
Poly(dG)·poly(dC), 250
Poly(rG), 250
Poly(rG)·poly(rC), 250
Poly(rU), 250
Population inversion, 28
Porphyrin, 221, 224
Potential energy distribution, 63, 71
Primary structure, of peptides, 206

Principal axes of polarizability, 121
Probe pulse, or probe beam, 130
Prolate top, 36
Prolyl prolyl proline, 248
Proper rotation, 91
Propylene oxide, 243
Prosthetic group, 220, 221
Protein, 205, 249
Pump pulse, or pump beam, 130
Purple membrane, 225
Pyroelectric detectors, 160

Q-branch, 40
Quantum efficiency, 168

Raman hypochromism, 220
Raman induced Kerr effect spectroscopy
 (RIKES), 140
Raman intensities, 127
Raman optical activity (ROA), 127, 236,
 256–260
Raman spectroscopy, 109, 110, 119, 204
Random coil, peptide structure, 246
Rayleigh scattering, 111, 114, 119, 147
R-branch, 40
Recursion formula (Hermite polynomials),
 23, 26
Redox proteins, 221
Reduced mass, 25, 63, 64
Reducible representation, 100
Reduction formula, 101
Redundancy, 75
Redundant coordinates, 75, 190, 198
Reflection, 92
Refractive index, 113, 115, 164, 239
Representation, 95
Resolving power, 148
Resonance Raman spectroscopy, 3, 112, 117,
 128, 204
 peptides and proteins, 220
Retardation, 239
Retinal, 202, 225
Reticon detector, 169
Rhodopsin, 132, 202, 225
Ribonuclease, 249
Ribonucleic acid (RNA), 226
Ribose vibration, 230
Right angle scattering, 124, 167
RIKES, *see* Raman induced Kerr effect
 spectroscopy
RNA, *see* Ribonucleic acid
ROA, *see* Raman optical activity
Rocking vibration, 183
Rotation reflection axis, 92
Rotational constant, 10, 36, 181

284 INDEX

Rotational energy, 33
Rotational Hamiltonian, 34
Rotational quantum number, 10
Rotational spectroscopy, 9, 33, 183
Rotational-vibrational spectra, 17, 39, 181
Rotatory polarizability, 256, 257
Rotatory strength, 237, 238, 246

Savitsky–Golay algorithm, 218
Scattering processes, 110
Scattering tensor, 128
Second derivative spectrum, 218
Secondary structure, of peptide, 207, 215–220, 244
Selection rules, 20
  hyper Raman effect, 140
  Raman effect, 116, 117
  rotational spectrum, 37
  vibrational spectrum, 54, 56, 103
Semiconductor detectors, see Detectors, infrared
SERS, see Surface enhanced Raman spectroscopy
Similarity transformation, 121
Simplex matrix, 173
Site interactions, in solids, 164
Site specific mutagenesis, 218
Solid angle, 121
Solvent interaction, 218
Soret band, 222
Space fixed coordinates, 119
Spherical harmonic functions, 10, 35
Spherical polar coordinates, 35
Spherical top rotor, 35, 181
S–S linkage, see Sulfur-sulfur linkages
Stationary states, 12, 19
Stefan–Boltzmann law, 159
Stimulated absorption, 32
Stimulated emission, 13, 32
Stimulated Raman effect, 112, 134, 138
Stokes (Raman) scattering, 111
Stress birefringence, 239
Subtractive dispersion, 168
Sulfur-sulfur linkage, 220
Surface enhanced Raman spectroscopy (SERS), 142
Symmetric top rotor, 36, 38
Symmetry coordinates, 69, 72, 75, 106
Symmetry group, 91
Symmetry operation, 91
Symmetry of vibration, 90

Taylor expansion, 48, 113
Tetrahedral molecules, 180
Tetranucleotide, 253

Teller–Redlich rule, 63
Tensor invariants, 121
Thallium halide, see KRS-5
Thermal detectors, 160. See also Detectors, infrared
Thermocouples, 160
Thiirane, 242, 260
Third order susceptibility, 134
Thymine, 227
Time dependent Schrödinger equation, 7, 8
Time independent Schrödinger equation, 8, 9
Time resolved Resonance Raman spectroscopy, 3, 112, 130
Toluene, 57
Torsional barrier, 191
Torsional vibration, 183, 191, 249
Trace, 99
$trans$-dideuteriocyclobutane, 242
$trans$-dideuteriocycylopropane, 242
Transmission spectroscopy, 163
Transition moment, 14, 19, 58
Transition moment, of C=O group, 216
Trihalomethane, 189
Trivial solution, secular determinant, 51
Two-photon spectroscopy, 117

UBFF, see Urey–Bradley force field
U-matrix, 107
Umbrella mode, 184, 186
Unit cell, 164
Unpolarized light, see Polarization
Uracil, 227
Urey–Bradley force field (UBFF), 75, 76, 85, 198

Valence force field, 81
Valence shell electron pair repulsion, 2
van der Waals forces, 21
VCD, see Vibrational circular dichroism
Vector potential, 29
Vector product, 238
Vibrational circular dichroism (VCD), 3, 236–256
Vibrational energy, polyatomics, 56
Vibrational intensity (absorption), 86
Vibrational optical activity, 236
Vibrational-rotational coupling constant, 43
Vibrational-rotational interaction, 43
Vibrational Schrödinger equation, 21, 54
Vibrational transition moment, 26
Vibronic states, 116, 172, 222
Vidicon detector, 169
Virtual state, 110, 112, 116
Visual pigment, 204

Wagging motion, 178
Water, normal mode calculation, 78, 176
Wave equation, 6
Wave vector, 30, 135
Wavenumber, 15, 24, 111

YAG laser, 173

Z-form of nucleic acid, 229, 231, 250
Z-matrix, 71
ZnSe, as an optical material, 158, 166, 239
Zwitterion, 194, 232

DATE DUE